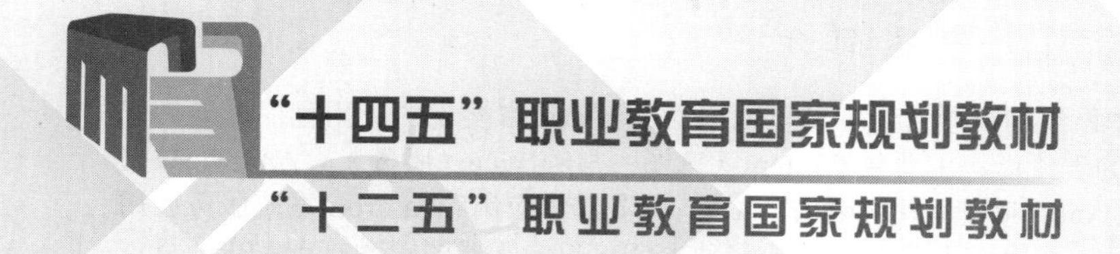

Android高级应用编程实战

（第二版）

李华忠　周彦兵　梁永生　主　编
汪　洋　卢　鑫　王炫盛　高　波　副主编

中国铁道出版社有限公司
CHINA RAILWAY PUBLISHING HOUSE CO., LTD.

内 容 简 介

本书按照Android平台的技术体系结构和项目内容，以面向对象Java语言实现的应用程序框架为基础编写而成。全书共分9章，前6章为理论篇，主要包括Intent（意图）与Service（服务）、Android数据永久存储应用、Android网络应用、Android调用外部数据、Android多媒体应用和Android系统服务应用等核心理论知识；后3章为综合项目实训篇，主要包括基于移动端GPS和传感器的运动打卡APP项目、Struggle车牌识别系统APP项目和基于Android智能仓储系统项目三个综合实训项目，综合应用了本书介绍的核心知识和关键技术。

本书编写理念遵循了基于工作过程导向、项目式学习、翻转课堂和线上线下混合式教学模式等现代新职教理念，实施了“项目引领、价值塑造、成果导向、精准培养”的专业核心能力培养策略，符合教学规律和课堂教学要求，转变教学设计思路，创新课堂教学实践，体现了二十大新精神和新发展理念，很好地反映了“以中国式现代化全面推进中华民族伟大复兴”新时代背景下，嵌入式和移动互联等行业出现的Android方面的新知识、新技术、新方法和新应用，打造“专业思政”路径，做到“三全育人”，解决了高校Android课程教学面临的迫切问题，既适合作为高等职业院校Android高级应用程序设计的教材，也适合作为移动开发爱好者的自学参考书。

图书在版编目（CIP）数据

Android高级应用编程实战/李华忠，周彦兵，梁永生主编. —2版. —北京：中国铁道出版社有限公司，2021.9（2024.7重印）
“十二五”职业教育国家规划教材　经全国职业教育教材审定委员会审定
ISBN 978-7-113-27717-8

Ⅰ.①A…　Ⅱ.①李…　②周…　③梁…　Ⅲ.①移动终端-应用程序-程序设计-高等职业教育-教材　Ⅳ.①TN929.53

中国版本图书馆CIP数据核字（2021）第016228号

书　　名：Android 高级应用编程实战
作　　者：李华忠　周彦兵　梁永生

策　　划：王春霞　　　　编辑部电话：（010）63551006
责任编辑：王春霞　彭立辉
封面设计：付　巍
封面制作：刘　颖
责任校对：焦桂荣
责任印制：樊启鹏

出版发行：中国铁道出版社有限公司（100054，北京市西城区右安门西街8号）
网　　址：https://www.tdpress.com/51eds/
印　　刷：三河市宏盛印务有限公司
版　　次：2015年1月27日第1版　2021年9月第2版　2024年7月第2次印刷
开　　本：880 mm×1 230 mm　1/16　印张：22.5　字数：590千
书　　号：ISBN 978-7-113-27717-8
定　　价：59.00元

前 言

Android是一个以Linux为基础的完整、开放、免费的手机平台。它由应用程序、应用程序框架、系统库、Android运行时以及Linux内核5部分组成。本教材以面向对象Java语言实现的Android平台技术应用程序框架为基础编写，易学、易用，极大地降低了开发移动互联应用程序的难度，大大提高了APP应用程序开发的效率。

目前，我国很多院校的嵌入式技术应用、软件技术、移动互联应用技术、工业机器人技术、工业互联网技术、无人机应用技术和物联网应用技术等专业，都将“Android高级应用”作为专业核心课。为了帮助院校老师比较全面、系统地讲授该课程，使学生能够熟练使用Android高级技术，我们几位长期在院校从事Android教学的教师和企业工程师，共同编写了这本《Android高级应用编程实战》教材。

本书编写理念遵循了基于工作过程导向、成果导向教育、项目式学习、翻转课堂和线上线下混合式教学模式等现代新职教理念，转变教学设计思路，创新课堂教学实践，实施了“项目引领、价值塑造、成果导向、精准培养”的专业核心能力培养策略，探索了“由知识第一，向能力第一转换”的专业课程人才培养模式，符合教学规律和课堂教学要求，体现了二十大新精神，有机地融入了中国式现代化、建设现代化产业体系、高水平科技自立自强和绿色生产生活方式等新发展理念，很好地反映了“以中国式现代化全面推进中华民族伟大复兴”时代背景下，在嵌入式和移动互联等行业出现的新知识、新技术、新方法和新应用，打造“专业思政”路径，做到“三全育人”，解决了高校Android课程教学面临的迫切问题。

本书全面贯彻落实了二十大重要会议精神，坚持“立德树人”教育根本任务，遵循以学习者为中心的自主学习、知识建构内化和巩固拓展等方面关键共性理论和知识技术，采用了新技术框架、组件，修订现有的理论知识体系。在教材形式上采用了嵌入微课（微视频）的一体化新形态，学生课前完成知识学习，课上实现知识内化，课后知识拓展。借助讲练结合、探究式教学法、自主学习法、案例式教学法等，开展项目实践、案例讨论、课内实训等系列“活动设计”，实现了“价值引领”和“知识内化”的统一。

本书内容的取舍和排序，一方面，遵循以工作需求为目标原则，务求反映当前Android高级应用编程开发的主流技术和主流开发工具，另一方面，重视软件工程的标准规范和业内工作过程的即成约定，使学生的学习内容与目标工作岗位能力要求无缝对接。本书采用了“项目引领，任务驱动”的教学模式。结合二十大精神和新发展理念，以“贴近生活，易于理解，实用性强”为原则，选取了“基于移动端GPS和传感器的运动打卡APP项目”、“Struggle车牌

识别系统 APP 项目”和“基于 Android 智能仓储系统项目”为主线项目，按照 Android 高级应用编程项目从设计到开发、实现流程，组织章节的演进。本书的体系结构，按照 Android 平台的技术体系结构和项目内容，如 Intent 与 Service、Android 数据永久存储、Android 网络、Android 调用外部数据、Android 多媒体和 Android 系统服务等项目做了精心的设计，设计了多个学习实践案例。每个案例又结合知识体系和实践技能细化为若干个针对具体知识点的学习实践案例，由浅入深，实用性强。最后，结合移动互联应用实际情况，安排了三个综合实训项目，在提高学生应用技能的同时，强化项目驱动，实施“工学结合”，提高理论教学和实践教学质量，充分满足了高职院校对教学和学生自学的需求。

本书配备了 PPT 课件、微课、源代码、习题答案、教学大纲、课程设计等丰富的教学资源，任课教师可到中国铁道出版社有限公司网站（https://www.tdpress.com/51eds/）免费下载使用。本书的教学参考总学时为 56 学时，其中讲授环节 28 学时，实践环节 28 学时。各章的参考学时参见下面的学时分配表。

章　节	课程内容	学时分配	
		讲　授	实　训
第 1 章	Intent（意图）与 Service（服务）	3	3
第 2 章	Android 数据永久存储应用	4	4
第 3 章	Android 网络应用	3	3
第 4 章	Android 调用外部数据	3	3
第 5 章	Android 多媒体应用	4	4
第 6 章	Android 系统服务应用	3	3
第 7 章	基于移动端 GPS 和传感器的运动打卡 APP 项目	4	4
第 8 章	Struggle 车牌识别系统 APP 项目	4	4
第 9 章	基于 Android 智能仓储系统项目（可选）		
课时总计		28	28

本书由深圳信息职业技术学院李华忠、副校长周彦兵和深圳技术大学副校长梁永生任主编，汪洋、卢鑫、王炫盛、高波任副主编，梁艳玲、唐海峰参与编写。深圳市越疆科技股份有限公司唐海峰、李晓亮深度参与了综合项目实训篇教材建设工作。感谢深圳市大雅新科技有限公司刘业涛、刘立明及深圳市盛泰奇科技有限公司黄华林、黄燕玲对本教材编写提出了很多宝贵意见。

由于 Android 开发技术发展日新月异，加之编者水平有限，书中难免存在疏漏和不妥之处，敬请广大读者批评指正。

编　者

2022 年 11 月

目 录

理论篇

综合项目实训篇

理论篇

本篇主要系统讲解 Android 高级应用编程项目中涉及的具有共性、可重用性和普遍性的一些核心理论知识。每章首先明确各共性知识的学习目标，通过学习导入，循序渐进地引导学生做必要的技术准备；然后通过典型案例讲解技术实施步骤和关键实现代码；最后，通过知识扩展和强化练习，夯实学生的理论和实践基本素养和技能。

本篇主要包含以下 6 章：

第 1 章　Intent（意图）与 Service（服务）

第 2 章　Android 数据永久存储应用

第 3 章　Android 网络应用

第 4 章　Android 调用外部数据

第 5 章　Android 多媒体应用

第 6 章　Android 系统服务应用

第 1 章

Intent（意图）与 Service（服务）

学习目标

视频

Intent（意图）与Service（服务）

- 理解Intent对于Android应用的作用。
- 理解Intent元素的组成及其作用。
- 理解意图过滤器（IntentFilter）的作用。
- 理解Service对于Android应用的作用。
- 理解Activity和Service的生命周期。
- 掌握使用Intent启动系统组件的方法。
- 掌握Service启动到销毁过程的实现。
- 掌握Service与Activity通信的实现。

1.1 学习导入

视频

Android意图（Intent）概念

1.1.1 Intent 的概念

Intent（意图）是一种运行时绑定（Runtime Binding）机制，它能在程序运行时动态地连接两个不同的组件。即Intent提供了一种通用的消息系统，它允许在应用程序与其他应用程序间传递Intent来执行动作和产生事件。使用Intent可以激活Android应用的3个核心组件：Activity（活动）、Service（服务）和广播接收器（Broadcast Receiver）。

（1）通过startActivity( )方法或startActivityForResult( )方法启动一个Activity。

（2）通过startService( )方法启动一个服务，或通过bindService( )方法绑定后台服务。

（3）通过广播方法（sendBroadcast(), sendOrderedBroadcast(), sendStickyBroadcast()）发给广播接收器（BroadcastReceiver）。

1.1.2　Service 的概念

视频

Android服务
Service概念

Service（服务）是运行在后台的线程，级别与Activity差不多。既然Service是运行在后台的线程，那么它就是不可见的，没有界面。用户可以启动一个Service播放背景音乐，或者记录地理信息位置的改变，或者启动一个服务一直监听某种事件发生。

Service和其他组件一样，都运行在主线程中，因此不能用它做耗时的请求或者动作。用户可以在服务中开一个线程，在线程中做耗时动作。

1.1.3　Android 平台应用开发技术回顾（Android 四大组件技术）

视频

Android活动
Activity

Android四大基本组件分别是Activity（活动）、Service（服务）、Content Provider（内容提供者）和BroadcastReceiver（广播接收器）。

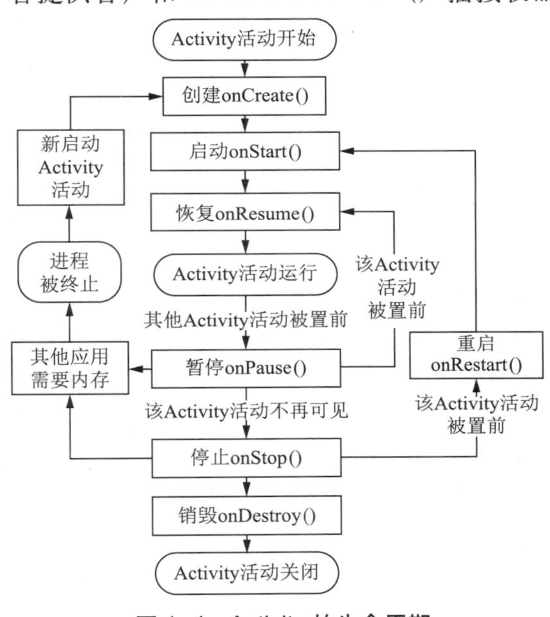

图 1-1　Activity 的生命周期

（1）Activity：在应用程序中，一个Activity通常就是一个单独的屏幕，它既可显示一些控件，也可监听并处理用户的事件并做出响应。Activity间通过Intent进行通信。Activity的生命周期如图1-1所示。

视频

Android Studio
开发环境安装

（2）Service：一个Service是一段长生命周期的没有用户界面的程序，可用来开发后台监控类程序。

（3）Content Provider：将应用程序的指定数据集提供给其他应用程序。这些数据可存储在文件系统、SQLite数据库或任何其他合适的数据存储媒介中，而其他应用程序可通过ContentResolver类从该内容提供者中获取或存入数据。

视频

Eclipe Android
开发环境安装

（4）BroadcastReceiver：它是一种被广泛运用在应用程序间传输信息的机制，通过广播接收器对发送出来的广播进行过滤接收并响应。

1.2　技术准备

Intent 协助应用间交互与通信；负责对应用中一次操作的动作、数据进行描述；Android根据Intent描述，负责找到对应组件，将 Intent传递给调用组件完成组件调用；负责应用程序内部Activity/Service间交互；可将Intent理解为不同组件间的通信媒介，专门提供组件互相调用相关信息。

视频

显示意图Intent
应用实战

1.2.1　Intent 的应用

1. 显式Intent

显式Intent是指在调用中通过代码直接指定哪个组件应该处理该Intent，但是由于开发者往往并不清楚应用程序组件名称，因此显式Intent大多用在应用程序内部传递消息。在调用过程中通过指定接收器的Class或ComponentName实现。要显式地调用一个Intent，可使用Intent.setClass()形式，通过该方法可直接传递Activity或Service类的引用处理Intent，从而简化整个Android的Intent解析过程。

下面示例演示了如何通过显式Intent启动另一个Activity。

```
Intent intent=new Intent();              //构造意图对象
//显示方式声明Intent，直接启动SecondActivity
intent.setClass(MainActivity.this, SecondActivity.class);
startActivity(intent);                   //启动Activity
```

视频

Android Intent意图通信机制

2. 隐式Intent

隐式Intent是指在调用过程中由Android系统决定哪个组件应该处理该Intent。在实现过程中通过一个Intent解析过程实现，该解析过程使用了Action（动作）、Data（数据）和Category（类别）。由于隐式Intent没有明确的目标组件名称，所以要由Android寻找与Intent请求意图最匹配的组件。具体方法是：Android将Intent的请求内容和IntentFilter（意图过滤器）进行比较，IntentFilter中包含系统所有可能待选组件。如果IntentFilter中某一组件匹配隐式Intent请求的内容，Android就选择该组件作为该隐式Intent的目标组件。

视频

隐式意图Intent应用实战

3. IntentFilter（意图过滤器）

在显式Intent和隐式Intent中，已经讲述了IntentFilter的作用是告诉Android系统，应用程序本身所能处理和响应的隐式Intent请求的内容。需要在Android应用程序的系统控制文件AndroidManifest.xml中添加<intent-filter>标签对IntentFilter进行声明。

每个<intent-filter>元素被解析成一个IntentFilter对象。当将一个apk包安装到Android系统中时，其中的组件就会向平台注册。一旦Android系统建立了一个IntentFilter的注册表项，系统就会知道如何把收到的Intent请求映射到已注册的Activity、Service、BraodcastReceiver和Content Provider。

请求Intent时，Android平台使用Intent的Action、Data和Category作为标准，通过已经注册的IntentFilter开始解析过程。Intent和IntentFilter匹配的原则如下：

（1）动作和类别必须匹配。

（2）如果指定数据类型，则数据类型必须匹配，或者数据方案、授权和路径的组合必须匹配。

一个隐式Intent请求要想顺利传递到目标组件，必须通过Action、Data和Category这三方面的检查，若三方面都不匹配，系统将不会把该隐式Intent传递给目标组件。

4. Intent元素

Intent元素包括：Action（动作）、Data（数据）、Category（类别）、Type（数据类型）、Component（组件）和Extras（附加信息）。

（1）Action：用来指明要实施的动作是什么，例如，拍照标准意向行动MediaStore.ACTION_IMAGE_CAPTURE、摄像MediaStore.ACTION_VIDEO_CAPTURE、ACTION_VIEW向用户显示数据、提供对给定数据的显式可编辑访问等。详细使用信息可查阅Android SDK->reference中的Android.content.intent类，里面的constants中定义了所有的Action。Action常用的值如下：

- ACTION_MAIN：Android应用入口，每个Android应用必须且只能包含一个此类型的Action声明。

- ACTION_VIEW：系统根据不同的Data类型，通过已注册的对应应用显示数据。
- ACTION_EDIT：系统根据不同的Data类型，通过已注册的对应应用编辑显示数据。
- ACTION_DIAL：打开系统默认拨号程序，如果Data中设置了电话号码，则自动在拨号程序中输入此号码。
- ACTION_CALL：直接呼叫Data中所带的号码。
- ACTION_ANSWER：接听来电。
- ACTION_SEND：由用户指定发送方式进行数据发送操作。
- ACTION_SENDTO：系统根据不同的Data类型，通过已注册的对应应用进行数据发送操作。
- ACTION_BOOT_COMPLETED：Android系统在启动完毕后发出带有此Action的广播。
- ACTION_TIME_CHANGED：Android系统的时间发生改变后发出带有此Action的广播。
- ACTION_PACKAGE_ADDED：Android系统安装了新的应用之后发出带有此Action的广播。
- ACTION_PACKAGE_CHANGED：Android系统中已存在应用发生改变后发出带有此Action的广播。

下面示例演示了如何利用系统Intent Action，调用系统UI。

```
Uri uri=Uri.parse("https://www.sziit.edu.cn/");          //定义Uri对象
Intent intent=new Intent(Intent.ACTION_VIEW,uri);        //浏览网页
startActivity(intent);                                   //启动调用系统UI
```

（2）Data：要实施的具体数据，一般由一个Uri变量表示。Data属性用来向Action提供用于操作的数据。setData语句令Data接收一个Uri对象，Uri对象内容通常为这个格式：<scheme>://<host>:<port>/<path>，也就是：<协议>://<主机名>:<端口号>/<路径>。

（3）Category：指定将要执行的Action的其他一些额外信息，例如，LAUNCHER_CATEGORY表示Intent的接受者应该在Launcher中作为顶级应用出现；而ALTERNATIVE_CATEGORY表示当前的Intent是一系列可选动作中的一个，这些动作可在同一块数据上执行。详细使用方法可参考Android SDK–>reference中的Android.content.intent类。

（4）Type：显式地指定Intent的媒体类型（Multipurpose Internet Mail Extensions，MIME）。一般Intent的数据类型能够根据数据本身进行判定，但是通过设置这个属性，可强制采用显式指定的类型而不再进行推导。MIME Type包括许多文件类型，包括图片、视频、音频等，如纯文本（text/plain）、GIF图像（image/gif）、JPEG图像（image/jpeg）、MPEG动画（video/mpeg）、MP3音乐（audio/mp3）、类型为image的所有文件（image/*）、类型为audio的所有文件（audio/*）和类型为video的所有文件（video/*）。下面是通过setType获取手机内指定格式文件的例子：

```
Intent intent=new Intent(Intent.ACTION_GET_CONTENT);
intent.setType("image/*");//指定类型为image的所有文件
startActivity(intent);
```

（5）Component：指定Intent的目标组件的类名称。通常Android会根据Intent中包含的其他属性信息（如Action、Data/Type、Category）进行查找，最终找到一个与之匹配的目标组件。但是，如果指定component属性，将直接使用它指定的组件，而不再执行上述查找过程。指定了这个属性以后，Intent的其他所有属性都是可选的。

视频

Android UI设计

（6）Extras（附加信息）：是其他所有附加信息的集合。Extras属性主要用于传递目标组件所需要的额外的数据。通过putExtras()方法设置。使用Extras可以为组件提供扩展信息，例如，如果要执行“发送电子邮件”这个动作，可以将电子邮件的标题、正文等保存在Extras里，传给电子邮件发送组件。

5. Intent典型应用案例

（1）利用Intent实现Activity间互相切换。该程序提供两个按钮，分别位于两个Activity中，通过这两个按钮实现两个Activity之间的互相切换。该程序的界面布局如图1-2和图1-3所示。

图 1-2　Intent 案例主活动布局

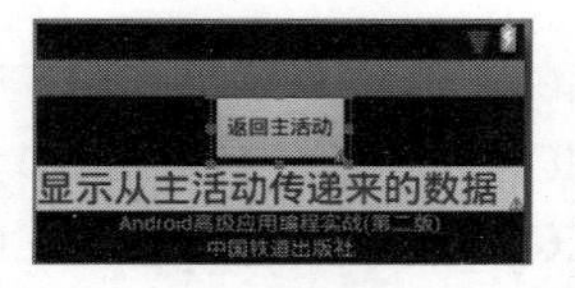

图 1-3　Intent 案例第二个活动布局

该程序的activity_main.xml主要布局代码如下：

```
<LinearLayout xmlns:android="http://schemas.android.com/apk/res/android"
    android:layout_width="fill_parent"
    android:layout_height="fill_parent"
    android:gravity="center_horizontal"
    android:orientation="vertical">
    <Button
        android:id="@+id/btnFirst"
        android:layout_width="wrap_content"
        android:layout_height="wrap_content"
        android:text="切换到第二个活动"
        android:textSize="30sp"/>
    <TextView
        android:id="@+id/DisplayData"
        android:layout_width="fill_parent"
        android:layout_height="wrap_content"
        android:background="#00FF00"
        android:gravity="center_horizontal"
        android:text="将数据从主活动传递第二个活动"
        android:textColor="#0000FF"
        android:textSize="20sp"/>
</LinearLayout>
```

activity_second.xml布局主要程序清单：

```
<LinearLayout xmlns:android="http://schemas.android.com/apk/res/android"
    android:layout_width="fill_parent"
    android:layout_height="fill_parent"
    android:orientation="vertical" >
```

```
    <Button
        android:id="@+id/btnSecond"
        android:layout_width="wrap_content"
        android:layout_height="wrap_content"
        android:layout_gravity="center"
        android:text="返回主活动" />
    <TextView
        android:id="@+id/DisplayData"
        android:layout_width="fill_parent"
        android:layout_height="wrap_content"
        android:background="#00FF00"
        android:text="显示从主活动传递来的数据"
        android:textColor="#0000FF"
        android:textSize="25sp" />
    <TextView
        android:layout_width="match_parent"
        android:layout_height="wrap_content"
        android:gravity="center_horizontal"
        android:text="@string/book_str" />
    <TextView
        android:layout_width="match_parent"
        android:layout_height="wrap_content"
        android:gravity="center_horizontal"
        android:text="@string/copyright_str" />
</LinearLayout>
```

该程序MainActivity类的主要Java代码如下：

```
//从活动Activity父类派生子类MainActivity，实现单击监听器接口
public class MainActivity extends Activity implements OnClickListener{
    private static final String TAG="MainActivity";  //定义私有静态字符串常量
    private Button btnFirst=null;  //声明按钮对象
    protected void onCreate(Bundle savedInstanceState){//重写onCreate()方法
        super.onCreate(savedInstanceState);  //调用基类onCreate()方法
        setContentView(R.layout.activity_main);  //设置手机主布局界面
        btnFirst=(Button)findViewById(R.id.btnFirst);  //从布局文件中查找按钮
        btnFirst.setOnClickListener(this);  //设置按钮单击事件监听器}
    public void onClick(View v){  //TODO 自动生成的方法存根
        switch(v.getId()){
        case R.id.btnFirst:
            Intent intent=new Intent();  //构建意图对象
            //显式方式声明Intent，直接启动SecondActivity，设置意图切换的目标类
            intent.setClass(MainActivity.this,SecondActivity.class);
            //注意putExtra()键"data01"与getStringExtra("data01")保持一致
```

```
            //通过意图传递数据（键值对）
            intent.putExtra("data01", "https://www.sziit.edu.cn");
            startActivity(intent);  //启动Activity
            break;}}}
```

程序SecondActivity类的主要Java代码清单如下：

```
//从活动Activity父类派生子类SecondActivity,实现单击监听器接口
public class SecondActivity extends Activity implements OnClickListener{
    private static final String TAG="SecondActivity";  //定义私有静态字符串常量
    private TextView textView=null;  //声明文本视图对象
    private Button btnSecond=null;  //声明按钮对象
    protected void onCreate(Bundle savedInstanceState){  //重写onCreate()方法
        super.onCreate(savedInstanceState);  //调用基类onCreate()方法
        setContentView(R.layout.activity_second);  //设置手机第二个布局界面
        btnSecond=(Button)findViewById(R.id.btnSecond);  //从布局中查找按钮
        btnSecond.setOnClickListener(this);  //设置按钮单击有名监听器
        textView=(TextView)findViewById(R.id.DisplayData);
        Intent intent=getIntent();  //活动意图对象
        //从意图对象中读取键"data01"所对应的值
        String value=intent.getStringExtra("data01");
        textView.setText("从意图获得的数据: "+value);  //将数据显示在TextView上
    public void onClick(View v){  //TODO 自动生成的方法存根
        switch(v.getId()){
        case R.id.btnSecond:
            Intent intent=new Intent();  //构建意图对象
            //设置意图切换的目标类
            intent.setClass(SecondActivity.this, MainActivity.class);
            startActivity(intent);  //启动Activity
            break;}}}
```

该程序运行效果如图1-4和图1-5所示。

（2）通过Intent调用系统现有照相机应用拍照。程序提供一个按钮，用户单击该按钮时会通过Intent调用系统现有照相机应用拍照，默认会把图片保存到系统图库的目录下。MediaStore.ACTION_IMAGE_CAPTURE为系统Intent的Action类型，从现有照相机应用请求一张相片。该程序的运行效果如图1-6所示。

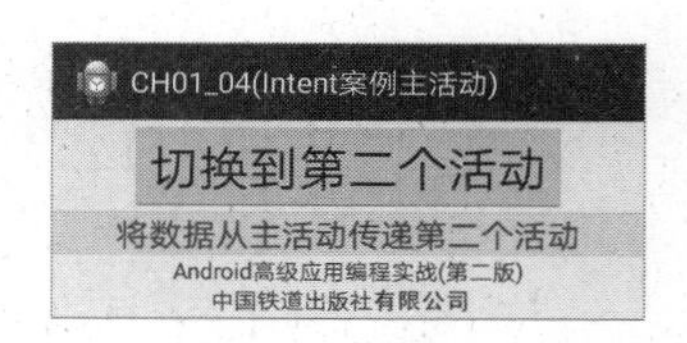

图1-4　Intent案例运行主界面

图1-5　Intent案例运行次界面

图1-6　通过Intent调用系统拍照

该程序的界面布局activity_main.xml代码如下：

```
<LinearLayout xmlns:android="http://schemas.android.com/apk/res/android"
    xmlns:tools="http://schemas.android.com/tools"
    android:layout_width="match_parent"
    android:layout_height="match_parent"
    android:orientation="vertical"
    tools:context=".MainActivity">
    <TextView
        android:layout_width="match_parent"
        android:layout_height="wrap_content"
        android:background="#0000FF"
           android:text="通过Intent调用系统现有照相机应用拍照案例:\nandroid.media.
action.IMAGE_CAPTURE"
        android:textColor="#FFFFFF"
        android:textSize="16sp"/>
    <Button
        android:id="@+id/btnTakePhoto"
        android:layout_width="match_parent"
        android:layout_height="wrap_content"
        android:text="拍照"/>
    <FrameLayout
        android:layout_width="match_parent"
        android:layout_height="match_parent">
        <ImageView
            android:id="@+id/ivPhoto"
            android:layout_width="match_parent"
            android:layout_height="match_parent"/>
    </FrameLayout>
</LinearLayout>
```

程序MainActivity.java的Java代码如下：

```
public class MainActivity extends Activity{
    private static final String TAG="MainActivity";
    private static final int TAKE_PHOTO=1;  //定义请求代码常量
    private Button btnTakePhoto=null;  //定义私有按钮对象
    private ImageView ivPhoto=null;  //定义私有图像视图对象
    protected void onCreate(Bundle savedInstanceState){
        super.onCreate(savedInstanceState);  //调用基类onCreate()方法
        setContentView(R.layout.activity_main);  //设置活动布局
        ivPhoto=(ImageView)findViewById(R.id.ivPhoto);
        btnTakePhoto=(Button)findViewById(R.id.btnTakePhoto);
        //设置按钮单击事件监听器
        btnTakePhoto.setOnClickListener(new OnClickListener(){
            public void onClick(View v){  //处理单击事件回调函数
```

```
                doImageCapture();}});}
            //启动系统现有照相机应用拍照
            private void doImageCapture(){
            //通过Intent调用系统现有照相机应用拍照，默认会把图片保存到系统图库的目录下
            startActivityForResult(new ntent(MediaStore.ACTION_IMAGE_CAPTURE),
            TAKE_PHOTO);}  //调用系统拍照功能
        protected void onActivityResult(int requestCode, int resultCode, Intent data)
    {super.onActivityResult(requestCode, resultCode, data);
            showPhoto(requestCode, resultCode, data);}
            //将相片显示到图像视图上，图片保存在系统默认相册路径
        private void showPhoto(int requestCode, int resultCode, Intent data){
            if(requestCode==TAKE_PHOTO){     //处理特定请求代码
                if(resultCode==RESULT_OK){  //处理正确结果代码
                //从系统返回相片，将相片显示在ImageView组件上
                    Bitmap bm=(Bitmap)data.getExtras().get("data");
                    ivPhoto.setImageBitmap(bm);}}}}
```

1.2.2 Service的应用

1. Service的两种模式

Service包括本地服务（Local Service）和远程服务（Remote Service）两种模式。

本地服务用于应用程序内部；远程服务用于Android系统内部的应用程序之间。前者用于实现应用程序自己的一些耗时任务（如查询升级信息），并不占用应用程序（如Activity所属线程），而是单独开启线程后台执行，这样用户体验比较好。

2. Service的生命周期

Service的生命周期（见图1-7）并不像Activity那么复杂，它只继承了onCreate()、onStart()和onDestroy()这3个方法，当第一次启动Service时，先后调用onCreate()和onStart()两个方法，当停止Service时，则执行onDestroy()方法，这里需要注意的是，如果Service已经启动了，当再次启动Service时，不会再执行onCreate()方法，而是直接执行onStart()方法。

（1）启动服务：Context.startService(Intent service, Bundle b)。

（2）绑定服务：Context.bindService(Intent service, ServiceConnection c,int flag)。

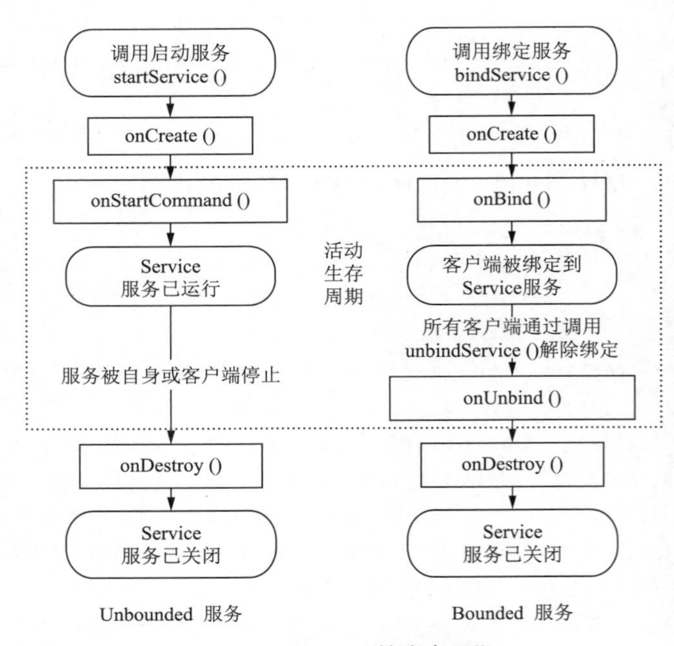

图1-7 Service的生命周期

（3）启动服务：Context.startService(Intent service, Bundle b)。

（4）绑定服务：Context.bindService(Intent service, ServiceConnection c, int flag)。

启动服务相当于告诉Android平台，在后台启动该Service并保持其运行，该服务与任何其他Activity或应用程序没有任何特定连接。启动服务和关闭服务流程如图1-8和图1-9所示。

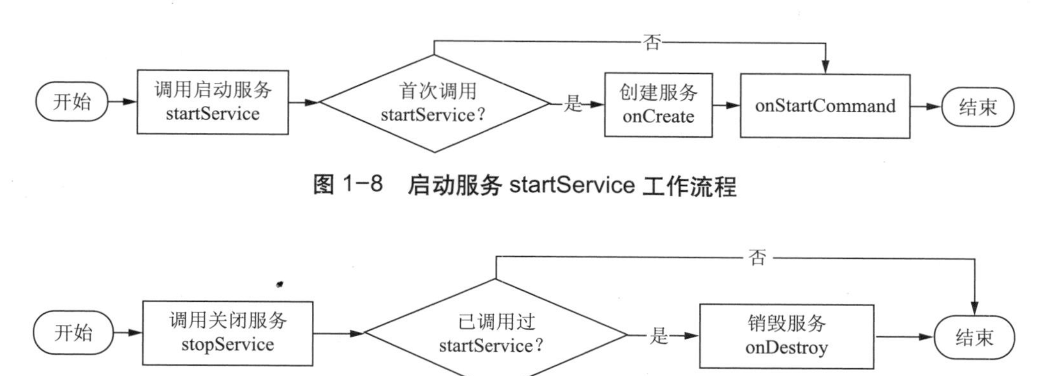

图 1-8　启动服务 startService 工作流程

图 1-9　关闭服务 stopService 工作流程

绑定服务是指获得远程对象的句柄并从Activity中调用已定义的方法。因为每个Android应用程序都在其自己的进程中运行，所以使用已绑定的Service可以在不同进程之间传递数据。绑定服务和解绑服务工作流程如图1-10和图1-11所示。

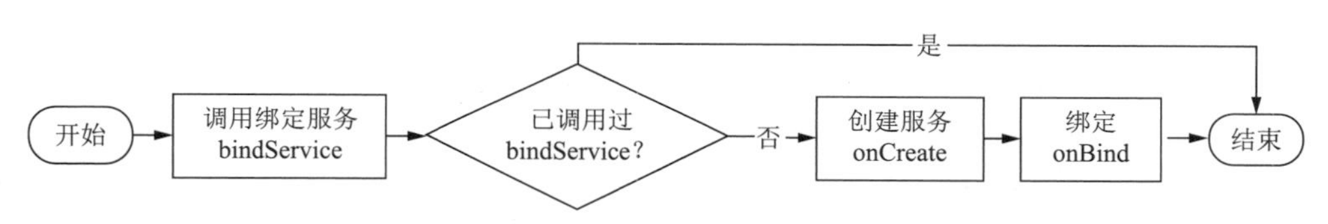

图 1-10　绑定服务 bindService 工作流程

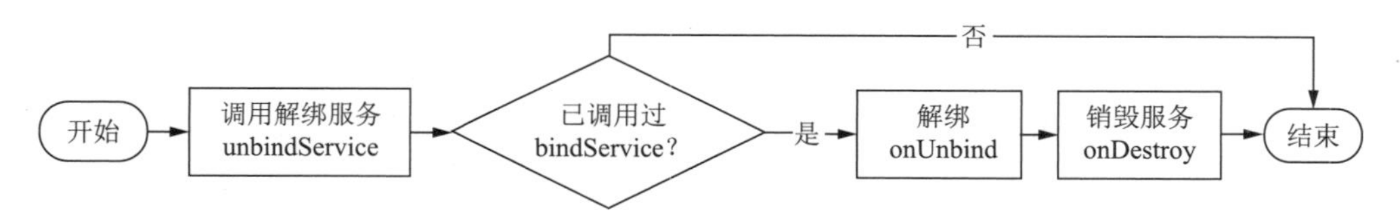

图 1-11　解绑服务 unbindService 工作流程

（1）被启动服务的生命周期。如果一个Service被Context.startService()方法启动，那么不管是否有活动绑定到该Service，该Service都在后台运行。在这种情况下，如果需要该Service的onCreate()方法将被调用，然后是onStart()方法。如果一个Service被启动不止一次，则onStart()方法将被多次调用，但是不会创建该Service的其他实例。该Service将一直在后台运行，直到被Context.stopService()方法或其自己的stopSelf()方法显式地停止该服务为止。

（2）被绑定服务的生命周期。如果一个Service被某个Activity调用Context.bindService()方法绑定，则只要该连接被建立，则该服务将一直运行。Activity可以使用Context建立到服务的连接，同时也要负责关闭该连接。当Service只是以这种方式被绑定而未被启动时，其onCreate()方法被调用，但是onStart()方法不被调用。在这种情况下，当绑定被解除时，平台可以停止和清除该

Service。

（3）被启动又被绑定服务的生命周期。如果一个Service既被启动又被绑定，则该Service基本上将在后台保持运行，与被启动服务的生命周期类似，唯一的区别在于生命周期本身。因为既启动服务又绑定服务，所以onStart()和onCreate()两个方法都将被调用。

（4）当服务被停止时清除服务。当一个Service被停止时，onDestory()方法或者在服务被启动以后被显式地调用，或者当不再有绑定的连接（没有被启动）时被隐式地调用。在onDestory()方法中，每个Service都应该执行最终的清除功能，如停止任何已生成的进程等。

视频

Android服务Service典型应用实战

（5）Service与Activity通信。通常每个应用程序都在它自己的进程内运行，但有时需要在进程间传递对象，可以通过应用程序UI的方式写一个运行在不同进程中的Service。在Android平台中，一个进程通常不能访问其他进程中的内存区域。所以，它们需要把对象拆分成操作系统能理解的简单形式，以便伪装成对象跨越边界访问。

3. 启动Service的两种方法

绑定Activity和Service的方式启动服务，通过bindService()方法可以将Activity和Service绑定。bindService()方法的定义如下：

```
Boolean bindService(Intent service,ServiceConnection conn,int flags)
```

该方法的第一个参数表示与服务类相关联的Intent对象，第二个参数是一个ServiceConnection类型的变量，负责连接Intent对象指定的服务。通过ServiceConnection对象可以获得连接成功或失败的状态，并可以获得连接后的服务对象。第三个参数是一个标志位，一般设为Context.BIND_AUTO_CREATE。

下面编写CustomService类，在该类中增加了几个与绑定相关的事件方法。

```
public class CustomService extends Service{
    private static final String TAG="CustomService";
    public class LocalBinder extends Binder{
        String strDataFromService="这是来自后台服务的数据！";
        public void startUploadData(){Log.d(TAG, "startUploadData()");
            //执行具体数据上传任务}
        CustomService getService(){
            Log.d(TAG, "getService()");return CustomService.this;}}
    private final IBinder mBinder=new LocalBinder();
    public IBinder onBind(Intent intent){  //重写onBind()方法
        Log.d(TAG, "onBind()");return mBinder;}
    public void onCreate(){  //重写服务onCreate()方法
        super.onCreate();  //调用服务基类onCreate()方法
        Log.d(TAG, "onCreate()");}
    public void onDestroy(){  //重写服务onDestroy()方法
        super.onDestroy();  //调用服务基类onDestroy()方法
    Log.d(TAG, "onDestroy()");}
    public void onStart(Intent intent, int startId){  //重写服务onStart()方法
        super.onStart(intent, startId);  //调用服务基类onStart()方法
        Log.d(TAG, "onStart()");}
```

```
    public int onStartCommand(Intent intent, int flags, int startId){
        //重写onStartCommand()方法
        Log.d(TAG, "onStartCommand()");
        return super.onStartCommand(intent, flags, startId);}
    public boolean onUnbind(Intent intent){  //重写onUnbind()方法
        Log.d(TAG, "onUnbind()");
        return super.onUnbind(intent);}}
```

现在定义一个CustomService变量和一个ServiceConnection变量，代码如下：

```
private CustomService mService;
private ServiceConnection sc=new ServiceConnection()
{  //连接服务失败后，该方法被调用
    public void onServiceDisconnected(ComponentName name){
        //只在service因异常而断开连接时，该方法才会用到
        sc=null;}
    //成功连接服务后，该方法被调用。在该方法中可以获得CustomService对象
    public void onServiceConnected(ComponentName name,IBinder service)
    {   //获得CustomService对象
         mService=((CustomService.LocalBinder)service).getService();
         String recStr=((CustomService.LocalBinder)service)
.strDataFromService;}};
```

最后使用bindService()方法绑定Activity和Service，代码如下：

```
bindService(intent,sc,Context.BIND_AUTO_CREATE);  //绑定服务
```

如果想解除绑定，可以使用下面的代码：

```
unbindService(sc);  //解绑服务
```

在CustomService类中定义了一个LocalBinder类，该类实际上是为了获得CustomService对象的实例。在ServiceConnection接口的onServiceConnected()方法的第二个参数是一个IBinder类型的变量，将该参数转换成CustomService.LocalBinder对象，并使用CustomService类中的getService()方法获得CustomService对象。在获得CustomService对象后，就可以在Activity中随意操作CustomService。

直接启动Service的调用方法startService()定义如下：

```
public ComponentName startService(Intent service)
```

在Activity中启动Service，先编写一个服务类CustomService并在AndroidManifest.xml文件中配置CustomService。该CustomService类代码如下：

```
public class CustomService extends Service{
    private static final String TAG="CustomService";
    public void onCreate(){
        super.onCreate();
        Toast.makeText(this, "服务已被创建...", Toast.LENGTH_LONG).show();
```

```
        Log.d(TAG, "onCreate()->服务已被创建...");}
    public void onStart(Intent intent, int startId){
        super.onStart(intent, startId); Log.d(TAG, "onStart()");}
    public int onStartCommand(Intent intent, int flags, int startId){
        Toast.makeText(this, "服务已经启动...", Toast.LENGTH_SHORT).show();
        Log.d(TAG, "onStartCommand()->服务已经启动...");
        return super.onStartCommand(intent, flags, startId);}
    public void onDestroy(){
        super.onDestroy();
        Toast.makeText(this, "服务已被删除...", Toast.LENGTH_SHORT).show();
        Log.d(TAG, "onDestroy()->服务已被删除...");}
    public IBinder onBind(Intent arg0){
        Log.d(TAG, "onBind()");return null;}}
```

下面完成启动服务关键一步就是建立Activity，调用startService()方法，其代码如下：

```
public class MainActivity extends Activity{
    private Intent intent;
    protected void onCreate(Bundle savedInstanceState){
        super.onCreate(savedInstanceState);
        setContentView(R.layout.activity_main);  }
    public void doClick(View view){
        switch(view.getId()){
        case R.id.btnStart:
            intent=new Intent(MainActivity.this,CustomService.class);
            startService(intent);break;  //启动服务
        case R.id.btnStop:
            stopService(intent);break;  //关闭服务}}}
```

如果以上两个案例安装后，绑定服务或启动服务出现异常，最大的可能是没有在AndroidManifest.xml文件中配置Service。配置Service的代码如下：

```
  <service android:name=".CustomService"></service>
```

4. Service典型应用案例

（1）启动服务Service的两种方法。程序中包括4个按钮，分别用于启动Service、关闭Service、绑定Service和解绑Service。通过该实例，让用户掌握启动服务Service的两种具体调用方法。该程序的界面布局大纲如图1-12所示，其运行结果如图1-13所示。该程序的界面布局代码activity_main.xml如下：

```
<LinearLayout mlns:android="http://schemas.android.com/apk/res/android"
    android:layout_width="fill_parent"
    android:layout_height="fill_parent"
    android:orientation="vertical">
    <LinearLayout
        android:layout_width="match_parent"
```

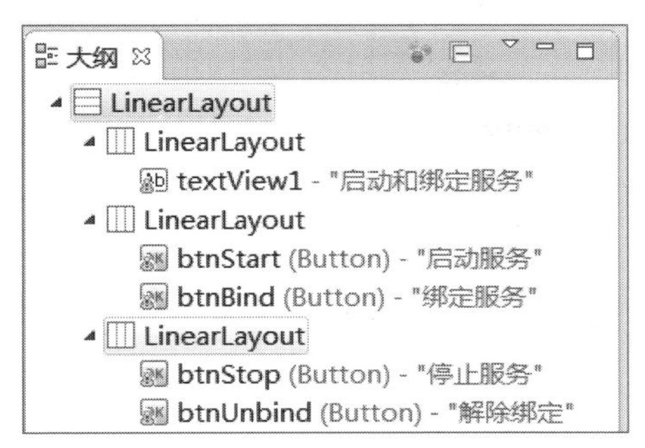

图 1-12 启动 Service 两种方法布局大纲

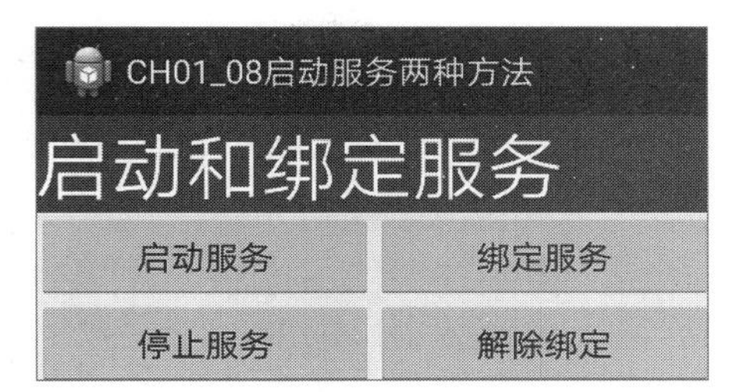

图 1-13　启动 Service 两种方法运行结果

```
        android:layout_height="wrap_content">
        <TextView
            android:id="@+id/textView1"
            android:layout_width="match_parent"
            android:layout_height="wrap_content"
            android:background="#0000FF"
            android:text="启动和绑定服务"
            android:textColor="#FFFFFF"
            android:textSize="40sp"/>
    </LinearLayout>
    <LinearLayout
        android:layout_width="match_parent"
        android:layout_height="wrap_content"
        android:orientation="horizontal">
        <Button
            android:id="@+id/btnStart"
            android:layout_width="wrap_content"
            android:layout_height="wrap_content"
            android:onClick="doClick"
            android:layout_weight="1"
            android:text="启动服务"/>
        <Button
            android:id="@+id/btnBind"
            android:layout_width="wrap_content"
            android:layout_height="wrap_content"
            android:onClick="doClick"
            android:layout_weight="1"
            android:text="绑定服务"/>
    </LinearLayout>
    <LinearLayout
        android:layout_width="match_parent"
        android:layout_height="wrap_content"
```

```
        android:orientation="horizontal">
        <Button
            android:id="@+id/btnStop"
            android:layout_width="wrap_content"
            android:layout_height="wrap_content"
            android:layout_weight="1"
            android:onClick="doClick"
            android:text="停止服务"/>
        <Button
            android:id="@+id/btnUnbind"
            android:layout_width="wrap_content"
            android:layout_height="wrap_content"
            android:layout_weight="1"
            android:onClick="doClick"
            android:text="解除绑定"/>
    </LinearLayout>
</LinearLayout>
```

从服务基类派生CustomService服务子类的Java代码如下：

```
public class CustomService extends Service{
    public static final String TAG="CustomService";
    private CustomBinder mBinder=new CustomBinder();
    class CustomBinder extends Binder{
        public void startUploadData(){
            Toast.makeText(CustomService.this, "startUploadData()被执行!",
        Toast.LENGTH_SHORT).show();Log.d(TAG, "startUploadData()");
            //执行具体数据上传任务}}
    public void onCreate(){  //重写服务onCreate()方法
        super.onCreate();  //调用服务基类onCreate()方法
        Toast.makeText(CustomService.this, "onCreate()被执行",
            Toast.LENGTH_SHORT).show();
        Toast.makeText(CustomService.this, "CustomService线程is "
+Thread.currentThread().getId(), Toast.LENGTH_SHORT).show();
Log.d(TAG, "onCreate()");}
    public int onStartCommand(Intent intent, int flags, int startId){  //重写
                                                    //onStartCommand()方法
        Toast.makeText(CustomService.this, "onStartCommand()被执行",
            Toast.LENGTH_SHORT).show();Log.d(TAG, "onStartCommand()");
        return super.onStartCommand(intent, flags, startId);  //调用服务基类
                                                    //onStartCommand()方法
    }
    public void onDestroy(){  //重写服务onDestroy()方法
        super.onDestroy();  //调用服务基类onDestroy()方法
```

```
        Toast.makeText(CustomService.this, "onDestroy()被执行",
            Toast.LENGTH_SHORT).show();Log.d(TAG, "onDestroy()");}
    public IBinder onBind(Intent intent){  //重写onBind()方法
        Toast.makeText(CustomService.this, "onBind()...", Toast.LENGTH_LONG)
    .show(); Log.d(TAG, "onBind()");
        return mBinder;}}
```

从活动基类派生子类MainActivity程序清单：

```
public class MainActivity extends Activity{
    private static final String TAG="MainActivity";
    private CustomBinder mBinder;
    //创建服务连接器对象
    private ServiceConnection sc=new ServiceConnection(){
        public void onServiceDisconnected(ComponentName name){
          sc=null;
             Toast.makeText(MainActivity.this, "onServiceDisconnected:
ServiceConnection --->"+sc,Toast.LENGTH_LONG).show();}
        public void onServiceConnected(ComponentName name, IBinder service)
{ mBinder=(CustomBinder)service;
            mBinder.startUploadData();} };
    /** 当活动第一次创建时被调用 */
    protected void onCreate(Bundle savedInstanceState){  //重写onCreate()方法
        super.onCreate(savedInstanceState);  //调用基类onCreate()方法
        setContentView(R.layout.activity_main);  //设置活动布局(UI)
        Toast.makeText(MainActivity.this, "MainActivity线程id是"+
Thread.currentThread().getId(),Toast.LENGTH_SHORT).show();
        Log.d(TAG, "onCreate()");  //写调试日志}
    public void doClick(View v){
        switch(v.getId()){
        case R.id.btnStart: //启动服务按钮ID
          Log.d("MainActivity","单击启动服务按钮");
            Intent startIntent=new Intent(this, CustomService.class);
            startService(startIntent); //调用启动服务方法
            break;
        case R.id.btnStop:  //停止服务按钮ID
            Log.d("MainActivity", "单击停止服务按钮");
            Intent stopIntent=new Intent(this, CustomService.class);
            stopService(stopIntent);  //调用停止服务方法
            break;
        case R.id.btnBind:  //绑定服务按钮ID
          Log.d("MainActivity","单击绑定服务按钮");
            Intent bindIntent=new Intent(this,CustomService.class);
            bindService(bindIntent,sc,BIND_AUTO_CREATE);  //调用绑定服务方法
```

```
            break;
        case R.id.btnUnbind:  //解除绑定服务按钮ID
            Log.d("MainActivity","单击解除绑定按钮！");
            unbindService(sc);  //调用解除绑定服务方法
            break;}}}
```

（2）启动Service播放背景音乐。程序通过在Activity中启动Service在后台播放背景音乐，程序的界面布局大纲如图1-14所示。通过DDMS可以观察到服务启动和停止过程。

图 1-14　启动 Service 播放背景音乐的界面布局大纲

程序的界面布局activity_main.xml代码如下：

```
<LinearLayout xmlns:android="http://schemas.android.
com/apk/res/android"
    android:layout_width="fill_parent"
    android:layout_height="fill_parent"
    android:orientation="vertical">
    <TextView
        android:layout_width="match_parent"
        android:layout_height="wrap_content"
        android:layout_marginTop="10dp"
        android:gravity="center"
        android:text="播放音乐服务"
        android:textColor="#FFFFFF"
        android:background="#0000FF"
        android:textSize="36sp"/>
    <LinearLayout
        android:layout_width="match_parent"
        android:layout_height="wrap_content"
        android:orientation="horizontal">
        <Button
            android:id="@+id/btnPlay"
            android:layout_width="wrap_content"
            android:layout_height="wrap_content"
            android:background="#00FF00"
            android:layout_weight="1"
            android:textSize="18sp"
            android:text="播放音乐"/>
        <Button
            android:id="@+id/btnPause"
            android:layout_width="wrap_content"
            android:layout_height="wrap_content"
            android:background="#00FF00"
            android:layout_weight="1"
            android:textSize="18sp"
```

```
            android:text="暂停音乐"/>
        <Button
            android:id="@+id/btnStop"
            android:layout_width="wrap_content"
            android:layout_height="wrap_content"
            android:background="#00FF00"
            android:layout_weight="1"
            android:textSize="18sp"
            android:text="停止音乐"/>
    </LinearLayout>
    <LinearLayout
        android:layout_width="match_parent"
        android:layout_height="wrap_content"
        android:orientation="horizontal">
    <Button
        android:id="@+id/btnClose"
        android:layout_width="wrap_content"
        android:layout_height="wrap_content"
        android:background="#00FF00"
        android:textSize="18sp"
         android:layout_weight="1"
        android:text="关闭服务"/>
    <Button
        android:id="@+id/exitBtn"
        android:layout_width="wrap_content"
        android:layout_height="wrap_content"
        android:background="#00FF00"
        android:textSize="18sp"
         android:layout_weight="1"
        android:text="退出程序"/>
     </LinearLayout>
</LinearLayout>
```

播放后台背景音乐服务子类MusicService程序的Java代码如下：

```
public class MusicService extends Service{
    private MediaPlayer mediaPlayer;
    public IBinder onBind(Intent arg0){
        return null;}
    public void onCreate(){
     if(mediaPlayer==null){
        mediaPlayer=MediaPlayer.create(this, R.raw.night);
        mediaPlayer.setLooping(false);}}
    public void onDestroy(){
```

```
        Toast.makeText(this, "多媒体播放器已被停止!", Toast.LENGTH_SHORT).show();
        if(mediaPlayer!=null){
            mediaPlayer.stop();
            mediaPlayer.release();}}
    public int onStartCommand(Intent intent, int flags, int startId){
       if(intent!=null){
       Bundle bundle=intent.getExtras();
       if(bundle!=null){
          int op=bundle.getInt("op");
          switch(op){
          case 1:play();    break;
          case 2:pause();   break;
          case 3:stop();    break;}}}
       return super.onStartCommand(intent, flags, startId);}
    public void play(){
        if(!mediaPlayer.isPlaying()){
           mediaPlayer.start();}}
    public void pause(){
        if(mediaPlayer!=null && mediaPlayer.isPlaying()){
           mediaPlayer.pause();}}
    public void stop(){
       if(mediaPlayer!=null){
            mediaPlayer.stop();
            //在调用stop()后如果需要再次通过start()进行播放,需要调用prepare()函数
            try{mediaPlayer.prepare();
        } catch(IOException ex){ex.printStackTrace();}}}}
```

启动服务MusicService的MainActivity类程序清单如下:

```
public class MainActivity extends Activity implements OnClickListener{
    private Button btnPlay;     private Button btnPause;
    private Button btnStop;     private Button btnClose;
    private Button exitBtn;     private Intent intent;
    public void onCreate(Bundle savedInstanceState){
        super.onCreate(savedInstanceState);
        setContentView(R.layout.activity_main);findView();  //初始化组件
        setClickListener();  //设置组件单击事件监听器}
    private void findView(){  //初始化组件
        btnPlay=(Button)findViewById(R.id.btnPlay);
        btnPause=(Button)findViewById(R.id.btnPause);
        btnStop=(Button)findViewById(R.id.btnStop);
        btnClose=(Button)findViewById(R.id.btnClose);
        exitBtn=(Button)findViewById(R.id.exitBtn);}
    private void setClickListener()  //设置组件单击事件监听器
```

```
        btnPlay.setOnClickListener(this);
        btnPause.setOnClickListener(this);
        btnStop.setOnClickListener(this);
        btnClose.setOnClickListener(this);
        exitBtn.setOnClickListener(this);}
    public void onClick(View v){
        int op=-1;intent=new Intent("MusicService");
        switch(v.getId()){
        case R.id.btnPlay: op=1;break;   //播放音乐
        case R.id.btnPause:op=2;break;   //暂停音乐
        case R.id.btnStop: op=3;break;   //停止音乐
        case R.id.btnClose:stopService(intent);break;  //停止服务
        case R.id.exitBtn:this.finish();break;  //关闭活动}
        Bundle bundle=new Bundle();
        bundle.putInt("op", op);intent.putExtras(bundle);
        startService(intent); //启动服务}
    public void onDestroy(){super.onDestroy();
        if(intent != null){stopService(intent);//关闭服务}}}
```

该程序的运行结果如图 1-15 所示。

图 1-15　启动 Service 播放背景音乐运行效果

1.3 案例

1.3.1　Android 应用程序闪屏页面

Android应用程序通常在启动时会显示闪屏页面，显示闪屏页面有几个好处：显示与程序相关的一些介绍，在闪屏页面显示过程中，初始化程序所需要的资源，避免在初始化资源过程中出现卡顿等现象，改善用户体验。可以采用Intent和AnimationUtils实现闪屏功能。程序的界面布局activity_main.xml代码如下：

```
<?xml version="1.0"encoding="utf-8"?>
<LinearLayout xmlns:android="http://schemas.android.com/apk/res/android"
    android:layout_width="fill_parent"
    android:layout_height="fill_parent"
```

```
    android:layout_gravity="center_vertical"
    android:background="@drawable/splash"
    android:orientation="vertical"
    android:paddingBottom="7.0dip"
    android:paddingLeft="7.0dip"
    android:paddingRight="7.0dip"
    android:paddingTop="7.0dip">
</LinearLayout>
```

闪屏SplashActivity类的Java代码如下：

```
//从活动Activity父类派生子类SplashActivity
public class SplashActivity extends Activity{
    //Handler对象主要接收子线程发送的数据，并用此数据配合主线程更新UI
    private Handler mHandler=new Handler();
    private final static int SPLASH_TIME=500;  //定义闪屏持续时间
    private static final String TAG="SplashActivity";
    public void onCreate(Bundle savedInstanceState){  //重写onCreate()方法
        super.onCreate(savedInstanceState);  //调用基类方法
        Log.d(TAG, "SplashActivity->onCreate!");
        //setFullScreen();
        View view=View.inflate(this,R.layout.activity_splash, null);
        setContentView(view);  //加载闪屏页面的布局文件
        Animation animation=AnimationUtils.loadAnimation(this, R.anim.myanim);
        //定义闪屏动画
        view.startAnimation(animation);
        animation.setAnimationListener(new AnimationListener(){  //设置监听器
            public void onAnimationStart(Animation arg0){}
            public void onAnimationRepeat(Animation arg0){}
            public void onAnimationEnd(Animation arg0){  //处理动画播放结束方法
                mHandler.postDelayed(new Runnable(){  //添加到消息队列
                    public void run(){  //跳转到主界面
                        goMain();}}, SPLASH_TIME);}}); }
    /***跳转到主窗体**/
    private void goMain(){
        startActivity(new Intent(this,MainActivity.class));
        Log.d(TAG, "跳转到主窗体!");};
    /***设置全屏**/
    private void setFullScreen(){getWindow().setFlags(WindowManager.LayoutParams.
FLAG_FULLSCREEN,WindowManager.LayoutParams.FLAG_FULLSCREEN);Log.d(TAG,
"设置全屏!");}
    /***退出全屏**/
    private void quitFullScreen(){
        final WindowManager.LayoutParams attrs=getWindow().getAttributes();
```

```
        attrs.flags &=(~WindowManager.LayoutParams.FLAG_FULLSCREEN);
        getWindow().setAttributes(attrs);getWindow().clearFlags(WindowManager.
LayoutParams.FLAG_LAYOUT_NO_LIMITS);
        Log.d(TAG, "退出全屏!");}}
```

从活动Activity父类派生子类MainActivity程序清单如下：

```
public class MainActivity extends Activity{
    private static final String TAG="MainActivity";
    protected void onCreate(Bundle savedInstanceState){  //重写onCreate()方法
        super.onCreate(savedInstanceState);  //调用基类onCreate()方法
        setContentView(R.layout.activity_main);  //设置屏幕布局
        Log.d(TAG, "MainActivity->onCreate!"); }}
```

该程序的运行结果如图1-16所示。

图 1-16　闪屏页面运行结果

1.3.2　服务器 / 客户端通信中的心跳包功能

在开发一些服务器/客户端程序时，需要定时与服务器通信或定时执行一些本地任务，在程序退出时，如果有消息更新，需要及时唤醒程序并对程序的界面进行更新，这就需要使用到Android中的Service服务来实现此功能。程序的界面布局大纲如图1-17所示。程序的界面布局代码activity_main.xml如下：

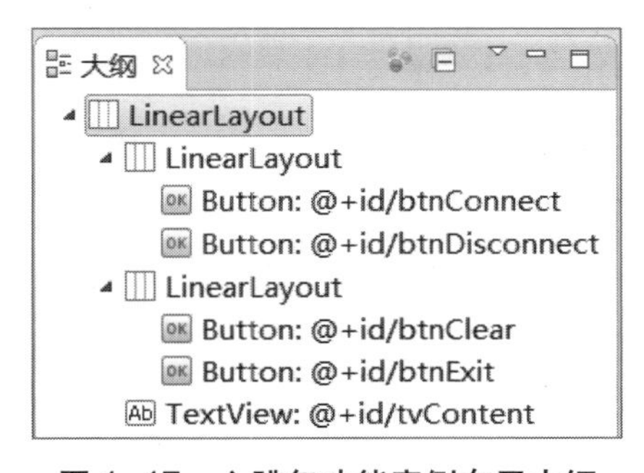

图 1-17　心跳包功能案例布局大纲

```
<LinearLayout xmlns:android="http://schemas.android.com/apk/res/android"
    xmlns:tools="http://schemas.android.com/tools"
    android:layout_width="match_parent"
    android:layout_height="match_parent"
    android:orientation="vertical"
    android:paddingBottom="@dimen/activity_vertical_margin"
    android:paddingLeft="@dimen/activity_horizontal_margin"
    android:paddingRight="@dimen/activity_horizontal_margin"
    android:paddingTop="@dimen/activity_vertical_margin"
    tools:context=".MainActivity">
    <LinearLayout
        android:layout_width="match_parent"
        android:layout_height="wrap_content"
        android:orientation="horizontal">
        <Button
            android:id="@+id/btnConnect"
            android:layout_width="wrap_content"
            android:layout_height="wrap_content"
            android:text="@string/connect"/>
        <Button
            android:id="@+id/btnDisconnect"
            android:layout_width="wrap_content"
            android:layout_height="wrap_content"
            android:text="@string/disconnect"/>
    </LinearLayout>
    <LinearLayout
        android:layout_width="match_parent"
        android:layout_height="wrap_content"
        android:orientation="horizontal">
        <Button
            android:id="@+id/btnClear"
            android:layout_width="wrap_content"
            android:layout_height="wrap_content"
            android:layout_weight="1"
            android:text="@string/clear"/>
        <Button
            android:id="@+id/btnExit"
            android:layout_width="wrap_content"
            android:layout_height="wrap_content"
            android:layout_weight="1"
            android:text="@string/exit"/>
    </LinearLayout>
    <TextView
```

```
            android:id="@+id/tvContent"
            android:layout_width="match_parent"
            android:layout_height="wrap_content"
            android:text=""/>
    </LinearLayout>
```

从活动Activity父类派生子类MainActivity程序的Java代码如下：

```
public class MainActivity extends Activity{
    //定义3个私有按钮控件对象
    private Button mConnect; //定义连接按钮对象
    private Button mDisConnect;  //定义断开按钮对象
    private Button mClear;  //定义清除按钮对象
private Button mExit; //定义退出按钮对象
     private static final String TAG="MainActivity";
    //活动加载时首先调用
    protected void onCreate(Bundle savedInstanceState){  //重写onCreate()方法
        Log.d(TAG, "onCreate!");
        super.onCreate(savedInstanceState);  //调用基类onCreate()方法
        setContentView(R.layout.activity_main);  //设置屏幕布局}
    protected void onResume(){  //重写onResume()方法
        Log.d(TAG, "onResume!");super.onResume();  //调用基类onResume()方法
        mConnect=(Button)findViewById(R.id.btnConnect);  //查找连接按钮对象
        mConnect.setOnClickListener(mOnclickListener);  //设置按钮单击监听器
        mDisConnect=(Button)findViewById(R.id.btnDisconnect);  //查找按钮
        mDisConnect.setOnClickListener(mOnclickListener);  //设置单击监听器
        mClear=(Button)findViewById(R.id.btnClear);  //查找清除按钮对象
        mClear.setOnClickListener(mOnclickListener);
        mExit=(Button)findViewById(R.id.btnExit);
        mExit.setOnClickListener(mOnclickListener);
        //显示当前更新系统时间
    addText(getResources().getString(R.string.updateTime)+":"+getCurrentTime()+
"\r\n");}
/***按钮点击事件监听类**/
private final OnClickListener mOnclickListener=new OnClickListener(){
    public void onClick(View v){  //处理按钮单击响应方法
        switch(v.getId()){  //查询被单击的控件id
            case R.id.btnConnect:  //若为连接按钮
            startService(new Intent(MainActivity.this,HeartService.class));
           addText(getResources().getString(R.string.connect));
            break;
        case R.id.btnDisconnect:  //若为断开按钮
            stopService(new Intent(MainActivity.this, HeartService.class));
            addText(getResources().getString(R.string.disconnect));
```

```
                break;
            case R.id.btnClear:  //若为清除按钮
                clearText();break; //清除
                case R.id.btnExit:  //若为退出按钮
                finish();break; //关闭程序
        }}};
        /***添加内容信息**/
        private void addText(final String str){
            new Handler().post(new Runnable(){
                public void run(){
                    TextView tvContent=(TextView)findViewById(R.id.tvContent);
        tvContent.setText(String.format("%s\r\n%s",tvContent.getText().toString(),
str));}});}
        /***清空内容信息**/
        private void clearText(){
            new Handler().post(new Runnable(){  //发送到消息对象
                public void run(){
                    TextView tv_content=(TextView)findViewById(R.id.tvContent);
                    tv_content.setText("");}});}
        /***获取系统当前时间**/
        private String getCurrentTime(){
            SimpleDateFormat formatter= new SimpleDateFormat("yyyy/MM/dd HH:mm:ss");
            Date curDate=new Date(System.currentTimeMillis());
            String str=formatter.format(curDate);
            return str;}}
```

推送广播接收器 HeartReceiver 程序清单如下：

```
public class HeartReceiver extends BroadcastReceiver{
    private static final String TAG="HeartReceiver";
    public void onReceive(Context context, Intent intent){
        String action=intent.getAction();
        Log.d(TAG, "action"+action);
        if(Const.ACTION_START_HEART.equals(action)){
            Log.d(TAG, "Start heart");
    }else if(Const.ACTION_HEARTBEAT.equals(action)){
        intent.setClass(context, MainActivity.class);
        intent.addFlags(Intent.FLAG_ACTIVITY_NEW_TASK);
        context.startActivity(intent);
        Log.d(TAG, "Heartbeat");
            //在此完成心跳需要完成的工作，比如请求远程服务器……
    } else if(Const.ACTION_STOP_HEART.equals(action)){
        Log.d(TAG, "Stop heart");}}}
```

心跳包服务HeartService程序清单如下：

```
public class HeartService extends Service{
    private static final String TAG="HeartService";
    /***心跳间隔一分钟*/
    private static final long HEARTBEAT_INTERVAL=6*1000L;
    private AlarmManager mAlarmManager;
    private PendingIntent mPendingIntent;
    public IBinder onBind(Intent intent){
        return null;}
    public void onCreate(){
        super.onCreate();
        mAlarmManager=(AlarmManager)getSystemService(ALARM_SERVICE);
        mPendingIntent=PendingIntent.getBroadcast(this, 0, new Intent(
        Const.ACTION_HEARTBEAT), PendingIntent.FLAG_UPDATE_CURRENT);}
    public int onStartCommand(Intent intent, int flags, int startId){
        Log.d(TAG, "onStartCommand");
        //发送启动推送任务的广播
        Intent startIntent=new Intent(Const.ACTION_START_HEART);
        sendBroadcast(startIntent);
        //启动心跳定时器
        long triggerAtTime=SystemClock.elapsedRealtime()+
 HEARTBEAT_INTERVAL;
        mAlarmManager.setInexactRepeating(AlarmManager.ELAPSED_REALTIME,
 triggerAtTime, HEARTBEAT_INTERVAL, mPendingIntent);
        return super.onStartCommand(intent, flags, startId);}
    public void onDestroy(){
        Intent startIntent=new Intent(Const.ACTION_STOP_HEART);
        sendBroadcast(startIntent);
        mAlarmManager.cancel(mPendingIntent); //取消心跳定时器
        super.onDestroy();}}
```

Const.java程序清单如下：

```
public interface Const{
    int PUSH_MSG=100;
    String ACTION_START_HEART="sziit.lhz.heart.
intent.STARTPUSH";
    String ACTION_HEARTBEAT="sziit.lhz.heart.
intent.HEARTBEAT";
    String ACTION_STOP_HEART="sziit.lhz.heart.
intent.STOPPUSH";}
```

该程序的运行结果如图1-18所示。

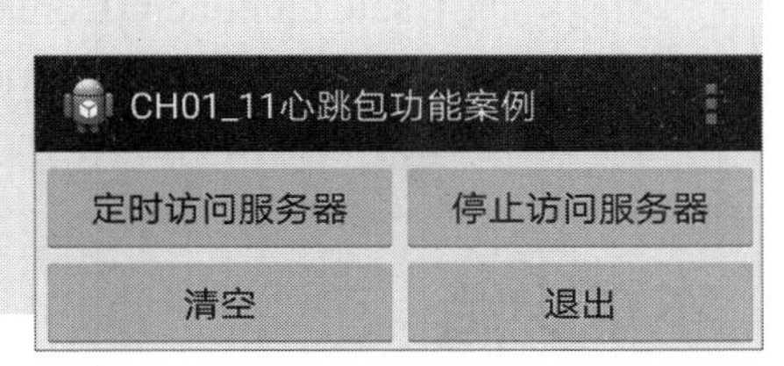

图 1-18　心跳包功能案例程序运行结果

1.4 知识扩展

1.4.1 BroadcastReceiver（广播接收器）

BroadcastReceiver是Android系统的四大组件之一，这种组件本质上是一种全局的监听器，用于监听系统全局的广播消息。BroadcastReceiver启动的步骤如下：

（1）创建需要启动的BroadcastReceiver的Intent。

（2）调用Content的sendBroadcast()或sendOrderedBroadcast()方法来启动指定的BroadcastReceiver。

1.4.2 数据绑定 Bundle 的主要功能函数

Bundle类是一个final类，其功能是应用于Activity之间相互传递值。其主要功能函数如下：

1. public void clear()

功能：从这种捆绑的映射中移除所有元素。

2. public boolean containsKey(String key)

功能：如果给定的键是在这种捆绑的映射中，则返回true。

参数：一个字符串键。

3. public Object get(String key)

功能：返回给定键的对象入口。

参数：一个字符串键。

4. public boolean getBoolean(String key)

功能：返回给定键的密钥，或值为相关联的默认值，如果没有所需的类型的映射关系，则给定键的存在值。

参数：一个字符串键。

5. public Bundle getBundle(String key)

功能：返回给定键的密钥，或者为null，如果没有所需的类型映射为给定的密钥或一个空值是明确与该键关联的存在值。

参数：一个字符串键。

6. public String getString(String key)

功能：返回值与给定的键，或NULL。

参数：一个字符串键。

7. public void putString(String key, String value)

功能：插入一个字符串值到这种捆绑的映射中，取代任何现有的值。

1.4.3 Intent 的主要功能函数

1. public Intent setClass(Context packageContext, Class<?>cls)

功能：调用setComponent(ComponentName)返回一个类对象的名称。

2. public Intent putExtra(String name, String value)

功能：添加扩展数据到意图中。

本章小结

本章主要介绍了Android系统中Intent和Service的功能和用法，并通过几个实例让用户掌握

Intent和Service的使用。

强化练习

一、填空题

1. Android四大基本组件分别是（　　）、（　　）、（　　）和（　　）。
2. Service启动到销毁的过程包括（　　）、（　　）和（　　）。
3. Intent元素包括（　　）、（　　）、（　　）、（　　）、（　　）和（　　）。
4. Service继承了（　　）、（　　）和（　　）3个方法。
5. 启动Service的两种方法是（　　）和（　　）。

二、单选题

1. Activity的生命周期方法onRestart()是在（　　）方法执行完后调用。
   A. onStart()　　B. onPause()　　C. onResume()　　D. onStop()
2. Android启动服务startService()方法的参数为（　　）类。
   A. Service　　B. Intent　　C. Activity　　D. Object
3. 下列不属于service服务生命周期的方法为（　　）。
   A. onStop()　　B. onStart()　　C. onCreate()　　D. onDestroy()
4. Android中发送邮件时使用的action为（　　）。
   A. Intent.ACTION_SEND　　B. Intent.ACTION_EMAIL
   C. Intent.EMAIL_ACTION　　D. Intent.SEND_ACTION
5. Android中显示网页时使用的action为（　　）。
   A. Intent.VIEW_INTERNET　　B. Intent.INTERNET_VIEW
   C. Intent.VIEW_ACTION　　D. Intent.ACTION_VIEW
6. Android中service生命周期回调方法包含（　　）。
   A. onCreate(),onStartCommand(), onStart(), onDestroy()
   B. onCreate(),onStartCommand(), onRestart(), onDestroy()
   C. onCreate(),onPause(), onStart(), onDestroy()
   D. onCreate(),onResume(), onStart(), onDestroy()
7. Activity绑定Service采用的方法为（　　）。
   A. startService()　　B. bindService()　　C. createService()　　D. onBind()
8. Intent可在各个对象间进行信息传递，这些对象不包含（　　）。
   A. Sqlite　　B. Activity　　C. Broadcast Receiver　　D. Service

三、简答题

1. 什么是Intent（意图）?
2. 什么是Service（服务）?
3. 如何实现Android平台组件之间的沟通?
4. 如何处理耗时的后台程序?

第2章 Android数据永久存储应用

学习目标

视频

Android数据永久存储应用

- 理解SharedPreferences的概念和作用。
- 理解Android的文件IO。
- 理解SQLite数据库。
- 理解ContentProvider的功能与意义。
- 掌握SharedPreferences保存程序的参数、选项的方法。
- 掌握读、写SD卡上文件的方法。
- 掌握Android的API操作SQLite数据库的方法。
- 掌握ContentProvider类的作用和常用方法的实现。
- 掌握网络存储数据的方法。

2.1 学习导入

Android一共提供了以下五种不同的数据存储方式。

（1）使用SharedPreferences存储数据。

（2）文件存储数据。

（3）SQLite数据库存储数据。

（4）使用ContentProvider存储数据。

（5）网络存储数据。

2.1.1 SharedPreferences（偏好数据存储）的概念

SharedPreferences是Android平台上一个轻量级的存储类，基于XML存储key-value键值对数据，特别适合用于保存软件配置参数。使用SharedPreferences保存数据，其背后是用xml文件存放数据，文件存放在 /data/data/<package name>/shared_prefs 目录下，其以键值对的方式存储，使得用户可很方便地读取和存入。

视频

Android Shared-Preferences 存储

2.1.2　文件存储数据的概念

顾名思义就是将要保存的数据以文件的形式保存。当需要调用所保存的数据时只要读取这些文件即可。需要注意，在 Android 中文件是 Linux 下的形式。

2.1.3　SQLite 数据库存储数据的概念

Android 内嵌了功能比其他手持设备操作系统强大的关系型数据库 sqlite3, SQL 语句基本都可以使用，创建的数据可以用 adb shell 来操作。具体路径是 /data/data/package_name/databases。SQLite 是一个开源的关系型数据库，与普通关系型数据库一样，也具有 ACID（原子性、一致性、隔离性、持久性）的特性。

视频

Android文件操作存储

2.1.4　ContentProvider 存储数据的概念

Android 提供了 ContentProvider，一个程序可以通过实现一个 ContentProvider 的抽象接口将自己的数据完全暴露出去，而且 ContentProviders 以类似数据库中表的方式将数据暴露，也就是说 ContentProvider 就像一个“数据库”。那么外界获取其提供的数据，也就应该与从数据库中获取数据的操作基本一样，只不过是采用 URI 表示外界需要访问的“数据库”。

视频

Android SQLite数据库

2.1.5　网络存储数据的概念

Android 中的网络存储数据就是将数据通过网络保存在网络上。在实际使用中会使用 java.net.* 和 android.net.* 等类。

2.2　技术准备

视频

SharedPre-ferences概念及典型应用实战

2.2.1　SharedPreferences 存储数据

SharedPreferences 是 Android 平台上一个轻量级的存储类，主要用于保存一些常用的配置（如窗口状态），一般在 Activity 中重载窗口状态时使用 SharedPreferences 完成，它提供了 Android 平台常规的 Long（长整型）、Int（整型）、String（字符串型）的保存方式。

SharedPreferences 存储数据的本质是基于 XML 文件存储 key-value 键值对数据，通常用来存储一些简单的配置信息。其存储位置在 /data/data/<包名>/shared_prefs 目录下。

SharedPreferences 对象本身只能获取数据而不支持存储和修改，存储修改通过 Editor 对象实现。实现 SharedPreferences 存储的步骤如下：

（1）根据 Context 获取 SharedPreferences 对象，有两种方法获得 SharedPreference 对象：

- SharedPreferences sp=PreferenceManager.getDefaultSharedPreferences(MainActivity.this)。
- SharedPreferences sp=getSharedPreferences("pref", MODE_PRIVATE)。

（2）利用 edit() 方法获取 Editor 对象：Editor editor=sp.edit()。

（3）通过 Editor 对象存储 key-value 键值对数据：利用 Editor 接口的 put×××() 方法保存 key-value 键值对；get×××() 方法获得键 key 所对应的值 value。

（4）通过commit()方法提交数据：editor.commit()。

利用SharedPreferences存储数据的示例代码如下：

```
package sziit.lihz.CH02_01sharedpreferences;
public class MainActivity extends Activity{
    private TextView tv1, tv2;
    private SharedPreferences sp;
    private Editor editor;
    protected void onCreate(Bundle savedInstanceState){
        super.onCreate(savedInstanceState);
        setContentView(R.layout.activity_main);
        tv1=(TextView)findViewById(R.id.tv1);
        tv2=(TextView)findViewById(R.id.tv2);
        sp=getSharedPreferences("synopsis", MODE_PRIVATE);
        editor=sp.edit();  //获取Editor实例
        //使用editor进行不同类型的信息存储
        editor.putString("name","张三");
        editor.putString("city","深圳市");
        editor.putInt("age", 24);
        editor.putFloat("distance", 12.34F);
        editor.putBoolean("key", true);
        editor.commit(); //提交生效
        editor.remove("default");
        editor.commit();
        tv1.setText(sp.getString("name", "未知").toString()+","
+sp.getInt("age", 1)+","+sp.getString("city", "未知"));
        tv2.setText(sp.getFloat("distance", 0.0F)+","
+sp.getBoolean("key", true));}}
```

这段代码执行过后，即在/data/data/sziit.lihz.CH02_01sharedpreferences/shared_prefs目录下生成了一个synopsis.xml文件，一个应用可以创建多个这样的xml文件。

SharedPreferences对象与SQLite数据库相比，免去了创建数据库、创建表、写SQL语句等诸多操作，相对而言更加方便、简洁。但是，SharedPreferences也有其自身缺陷，例如其只能存储Boolean、Int、Float、Long和String五种简单的数据类型，无法进行条件查询等。所以，不论SharedPreferences的数据存储操作如何简单，它也只能是存储方式的一种补充，而无法完全替代如SQLite数据库等其他数据存储方式。

获得SharedPreferences对象的方法为getSharedPreferences(name,mode)：该方法的第一个参数name用于指定文件名称，名称不用带扩展名，扩展名会由Android自动加上。第二个参数mode指定文件操作模式，共有4种操作模式：Activity.MODE_PRIVATE默认操作模式，代表该文件是私有数据，只能被应用本身访问，在该模式下，写入的内容会覆盖原文件内容；如果想把新写入内容追加到原文件中，可以使用Activity.MODE_APPEND；Activity.MODE_WORLD_READABLE表示当前文件可以被其他应用读取和写入；Activity.MODE_APPEND以追加方式存储。SharedPrefences常用方法如表2-1所示，SharedPreferences.Editor接口常用方法如表2-2所示。

表 2-1 SharedPrefences 常用方法

编号	方 法	功 能 描 述
1	public SharedPreferences.Editor edit()	使其处于可编辑状态
2	public boolean contains(String key)	判断某一个 key 是否存在
3	public Map<String, ?>getAll()	取出全部数据
4	public putBoolean(String key, boolean defValue)	取出 boolean 型数据，并指定默认值
5	public float getFloat(String key, float defValue)	取出 float 型数据，并指定默认值
6	public int getInt(String key, int defValue)	取出 int 型数据，并指定默认值
7	public long getLong(String key, long defValue)	取出 long 型数据，并指定默认值
8	public String getString(String key, String defValue)	取出 String 型数据，并指定默认值

表 2-2 SharedPreferences.Editor 接口常用方法

编号	方 法	功 能 描 述
1	public SharedPreferences.Editor clear()	清除所有数据
2	public boolean commit()	提交更新的数据
3	public SharedPreferences.Editor remove(String key)	删除指定 key 的数据
4	public Editor putBoolean(String key, boolean value)	保存一个 boolean 型数据
5	public Editor putFloat(String key, boolean value)	保存一个 float 型数据
6	public Editor putInt(String key, int value)	保存一个 int 型数据
7	public Editor putLong(String key, long value)	保存一个 long 型数据
8	public Editor putString(String key, String value)	保存一个 String 型数据

SharedPreferences典型应用案例：存储用户登录信息

该案例主要实现存储用户登录用户名和密码，并能选择是否记住用户名。该程序的布局大纲如图2-1所示，运行结果如图2-2所示。

图 2-1 存储用户登录信息界面布局大纲

图 2-2 存储用户登录信息运行结果

该程序MainActivity类的Java核心代码如下：

```
public class MainActivity extends Activity{
    private EditText etusername, etpassword;
```

```
    private CheckBox chk;SharedPreferences sp; Editor editor;
    private String sp_name,sp_password;
    String name,String password;
    protected void onCreate(Bundle savedInstanceState){
        super.onCreate(savedInstanceState);
        setContentView(R.layout.activity_main);
        //获取界面元素
        etusername=(EditText)findViewById(R.id.etusername);
        etpassword=(EditText)findViewById(R.id.etpassword);
        chk=(CheckBox)findViewById(R.id.chkSaveName);
        //获取SharedPreferences和editor
        sp=getSharedPreferences("userinfo", MainActivity.MODE_PRIVATE);
        editor=sp.edit();
        sp_name=sp.getString("username", "admin").trim();
        sp_password=sp.getString("password", "666666").trim();
        name=etusername.getText().toString().trim();
        //如果没有填写用户名，则不能单击复选框
        if(name.equals(""))
        {chk.setChecked(false);}{
            chk.setChecked(true);etusername.setText(name);}}
    public void doClick(View v){
        switch(v.getId()){
        case R.id.btnLogin:doLogin();break;
        case R.id.btnCancel:doCancel();break;
        case R.id.btnExit:doExit();break;}}
    private void doLogin(){
        name=etusername.getText().toString().trim();
        password=etpassword.getText().toString().trim();
        if(sp_name.equals(name)&& sp_password.equals(password)){
            if(chk.isChecked()){  //如果选择了保存用户名
                editor.putString("username", name);
                editor.putString("password", password);editor.commit();
            }
        else{
            editor.remove("username");editor.remove("password");
            editor.commit();}
            Toast.makeText(MainActivity.this,"登录成功",Toast.LENGTH_LONG).show();
    } else{  //用户名或密码错误，禁止登录
  Toast.makeText(MainActivity.this,"登录被拒绝",Toast.LENGTH_LONG).show();
    }}
    private void doCancel(){
etusername.setText("");etpassword.setText("");chk.setChecked(false);}
    private void doExit(){finish();}}
```

观察SharedPreference的运行结果：在Android虚拟设备（模拟器）上运行该应用程序；采用DDMS将userinfo.xml文件导出，选择DDMS→File Explorer→data→data→当前应用程序包名→当前项目的userinfo.xml文件→图标Pull a file from the device，如图2-3所示；打开userinfo.xml文件可观察到保存的数据。

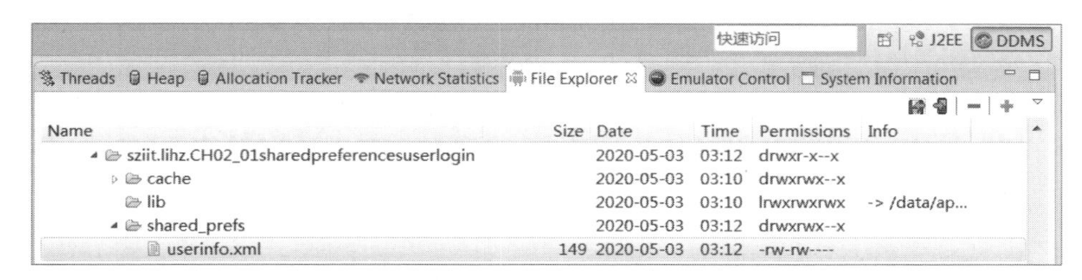

图 2-3　DDMS 导出 userinfo.xml 文件

2.2.2　文件存储数据

基于文件流的读取与写入是Android平台上的数据存取方式之一。在Android中，可通过Context.openFileInput和Context.openFileOutput分别获取FileInputStream和FileOutputStream。

public FileInputStream openFileInput(String name)方法打开应用程序私有目录下的指定私有文件以读入数据，返回一个FileInputStream对象。

视频

文件存储数据概念及应用实战

openFileOutput(String name,int mode)方法打开应用程序私有目录下的指定私有文件以写入数据，返回一个FileOutputStream对象，如果文件不存在就创建这个文件。openFileOutput()方法具体的实现过程与在J2SE（即Jave SE）环境中保存数据到文件中是一样的。文件可用来存放大量数据，如文本、图片、音频等。默认位置：/data/data/<包>/files/*.**。该方法的第一个参数name用于指定文件名称，其保存在data/data/<包>/files/下，通过选择Eclipse→Window→Show View→Other命令，在打开的对话框中展开android文件夹，选择下面的File Explorer视图，然后在File Explorer视图中展开/data/data/<包>/files目录就可以看到该文件。第二个参数mode用于指定操作模式，有4种模式，具体如下：

```
Context.MODE_PRIVATE=0
Context.MODE_APPEND=32768
Context.MODE_WORLD_READABLE=1
Context.MODE_WORLD_WRITEABLE=2
```

- Context.MODE_PRIVATE：为默认操作模式，代表该文件是私有数据，只能被应用本身访问，在该模式下，写入的内容会覆盖原文件的内容。如果想把新写入的内容追加到原文件中，可以使用Context.MODE_APPEND。
- Context.MODE_APPEND：该模式会检查文件是否存在，如果存在就向文件追加内容，否则就创建新文件。
- Context.MODE_WORLD_READABLE和Context.MODE_WORLD_WRITEABLE：用来控制其他应用是否有权限读/写该文件。

MODE_WORLD_READABLE：表示当前文件可以被其他应用读取。

MODE_WORLD_WRITEABLE：表示当前文件可以被其他应用写入。

私有文件只能被创建该文件的应用访问，如果希望文件能被其他应用读和写，可以在创建文件时，指定Context.MODE_WORLD_READABLE和Context.MODE_WORLD_ WRITEABLE权限。利用openFileInput()方法和openFileOutput()方法实现文件存储数据的示例代码如下。

从文件读数据示例：

```
public String readFiles(String fileName)throws IOException{   //读数据
    String res="";
    try{FileInputStream fin=openFileInput(fileName);
        int length=fin.available();byte[] buffer=new byte[length];
        fin.read(buffer);res=EncodingUtils.getString(buffer, "UTF-8");
        fin.close();
    }catch(Exception e){e.printStackTrace();}
    return res;}
```

将数据写入文件示例：

```
public void writeFiles(String content){   //保存文件内容
    FileOutputStream fos;
    try{fos=openFileOutput(fileName, MODE_PRIVATE);
    fos.write(content.getBytes());fos.close();
} catch(Exception e){e.printStackTrace();}}
```

Activity还提供了getCacheDir()和getFilesDir()方法：getCacheDir()方法用于获取/data/data/包>/cache目录；getFilesDir()方法用于获取/data/data/<包>/files目录。把文件存入SDCard。

使用Activity的openFileOutput()方法保存文件，文件存放在手持设备空间上。对于像视频这样的大文件，可以把它存放在内置SDCard或外置SDCard。创建SDCard可以在Eclipse创建模拟器的同时创建，也可以使用DOS命令进行创建。在程序中访问SDCard时，用户需要申请访问SDCard的权限。在AndroidManifest.xml中加入访问SDCard的权限如下：

```
<!-- 在SDCard中创建与删除文件权限 -->
<uses-permission android:name="android.permission.MOUNT_UNMOUNT_FILESYSTEMS"/>
<!-- 往SDCard写入数据权限 -->
<uses-permission android:name="android.permission.WRITE_EXTERNAL_STORAGE"/>
```

要往SDCard中存放文件，程序必须先判断手持设备是否装有SDCard，且可以进行读/写。

```
if(Environment.getExternalStorageState().equals(Environment.MEDIA_MOUNTED)){
    File sdCardDir=Environment.getExternalStorageDirectory();
    File saveFile=new File(sdCardDir, "test.txt");
    FileOutputStream outStream=new FileOutputStream(saveFile);
    outStream.write("test".getBytes());
    outStream.close();}
```

Environment.getExternalStorageState()方法用于获取SDCard的状态，如果手持设备装有SDCard，并且可以进行读/写，那么该方法返回的状态为Environment.MEDIA_MOUNTED。

Environment.getExternalStorageDirectory()方法用于获取SDCard的目录。当然，要获取SDCard

的目录，用户也可以编写如下代码：

```
File sdCardDir=new File("/sdcard");
File afile=new File(sdCardDir, "test.txt");
//上面两句代码可以合成一句
File afile=new File("/sdcard/test.txt");
```

文件存储数据典型应用案例：将内容写入SD卡文件

该案例利用Java的File类实现将输入内容写入指定SD卡文件中。界面设计组件包括Button、EditText、TextView。该程序的界面布局大纲如图2-4所示。实现SD卡读写的ExternalStorage类程序清单如下：

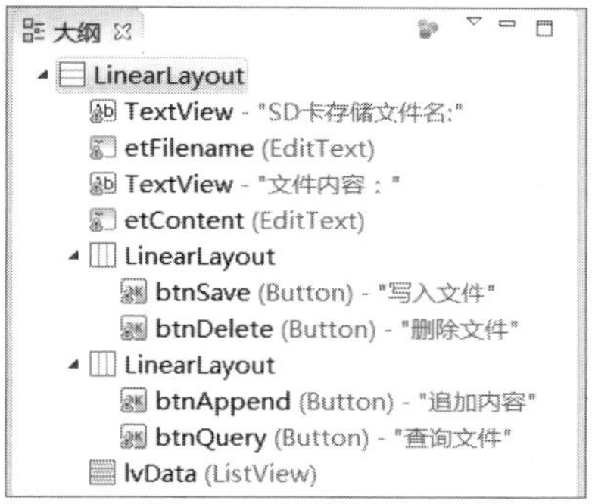

图 2-4　将内容写入 SD 卡文件布局大纲

```
public class ExternalStorage{
private Context context; //保存当前调用对象的Context
public static final String ENCODING="UTF-8";
//编码格式常量
public ExternalStorage(Context context){
    this.context=context;}
/***保存内容到内部存储器中******
*@param filename 文件名
*@param content  内容 ******/
public void saveToSdcard(String filename, String content)
    throws IOException{
    if(Environment.MEDIA_MOUNTED.equals(Environment
    .getExternalStorageState())){  //判断是否有可用的SDcard
    File file=new File(Environment.getExternalStorageDirectory(),
    filename);
    FileOutputStream fos=null;
    fos=new FileOutputStream(file);
    fos.write(content.getBytes());
    fos.close();}}
/***通过文件名获取内容********
*@param filename 文件名
*@return 文件内容*********/
public String getFromSdcard(String filename)throws IOException{
    String result="";
    try{FileInputStream fin=context.openFileInput(filename);
        int lenght=fin.available();  //获取文件长度
        byte[] buffer=new byte[lenght];
        fin.read(buffer);
        //将byte数组转换成指定格式的字符串
        result=EncodingUtils.getString(buffer, ENCODING);
} catch(Exception e){e.printStackTrace();}return result;}
```

```
/***以追加的方式在文件的末尾添加内容****
*@param filename文件名
*@param content 追加的内容*******/
public void append(String filename,String content)throws IOException{
    if(Environment.MEDIA_MOUNTED.equals(Environment
            .getExternalStorageState())){  //判断是否有可用的SDcard
        FileOutputStream fos=context.openFileOutput(filename,
                Context.MODE_APPEND);
        fos.write(content.getBytes());
        fos.close();}}
/***删除文件***************
*@param filename  文件名
*@return 是否成功******/
public boolean delete(String filename){
    return context.deleteFile(filename);}
/***获取外部存储路径下的所有文件名*
*@return 文件名数组*********/
public String[] queryAllFile(){
    if(Environment.MEDIA_MOUNTED.equals(Environment
            .getExternalStorageState())){  //判断是否有可用的SDcard
        return context.fileList();
} else{return null;}}}
```

调用ExternalStorage()函数的MainActivity程序Java代码如下：

```
public class MainActivity extends Activity implements OnClickListener{
    private static final String TAG="MainActivity";
    private EditText etFilename, etContent;
    private Button btnSave, btnQuery, btnDelete, btnAppend;
    private ListView lvData;
    protected void onCreate(Bundle savedInstanceState){
        super.onCreate(savedInstanceState);
        setContentView(R.layout.activity_main);
        initView(); setOnClickListener();}
    private void setOnClickListener(){
        btnSave.setOnClickListener(this);
        btnAppend.setOnClickListener(this);
        btnQuery.setOnClickListener(this);
        btnDelete.setOnClickListener(this);
        lvData.setOnItemClickListener(new OnItemClickListener(){
        public void onItemClick(AdapterView<?>parent, View view,
                int position, long id){
            showFileContent(parent, position, id);}});}
    private void initView(){
```

```
        lvData=(ListView)findViewById(R.id.lvData);
        etFilename=(EditText)findViewById(R.id.etFilename);
        etContent=(EditText)findViewById(R.id.etContent);
        btnSave=(Button)findViewById(R.id.btnSave);
        btnAppend=(Button)findViewById(R.id.btnAppend);
        btnQuery=(Button)findViewById(R.id.btnQuery);
        btnDelete=(Button)findViewById(R.id.btnDelete);}
    public void onClick(View v){
        switch(v.getId()){
        case R.id.btnSave:doSave();break;      //保存数据
        case R.id.btnAppend:doAppend();break;  //向文件添加数据
        case R.id.btnQuery:doQuery();break;    //查询文件
        case R.id.btnDelete:doDelete();break;  //删除文件}}
    private void doDelete(){  //删除文件
        String filename=etFilename.getText().toString();
        ExternalStorage es=new ExternalStorage(MainActivity.this);
        es.delete(filename);}
    private void doQuery(){  //查询文件
        ExternalStorage es=new ExternalStorage(MainActivity.this);
        String[] files=es.queryAllFile();
        ArrayAdapter<String>fileArray=new ArrayAdapter<String>(
        MainActivity.this, android.R.layout.simple_list_item_1, files);
        lvData.setAdapter(fileArray);}
    private void doAppend(){  //向文件添加数据
        String filename=null;String content=null;
        filename=etFilename.getText().toString();
        content=etContent.getText().toString();
        ExternalStorage es=new ExternalStorage(MainActivity.this);
        try{es.append(filename, content);
    } catch(IOException e){e.printStackTrace();}}
    private void doSave(){    //保存数据
        String filename=null;String content=null;
        filename=etFilename.getText().toString();
        content=etContent.getText().toString();
        ExternalStorage es=new ExternalStorage(MainActivity.this);
        try{es.saveToSdcard(filename, content);
    }catch(IOException e){e.printStackTrace();}}
private void showFileContent(AdapterView<?>parent, int position, long id){
        ListView lv=(ListView)parent;
        ArrayAdapter<String>adapter=(ArrayAdapter<String>)lv.getAdapter();
        String filename=adapter.getItem(position);
        etFilename.setText(filename);
        ExternalStorage es=new ExternalStorage(MainActivity.this);
```

```
        try{String content=es.getFromSdcard(filename);
            etContent.setText(content);
    } catch(IOException e){e.printStackTrace();}}
}
```

该程序的运行结果如图2-5所示。

2.2.3 SQLite 数据库存储数据

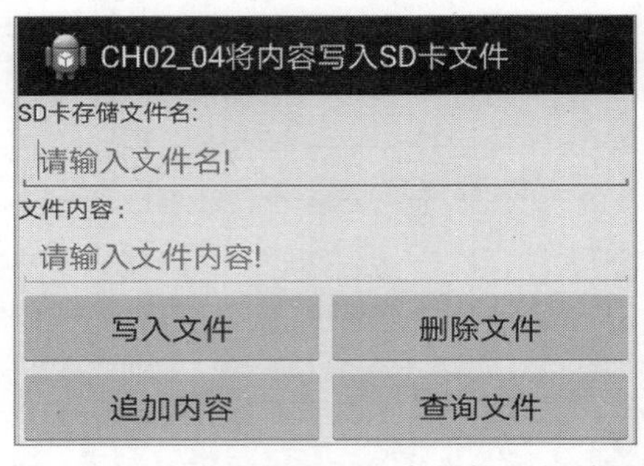

图 2-5　将内容写入 SD 卡文件运行结果

SQLite是轻量级嵌入式数据库引擎，它支持SQL语句，并且只利用很少的内存就有很好的性能。许多开源项目（Mozilla、PHP、Python）都使用了SQLite。SQLite 由以下几个组件组成：SQL 编译器、内核、后端以及附件。SQLite 通过利用虚拟机和虚拟数据库引擎（Virtual Datebase Engine，VDBE），使调试、修改和扩展 SQLite 的内核变得更加方便。其主要特点如下：

（1）面向资源有限的设备。

（2）没有服务器进程。

（3）所有数据存放在同一文件中跨平台。

（4）可自由复制。

SQLite 内部结构如图2-6所示。

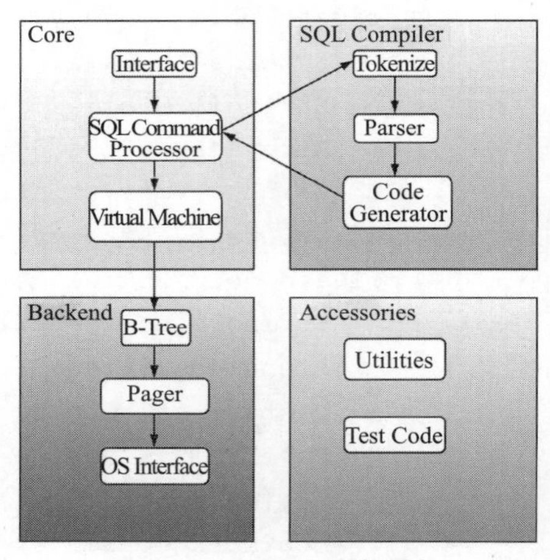

图 2-6　SQLite 内部结构

SQLite 基本上符合 SQL-92标准，和其他的主要 SQL 数据库没什么区别。它的优点就是高效，Android 运行时环境包含了完整的 SQLite。

Android 集成了SQLite数据库，所以每个Android应用程序都可以使用 SQLite 数据库。

对于熟悉 SQL 的开发人员来说，在 Android 开发中使用 SQLite 相当简单。但是，由于 JDBC 会消耗太多的系统资源，所以 JDBC 对于手持设备这种内存受限设备来说并不合适。因此，Android 提供了一些新的 API 来使用 SQLite 数据库。Android 开发中，程序员需要学会使用这些 API。

视频

SQLite数据库概念及应用实战

SQLite数据库存储在 /data/data/<包>/databases/ 下。Android 开发中使用 SQLite 数据库 Activites 可以通过 Content Provider 或者 Service 访问一个数据库。

SQLite数据库框架功能主要包括：自动建表，支持属性来自继承类，可根据注解自动完成建表，并且对于继承类中的注解字段也支持自动建表；自动支持增删改，增改支持对象化操作——增删改是数据库操作最基本单元，不用重复写这些增删改代码；查询方式灵活，支持Android框架提供的方式，也支持原生SQL方式；查询结果对象化，对于查询结果可自动包装为实体对象；查询结果灵活，支持对象化。

SQLite数据库框架包括SQLiteDatabase类和SQLiteDataHelper类。既可利用管理SQLite的类SQLiteDatabase创建、执行、删除SQL命令，也可利用SQLiteDatabase的帮助类SQLiteDataHelper管理数据库的创建和版本更新。使用SQLiteOpenHelper时，要建立一个类继承它，并重写

onCreate() 和 onUpgrade() 方法。

SQLiteDatabase 类是 Android 系统提供的用于管理和操作 SQLite 数据库的 API，一个 SQLiteDatabase 对象实例相当于一个 SQLite 数据库，并提供了创建、删除、执行 SQL 语句和其他常用数据库操作的方法。

SQLiteDatabase 类的实例代表了一个 SQLite 数据库操作，通过 SQLiteDatabase 类可执行 SQL 语句，以完成对数据表的增删改查（增加、删除、修改、查询）等操作，在该类之中定义了基本的数据库执行 SQL 语句的操作方法及其操作模式常量。SQLiteDatabase 封装了一些操作数据库 API。

（1）static SQLiteDatabase create(SQLiteDatabase.CursorFactory factory)：创建一个数据库，其中 factory 为可选的数据库游标工厂类，当查询被提交时，该对象会被调用来实例化一个游标。

（2）static SQLiteDatabase openDatabase(String path, CursorFactory factory, int flags)：根据提供的模式打开一个数据库。其中，path 用于打开或创建的数据库文件；factory 为可选的数据库游标工厂类，当查询被提交时，该对象会被调用来实例化一个游标，默认为 null；flags 控制数据库的访问模式。

（3）static SQLiteDatabase openOrCreateDatabase(File file, CursorFactory factory)：等同于 openDatabase(file.getPath(), factory, CREATE_IF_NECESSARY)。

（4）static SQLiteDatabase openOrCreateDatabase(String path, CursorFactory factory)：等同于 openDatabase(path,factory, CREATE_IF_NECESSARY)。

（5）void close()：关闭数据库。

（6）boolean deleteDatabase(String name)：删除指定的数据库。其中，name 为要关闭的数据库的名字。

（7）long insert(String table, String nullColumnHack, ContentValues values)：插入数据。其中，table 为表名；nullColumnHack 表示传入的 valuesnull 则列被设为 null；values 为所有要插入的数据。

（8）long insertOrThrow(String table, String nullColumnHack, ContentValues values)：插入数据，参数同 insert，但会抛出 SQLException 异常。

（9）int update(String table, ContentValues values, String whereClause, String[] whereArgs)：修改数据，其中，table 为表名，values 为更新数据，whereClause 指明 WHERE 子句，whereArgs 为 WHERE 子句参数，用于替换“?”。

（10）int delete(String table, String whereClause, String[] whereArgs)：删除数据。其中，table 为表名；whereClause 指明 WHERE 子句；whereArgs 为 WHERE 子句参数，用于替换“?”。

（11）Cursor query(boolean distinct, String table, String[] columns,String selection, String[] selectionArgs, String groupBy, String having, String orderBy, String limit)：查询数据表。其中，参数有 distinct(是否去掉重复行)、table(表名)、columns(列名)、selection(WHERE 子句)、selectionArgs(WHERE 条件)、groupBy(分组)、having(分组过滤)、orderBy(排序)、limit(LIMIT 子句。

（12）Cursor query(String table, String[] columns, String selection, String[] selectionArgs, String groupBy, String having, String orderBy)：查询数据表。

（13）Cursor rawQuery(String sql, String[] selectionArgs)：执行指定的 SQL 查询语句。

SQLiteDatabase 类也支持 SQL 命令来完成各种数据操作，常用方法包括：

（1）添加记录：db.execSQL("insert into users(userid,password)values(?,?)", new String[]{name,password})。

（2）修改记录：db.execSQL("update users set password=? where userid=?",new String[]{password,name})。

（3）删除记录：db.execSQL("delete from users where userid=?",new String[]{name})。

（4）查询：Cursor c=db.rawQuery("select id as _id,userid,password from users",null)。

SQLite可以解析大部分标准SQL语句：

```
select *from person
select *from person order by id desc
select name from person group by name having count(*)>1
select *from person limit 5 offset 3
select *from person limit 3,5 //获取5条记录，跳过前面3条记录
insert into person(name, age)values('张三',3)
update person set name='张三'where id=10
delete from person  where id=10
```

这里给出SQLiteDatabase简单示例。先调用openOrCreateDatabase()方法创建一个数据库并打开，可在/data/data/包名/databases目录下生成数据库user.db文件，然后执行SQL语句创建数据表，再向数据表中插入数据。运行应用程序，利用DDMS中文件浏览器从/data/data/包名/databases目录中导出被创建的数据库user.db。利用NaviCat工具打开SQLite数据库，查询数据表中数据。该案例的主要代码如下（代码详细见CH02_05）：

```
SQLiteDatabase db=openOrCreateDatabase("user.db", MODE_PRIVATE, null);
//执行SQL语句
db.execSQL("create table if not exists person(_id integer primary key
autoincrement,name text not null,"+"age integer not null, sex text not null)");
db.execSQL("insert into person(name, age, sex)values('张三', 16, '男')");
db.execSQL("insert into person(name,age,sex)values('李四',24,'男')");
db.execSQL("insert into person(name,age,sex)values('小倩',18,'女')");
db.execSQL("insert into person(name,age,sex)values('小琳',18,'女')");
Cursor c=db.rawQuery("select *from person", null);
if(c!=null){
while(c.moveToNext()){
    Log.i("info", "_id:"+c.getInt(c.getColumnIndex("_id")));
    Log.i("info", "name:"+c.getString(c.getColumnIndex("name")));
    Log.i("info", "age:"+c.getInt(c.getColumnIndex("age")));
    Log.i("info", "sex:"+c.getInt(c.getColumnIndex("sex")));}
    c.close();     //关闭数据集合游标}
    db.close();    //关闭数据库对象
```

Android提供SQLiteOpenHelper帮助用户创建数据库，用户只需继承SQLiteOpenHelper类即可轻松创建数据库。SQLiteOpenHelper类根据开发应用程序的需要，封装了创建和更新数据库使用的逻辑。

SQLiteOpenHelper的子类，至少需要实现以下3个方法：

（1）构造函数：调用父类 SQLiteOpenHelper 的构造函数。这个方法需要4个参数：上下文环

境（如一个Activity）、数据库名字、一个可选的游标工厂（通常是Null）、一个代表用户正在使用的数据库模型版本的整数。

（2）onCreate()方法：当数据库第一次被建立的时候被执行，如创建表、初始化数据等。它需要一个SQLiteDatabase对象作为参数，根据需要对这个对象填充表和初始化数据。

（3）onUpgrage()方法：更新数据库时执行，如删除旧表、创建新表。它需要3个参数：一个SQLiteDatabase对象、一个旧的版本号和一个新的版本号，这样用户就可以清楚如何把一个数据库从旧的模型转变到新的模型。

类SQLiteOpenHelper是一个帮助类，提供了方法来辅助创建和打开数据库，管理数据库的不同版本。使用时需要创建一个类，继承自类SQLiteOpenHelper，并重写其相对应的方法来实现创建、打开、更新数据库的操作。SQLiteOpenHelper类常用方法包括：onCreate()、onOpen()、onUpgrade()、getWritableDatabase()、getReadableDatabase()和close()等。

（1）void onCreate(SQLiteDatabase db)：在数据库第一次生成时会调用该方法，通常在该方法中生成数据表。

（2）void onUpgrade(SQLiteDatabase db, int oldVersion, int newVersion)：当数据库需要升级时，系统会自动调用该方法，通常在该方法里删除数据表，建立新数据表，并根据实际需求做其他操作。

（3）void onOpen(SQLiteDatabase db)：打开数据库时的回调函数，通常不会用到。

（4）SQLiteDatabase getReadableDatabase()：创建或打开数据库。

（5）SQLiteDatabase getWritableDatabase()：以读写方式创建或打开数据库。

下面的示例代码展示了如何继承SQLiteOpenHelper创建数据库：

```
public class DBHelper extends SQLiteOpenHelper{
public DBHelper(Context context,String name){
    super(context,name,null,1);  //factory为null,version=1}
public void onCreate(SQLiteDatabase db){
    db.execSQL("create table if not exists person(_id integer primary key
autoincrement, name text not null, age integer not null, sex text not null)");
    db.execSQL("insert into person(name,age,sex)values('张三',18,'男')");
    db.execSQL("insert into person(name,age,sex)values('李四',24,'男')");
    db.execSQL("insert into person(name,age,sex)values('小倩',18,'女')");
    db.execSQL("insert into person(name,age,sex)values('小琳',18,'女')");}
public void onUpgrade(SQLiteDatabase db, int oldVersion, int newVersion)
{db.execSQL("DROP TABLE IF EXISTS "+"person");
   onCreate(db);}}
```

接下来讨论具体如何创建表、插入数据、删除表等。调用getReadableDatabase()方法或getWriteableDatabase()方法，用户可以得到SQLiteDatabase实例，具体调用哪个方法，取决于用户是否需要改变数据库的内容：

```
DBHelper helper=new DBHelper(MainActivity.this, "student.db");
SQLiteDatabase db=helper.getWritableDatabase();
```

上面这段代码会返回一个SQLiteDatabase类的实例，使用这个对象，就可以查询或者修改数据库。当完成了对数据库的操作（如Activity已经关闭），需要调用 SQLiteDatabase 的 close()方法释放数据库连接。创建表和索引，需要调用 SQLiteDatabase 的 execSQL()方法来执行 DDL 语句。如果没有异常，这个方法没有返回值。例如，执行如下代码：

```
db.execSQL("create table if not exists person(id integer primary key
    autoincrement
```

这条语句会创建一个名为 person的表，表有一个列名为 id，并且是主键，这列的值是会自动增长的整数（例如，当插入一行时，SQLite会给这列自动赋值）。另外，还有3列：name（文本）、age（整数）和sex（文本）。SQLite 会自动为主键列创建索引。通常情况下，第一次创建数据库时创建了表和索引。

如果不需要改变表的 schema，不需要删除表和索引。删除表和索引，需要使用 execSQL()方法调用 DROP INDEX 和 DROP TABLE 语句。给表添加数据，上面的代码已经创建了数据库和表，现在需要给表添加数据。有两种方法可以给表添加数据。

像上面创建表一样，可以使用 execSQL()方法执行INSERT、UPDATE、DELETE等语句来更新表的数据。execSQL()方法适用于所有不返回结果的 SQL 语句。例如：

```
db.execSQL("insert into person(name,age,sex)values('张三',18,'男')");
```

另一种方法是使用 SQLiteDatabase 对象的insert()、update()、delete()方法。这些方法把 SQL 语句的一部分作为参数。例如：

```
ContentValues cv=new ContentValues();
cv.put("name", "张三");cv.put("age",24); cv.put("sex", "男");
db.insert("person", getNullColumnHack(), cv);
```

update()方法有4个参数，分别是表名，表示列名和值的 ContentValues 对象，可选的WHERE条件和可选的填充 WHERE 语句的字符串，这些字符串会替换 WHERE 条件中的“?”标记。update()根据条件更新指定列的值，所以用 execSQL()方法可以达到同样的目的。WHERE 条件和其参数和用过的其他 SQL APIs 类似。例如：

```
cv=new ContentValues();cv.put("name", "张三"); cv.put("age", "24");
String whereClause="id=?"; String[] whereArgs={String.valueOf(1)};
db.update("person", cv, whereClause, whereArgs);
```

delete()方法的使用和 update()方法类似，使用表名，可选的 WHERE 条件和相应的填充WHERE条件的字符串。查询数据库，类似INSERT、UPDATE、DELETE，有两种方法使用SELECT从SQLite数据库检索数据。

使用 rawQuery()方法直接调用SELECT语句；使用query()方法构建一个查询。例如：

```
cursor=db.query("person", new String[]{"id", "name", "age", "sex"},
  "age <= ?", new String[]{"24"}, null, null, null);
```

在上面的例子中，查询SQLite系统表（sqlite_master）检查person表是否存在。返回值是一个cursor对象，这个对象的方法可以迭代查询结果。如果查询是动态的，使用这个方法就会非常复杂。例如，如果需要查询的列在编译程序时不能确定，使用query()方法会方便很多。使用游标，不管如何执行查询，都会返回一个Cursor，这是 Android的SQLite 数据库游标。使用游标可以：

（1）通过使用 getCount() 方法得到结果集中有多少记录。

（2）通过 moveToFirst()、moveToNext() 和 isAfterLast() 方法遍历所有记录。

（3）通过 getColumnNames() 方法得到字段名。

（4）通过 getColumnIndex() 方法转换成字段号。

（5）通过 getString()、getInt() 等方法得到给定字段当前记录的值。

（6）通过 requery() 方法重新执行查询得到游标。

（7）通过 close() 方法释放游标资源。

下面代码遍历 person 表：

```
cursor=db.query("person", new String[]{"id", "name", "age", "sex"},
      "age <= ?", new String[]{"24"}, null, null, null);
while(cursor.moveToNext()){
   id=cursor.getInt(cursor.getColumnIndex("id"));
   name=cursor.getString(cursor.getColumnIndex("name"));
   age=cursor.getString(cursor.getColumnIndex("age"));
   sex=cursor.getString(cursor.getColumnIndex("sex"));}
cursor.close();
```

在 Android 中使用 SQLite 数据库管理工具。在其他数据库进行开发时，一般都使用工具来检查和处理数据库的内容，而不是仅仅使用数据库的 API。

使用 Android 模拟器，有两种可选方法来管理数据库。若模拟器绑定了 sqlite3 控制台程序，可使用 adb shell 命令来调用它。只要进入了模拟器的 shell，在数据库的路径执行 sqlite3 命令即可。数据库文件一般存放在 /data/data/<包>/databases/ 数据库。如果喜欢使用更友好的 DDMS 工具，可把数据库复制到开发机上，使用 NaviCat 客户端来操作它。这样，在一个数据库的副本上操作，如果想要修改能反映到设备上，需要把数据库备份回去。把数据库从设备上复制出来，可使用 adb pull 命令。存储一个修改过的数据库到设备上，使用 adb push 命令或使用 DDMS 调试工具。

SQLite 数据库存储数据典型应用案例：学生信息 CRUD

该案例实现的功能：定义 SQLiteOpenHelper 子类，重写 onCreate (SQLiteDatabase db) 方法，在 synopsis.db 数据库上创建 person，完成学生信息 CRUD 操作。该程序界面布局大纲如图 2-7 所示。

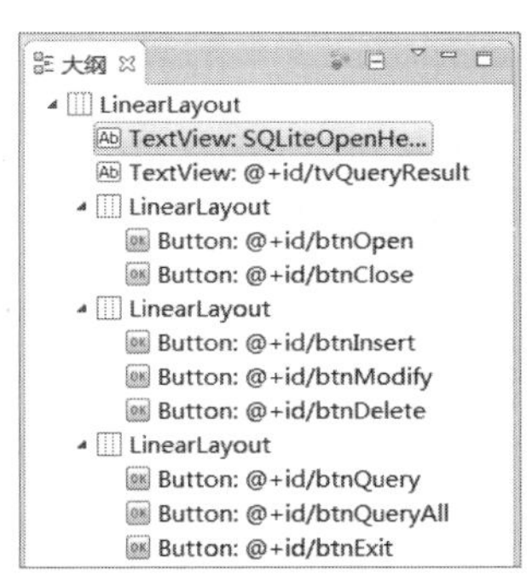

图 2-7 学生信息 CRUD 布局大纲

程序的界面布局 layout\ activity_main.xml 代码如下：

```
<LinearLayout xmlns:android="http://schemas.android.com/
apk/res/android"
     android:layout_width="fill_parent"
     android:layout_height="fill_parent"
     android:orientation="vertical">
     <TextView
         android:layout_width="fill_parent"
         android:layout_height="wrap_content"
         android:background="#0000FF"
         android:gravity="center_horizontal"
```

```
        android:text="SQLiteOpenHelper案例"
        android:textColor="#FFFFFF"
        android:textSize="20sp"/>
    <TextView
        android:id="@+id/tvQueryResult"
        android:layout_width="fill_parent"
        android:layout_height="wrap_content"
        android:text=""/>
    <LinearLayout
        android:layout_width="match_parent"
        android:layout_height="wrap_content"
        android:orientation="horizontal">
        <Button
            android:id="@+id/btnOpen"
            android:layout_width="wrap_content"
            android:layout_height="wrap_content"
            android:layout_weight="1"
            android:text="打开/创建数据库"/>
        <Button
            android:id="@+id/btnClose"
            android:layout_width="wrap_content"
            android:layout_height="wrap_content"
            android:layout_weight="1"
            android:enabled="false"
            android:text="关闭数据库"/>
    </LinearLayout>
    <LinearLayout
        android:layout_width="match_parent"
        android:layout_height="wrap_content"
        android:orientation="horizontal">
        <Button
            android:id="@+id/btnInsert"
            android:layout_width="wrap_content"
            android:layout_height="wrap_content"
            android:layout_weight="1"
            android:enabled="false"
            android:text="插入记录"/>
        <Button
            android:id="@+id/btnModify"
            android:layout_width="wrap_content"
            android:layout_height="wrap_content"
            android:layout_weight="1"
```

```
            android:enabled="false"
            android:text="修改记录"/>
        <Button
            android:id="@+id/btnDelete"
            android:layout_width="wrap_content"
            android:layout_height="wrap_content"
            android:layout_weight="1"
            android:enabled="false"
            android:text="删除记录"/>
    </LinearLayout>
    <LinearLayout
        android:layout_width="match_parent"
        android:layout_height="wrap_content"
        android:orientation="horizontal">
        <Button
            android:id="@+id/btnQuery"
            android:layout_width="wrap_content"
            android:layout_height="wrap_content"
            android:layout_weight="1"
            android:enabled="false"
            android:text="查询记录"/>
        <Button
            android:id="@+id/btnQueryAll"
            android:layout_width="wrap_content"
            android:layout_height="wrap_content"
            android:layout_weight="1"
            android:enabled="false"
            android:text="列举记录"/>
        <Button
            android:id="@+id/btnExit"
            android:layout_width="wrap_content"
            android:layout_height="wrap_content"
            android:layout_weight="1"
            android:enabled="false"
            android:text="关闭程序"/>
    </LinearLayout>
</LinearLayout>
```

实体类Person程序清单如下：

```
public class Person{
    private String name;
    private int age;
    private String sex;
```

```
    public String getSex(){return sex;}
    public void setSex(String sex){this.sex=sex;}
    public String getName(){return name;}
    public void setName(String name){this.name=name;}
    public int getAge(){return age;}
    public void setAge(int age){this.age=age;}}
public class DBHelper extends SQLiteOpenHelper{
    private static final String TAG="DBHelper";
    public static final int VERSION=1;
    public DBHelper(Context context, String name, CursorFactory factory,
    int version){  //构造函数
    super(context, name, factory, version);Log.d(TAG, "DBHelper()");}
    //第一次创建数据库时，调用该方法
    public void onCreate(SQLiteDatabase db){
    String sql="create table person(id INTEGER PRIMARY KEY AUTOINCREMENT, name
VARCHAR(20),age INTEGER,sex VARCHAR(10))";
    Log.d(TAG, "create Database!"); //输出创建数据库的日志信息
    //execSQL()函数用于执行SQL语句
    db.execSQL("DROP TABLE IF EXISTS person");  db.execSQL(sql);}}
```

MainActivity类程序的Java代码如下：

```
public class MainActivity extends Activity implements OnClickListener{
private TextView tvQueryResult;
private Button btnOpen, btnInsert, btnModify, btnDelete, btnQuery,
    btnQueryAll, btnClose, btnExit;
private DBHelper dbHelper;
private SQLiteDatabase db;
private ContentValues cv;
private Cursor cursor;
private int id;
private String name, age, sex;
protected void onCreate(Bundle savedInstanceState){
    super.onCreate(savedInstanceState);
    setContentView(R.layout.activity_main);
    findViews();setClickListener();
    //创建DBHelper对象
    dbHelper=new DBHelper(MainActivity.this, "synopsis.db", null, 1);
  cv=new ContentValues();}
  private void findViews(){
      tvQueryResult=(TextView)findViewById(R.id.tvQueryResult);
      btnOpen=(Button)findViewById(R.id.btnOpen);
      btnInsert=(Button)findViewById(R.id.btnInsert);
      btnModify=(Button)findViewById(R.id.btnModify);
```

```
        btnDelete=(Button)findViewById(R.id.btnDelete);
        btnQuery=(Button)findViewById(R.id.btnQuery);
        btnQueryAll=(Button)findViewById(R.id.btnQueryAll);
        btnClose=(Button)findViewById(R.id.btnClose);
        btnExit=(Button)findViewById(R.id.btnExit);}
    private void setClickListener(){
        btnOpen.setOnClickListener(this);
        btnInsert.setOnClickListener(this);
        btnModify.setOnClickListener(this);
        btnDelete.setOnClickListener(this);
        btnQuery.setOnClickListener(this);
        btnQueryAll.setOnClickListener(this);
        btnClose.setOnClickListener(this);
        btnExit.setOnClickListener(this);}

    private void doOpen(){
        db=dbHelper.getReadableDatabase();  //得到一个可读的数据库
        btnInsert.setEnabled(true);btnModify.setEnabled(true);
        btnDelete.setEnabled(true);btnQuery.setEnabled(true);
        btnQueryAll.setEnabled(true);btnClose.setEnabled(true);
        btnExit.setEnabled(true);}
    private void doInsert(){
        db=dbHelper.getWritableDatabase();  //得到一个可写的数据库
        Person person=new Person();
        person.setName("李四");person.setAge(21);person.setSex("男");
        //插入第一条记录
        db.execSQL("INSERT INTO person VALUES(NULL, ?, ?,?)", new Object[]
            {person.getName(), person.getAge(), person.getSex()});
        cv=new ContentValues();
        cv.put("name", "张三"); cv.put("age", 22);cv.put("sex", "男");
        db.insert("person", null, cv);   //插入第二条记录
        cv.put("name", "小倩");  cv.put("age", 23);cv.put("sex", "女");
        db.insert("person", null, cv);   //插入第三条记录
        db.close();  //关闭数据库}

    private void doUpdate(){
        db=dbHelper.getWritableDatabase();  //得到一个可写的数据库
        cv=new ContentValues();cv.put("name","张三");cv.put("age","24");
        String whereClause="id=?";
        String[] whereArgs ={String.valueOf(1)};
        db.update("person",  //参数person 是要更新的表名
            cv,              //参数cv 是一个ContentValeus对象
            whereClause,     //参数whereClause 是where子句
```

```
        whereArgs);       //参数whereArgs 是子句条件}
    private void doDelete(){
        String whereClause;
        db=dbHelper.getWritableDatabase();  //得到一个可写的数据库
        whereClause="age>=?";
        String[] whereArgs2={String.valueOf(0)};
        //调用delete方法，删除数据
        db.delete("person", whereClause, whereArgs2);}
    private void doQuery(){
        tvQueryResult.setText("");
        db=dbHelper.getReadableDatabase();  //得到一个可读的数据库
        cursor=db.query("person",new String[]{"id","name","age","sex"},
        "age <= ?", new String[]{"24"}, null, null, null);
        while(cursor.moveToNext()){
            id=cursor.getInt(cursor.getColumnIndex("id"));
            name=cursor.getString(cursor.getColumnIndex("name"));
            age=cursor.getString(cursor.getColumnIndex("age"));
            sex=cursor.getString(cursor.getColumnIndex("sex"));
            tvQueryResult.setText(tvQueryResult.getText().toString()+
            id+"," +name+","+age+","+sex+"\n");}
      cursor.close();
      db.close();  //关闭数据库}
    private void doQueryAll(){
        tvQueryResult.setText("");
        db=dbHelper.getReadableDatabase();  //得到一个可读的数据库
        cursor=db.query("person",
        new String[]{"id", "name", "age", "sex"}, "id >?",
        new String[]{"0"}, null, null, null);
      while(cursor.moveToNext()){
          id=cursor.getInt(cursor.getColumnIndex("id"));
          name=cursor.getString(cursor.getColumnIndex("name"));
          age=cursor.getString(cursor.getColumnIndex("age"));
          sex=cursor.getString(cursor.getColumnIndex("sex"));
          tvQueryResult.setText(tvQueryResult.getText().toString()+
          id+"," +name+","+age+","+sex+"\n");}
        cursor.close();
        db.close();  //关闭数据库}
    private void doClose(){
        db.close();  //删除数据库
        btnInsert.setEnabled(false); btnModify.setEnabled(false);
        btnDelete.setEnabled(false); btnQuery.setEnabled(false);
        btnQueryAll.setEnabled(false);btnExit.setEnabled(false);
        tvQueryResult.setText("");}
```

```
    private void doExit(){
        doClose();finish();}
    public void onClick(View v){
        switch(v.getId()){
        case R.id.btnOpen:doOpen(); break;            //打开或创建数据库
        case R.id.btnInsert: doInsert(); break;       //插入记录
        case R.id.btnModify: doUpdate(); break;       //修改记录
        case R.id.btnDelete: doDelete(); break;       //删除记录
        case R.id.btnQuery:doQuery();break;           //查询记录
        case R.id.btnQueryAll:doQueryAll();break;     //列举记录
        case R.id.btnClose:doClose();  break;         //关闭数据库
        case R.id.btnExit: doExit(); break;           //关闭应用}}}
```

该程序的运行结果图如图2-8所示。采用DDMS调试工具在File Explorer中从以下路径/data/data/sziit.lihz.CH02_07sqliteopenhelper/databases导出contacts.db数据库，再利用NaviCat打开SQLite数据库，观察其中的person数据表，如图2-9所示。

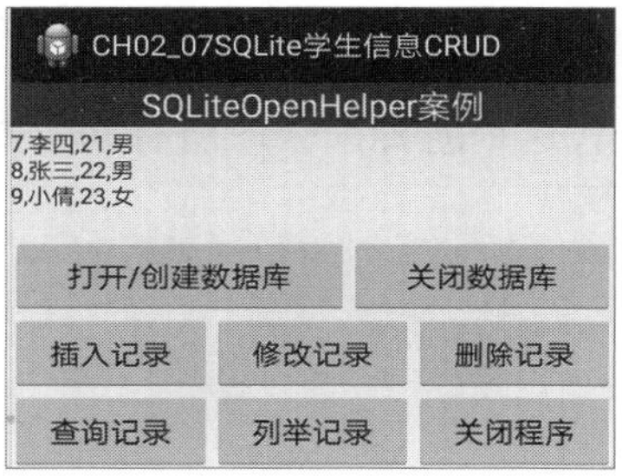

图 2-8　学生信息 CRUD 运行结果

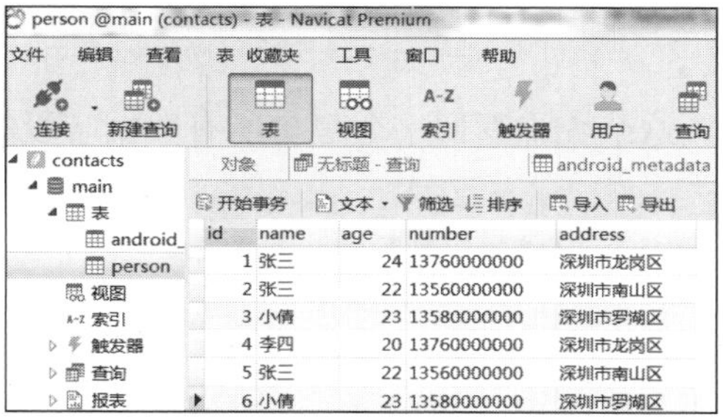

图 2-9　用 NaviCat 打开 SQLite 数据库 contacts.db 观察 person 表

2.2.4　ContentProvider 存储数据

一个ContentProvider类实现了一组标准的方法接口，从而能够让其他的应用保存或读取此ContentProvider的各种数据类型。也就是说，一个程序可通过实现一个ContentProvider的抽象接口将自己的数据暴露出去。外界根本看不到，也不用看到这个应用暴露的数据在应用当中是如何存储的，或者是用数据库存储还是用文件存储，还是通过网上获得，这些都不重要，重要的是外界可以通过这套标准及统一的接口和程序里的数据打交道，可以读取程序的数据，也可以删除程序的数据，当然，中间也会涉及一些权限的问题。

ContentProvider提供了一种多应用间数据共享的方式，例如，联系人信息可以被多个应用程序访问。ContentProvider是个实现了一组用于提供其他应用程序存取数据的标准方法的类。应用程序可以在ContentProvider中执行如下操作：查询数据、修改数据、添加数据、删除数据。标准的ContentProvider：Android提供了一些已经在系统中实现的标准ContentProvider，如联系人信息、图片库等，可以用这些ContentProvider访问设备上存储的联系人信息、图片等。

1. 查询记录

在ContentProvider中使用的查询字符串有别于标准的SQL查询。很多诸如select、add、delete、

modify等操作都使用一种特殊的URI（Uniform Resource Identifier，通用资源标识符）来进行，这种URI由三部分组成，“content://”代表数据的路径和一个可选的标识数据的ID。以下是一些示例URI：

```
content://media/internal/images        //这个URI将返回设备上存储的所有图片
content://contacts/people/             //这个URI将返回设备上的所有联系人信息
content://contacts/people/45           //这个URI返回联系人信息中ID为45的联系人记录
```

尽管这种查询字符串格式很常见，但是它看起来还是有点令人迷惑。为此，Android提供了一系列的帮助类（在android.provider包下），里面包含了很多以类变量形式给出的查询字符串，这种方式更容易让人理解。例如：

```
MediaStore.Images.Media.INTERNAL_CONTENT_URI
Contacts.People.CONTENT_URI
```

因此，如上面content://contacts/people/45这个URI就可以写成如下形式：

```
Uri person=ContentUris.withAppendedId(People.CONTENT_URI, 45);
```

然后执行数据查询：

```
Cursor cur=managedQuery(person, null, null, null);
```

这个查询返回一个包含所有数据字段的游标，可通过迭代这个游标来获取所有的数据：

```
private void displayRecords(){
    //该数组中包含了所有要返回的字段
    String columns[]=new String[]{People.NAME, People.NUMBER };
    Uri mContacts=People.CONTENT_URI;
    Cursor cur=managedQuery(mContacts,
          columns,                //要返回的数据字段
          null,                   //WHERE子句
          null,                   //WHERE 子句的参数
          null;                   //Order-by子句
      if(cur.moveToFirst()){
          String name=null;
          String phoneNo=null;
          do{                     //获取字段的值
          name=cur.getString(cur.getColumnIndex(People.NAME));
          phoneNo=cur.getString(cur.getColumnIndex(People.NUMBER));
             Toast.makeText(this,name+""+phoneNo, Toast.LENGTH_LONG).show();
 } while(cur.moveToNext()); }}
```

2. 修改记录

```
private void updateRecord(int recNo, String name){
    Uri uri=ContentUris.withAppendedId(People.CONTENT_URI, recNo);
    ContentValues values=new ContentValues();
    values.put(People.NAME, name);
```

```
    getContentResolver().update(uri, values, null, null);}
```

现在可以调用上面的方法来更新指定记录：

```
updateRecord(10, "XYZ");     //更改第10条记录的name字段值为XYZ
```

3. 添加记录

要增加记录，可以调用ContentResolver.insert()方法，该方法接受一个要增加的记录的目标URI，以及一个包含了新记录值的Map对象，调用后的返回值是新记录的URI，包含记录号。上面的例子中都是基于联系人信息簿这个标准的ContentProvider，现在继续来创建一个insertRecord()方法对联系人信息簿添加数据：

```
private void insertRecords(String name, String phoneNo){
ContentValues values=new ContentValues();
    values.put(People.NAME, name);
    Uri uri=getContentResolver().insert(People.CONTENT_URI, values);
    Log.d("ANDROID", uri.toString());
    Uri numberUri=Uri.withAppendedPath(uri,People.Phones.CONTENT_DIRECTORY);
    values.clear();
    values.put(Contacts.Phones.TYPE, People.Phones.TYPE_MOBILE);
    values.put(People.NUMBER, phoneNo);
    getContentResolver().insert(numberUri, values);}
```

这样就可以调用insertRecords(name,phoneNo)的方式来向联系人信息簿中添加联系人姓名和电话号码。

4. 删除记录

ContentProvider中的getContextResolver.delete()方法可以用来删除记录。

下面的记录用来删除设备上所有的联系人信息：

```
private void deleteRecords(){
    Uri uri=People.CONTENT_URI;
    getContentResolver().delete(uri, null, null);}
```

也可以指定WHERE条件语句来删除特定的记录：

```
getContentResolver().delete(uri, "NAME="+"'XYZ XYZ'", null);
```

这将会删除NAME为XYZ XYZ的记录。

5. 创建ContentProvider

至此已经知道如何使用ContentProvider，那么如何自己创建一个ContentProvider？要创建自己的ContentProvider，需要遵循以下几步：

（1）创建一个继承了ContentProvider父类的子类。

（2）定义一个名为CONTENT_URI，并且是public static final的Uri类型的类变量，用户必须为其指定唯一的字符串值，最好的方案是类的全名称。

（3）创建数据存储系统。大多数ContentProvider使用Android文件系统或SQLite数据库保存数据，但是也可以以任何想要的方式来存储。

（4）定义要返回给客户端的数据列名。如果正在使用Android数据库，则数据列的使用方式就和以往所熟悉的其他数据库一样。但是，必须为其定义一个名称为_id的列，它用来表示每条记录的唯一性。

（5）如果要存储字节型数据。（如位图文件等），那么保存该数据的数据列，其实是一个表示实际保存文件的URI字符串，客户端通过它来读取对应的文件数据。处理这种数据类型的ContentProvider需要实现一个名称为_data的字段，_data字段列出了该文件在Android文件系统上的精确路径。这个字段不仅是供客户端使用，而且也可以供ContentResolver使用。客户端可以调用ContentResolver.openOutputStream()方法处理该URI指向的文件资源。如果是ContentResolver本身，由于其拥有的权限比客户端要高，所以它能直接访问该数据文件。

（6）声明public static String型的变量，用于指定要从游标处返回的数据列。

（7）查询返回一个Cursor类型的对象。所有执行写操作的方法如insert()、update()以及delete()都将被监听。可以通过使用ContentResover().notifyChange()方法通知监听器关于数据更新的信息。

（8）在AndroidMenifest.xml中使用标签来设置ContentProvider。

（9）如果要处理的数据类型是一种比较新的类型，就必须先定义一个新的MIME类型，以供ContentProvider.geType(url)返回。MIME类型有两种形式：

一种是为指定的单个记录的MIME类型，还有一种是为多条记录的MIME类型。这里给出一种常用的格式：vnd.android.cursor.item/vnd.yourcompanyname.contenttype（单个记录的MIME类型）。例如，一个请求列车信息的URI如content://com.example.transportationprovider/trains/122可能就会返回typevnd.android.cursor.item/vnd.example.rail这样一个MIME（Multipurpose Internet Mail Extensions，多用途互联网邮件扩展）类型。

vnd.android.cursor.dir/vnd.yourcompanyname.contenttype（多个记录的MIME类型）。例如，一个请求所有列车信息的URI如content://com.example.transportationprovider/trains 可能就会返回vnd.android.cursor.dir/vnd.example.rail这样一个MIME类型。

6. ContentProvider存储数据典型应用案例：短信监听

该程序通过监听Uri(Content://sms)的数据改变即可监听用户短信的数据改变，并在监听器的onChange()方法里查询Uri(content://sms/outbox)的数据，这样即可获取用户正在发送的短信。MonitorSms程序的Java代码如下：

```
public class MonitorSms extends Activity {           //从活动基类派生子类
public void onCreate(Bundle savedInstanceState) { //子类重写onCreate()方法
    super.onCreate(savedInstanceState);              //调用基类onCreate()方法
    setContentView(R.layout.activity_main);          //设置活动界面布局
    //为content://sms的数据改变注册监听器
    getContentResolver().registerContentObserver(
        Uri.parse("content://sms"), true,
        new SmsObserver(new Handler()));}
    //提供自定义的ContentObserver监听器类
    private final class SmsObserver extends ContentObserver{
        public SmsObserver(Handler handler){super(handler);}
        public void onChange(boolean selfChange)
            {//查询发送箱中的短信(处于正在发送状态的短信放在发送箱)
            Cursor cursor=getContentResolver().query(
```

```
        Uri.parse("content://sms/outbox"),null,null,null,null);
    //遍历查询得到的结果集，即可获取用户正在发送的短信
    while(cursor.moveToNext()){
        StringBuilder sb=new StringBuilder();
        //获取短信的发送地址
        sb.append("address=").append(cursor
        .getString(cursor.getColumnIndex("address")));
        //获取短信的标题
        sb.append(";subject=").append(cursor
        .getString(cursor.getColumnIndex("subject")));
        //获取短信的内容
        sb.append(";body=").append(cursor
        .getString(cursor.getColumnIndex("body")));
        //获取短信的发送时间
        sb.append(";time=").append(cursor
    .getLong(cursor.getColumnIndex("date")));}}}}
```

别忘记了在AndroidManifest.xml文件中设置读短信READ_SMS的权限。

该程序的运行结果如图2-10所示。

图 2-10　短信监听案例运行主界面

2.2.5　网络存储数据

前面介绍的几种存储都是将数据存储在本地设备上，除此之外，还有一种存储（获取）数据的方式，通过网络来实现数据的存储和获取。可调用WebService返回的数据或者解析HTTP（Hypertext Transfer Protocol，超文本传输协议）实现网络数据交互。具体需要熟悉java.net.*、Android.net.*这两个包的内容，这里不再赘述，请大家参阅相关文档。下面是一个通过地区名称查询该地区的天气预报，以POST发送的方式发送请求到webservicex.net站点，访问WebService.webservicex.net站点上提供查询天气预报的服务。代码如下：

```
public class MyAndroidWeatherActivity extends Activity{
    //定义需要获取的内容来源地址
    private static final String SERVER_URL="http://www.webservicex.net/
WeatherForecast.asmx/GetWeatherByPlaceName";
    public void onCreate(Bundle savedInstanceState){
        super.onCreate(savedInstanceState);
        setContentView(R.layout.main);
        HttpPost request=new HttpPost(SERVER_URL);
        //根据内容来源地址创建一个HTTP请求，添加一个变量
        List<NameValuePair>params=new ArrayList<NameValuePair>();
```

```
        //设置一个地区名称，添加必需的参数
        params.add(new BasicNameValuePair("PlaceName", "ShenZhen"));
        try{//设置参数的编码
            request.setEntity(new UrlEncodedFormEntity(params, HTTP.UTF_8));
            //发送请求并获取反馈
            HttpResponse httpResponse=new DefaultHttpClient().execute(request);
            if(httpResponse.getStatusLine().getStatusCode()!= 404){
                //解析返回的内容
                String result=EntityUtils.toString(httpResponse.getEntity());
                System.out.println(result);}
        } catch(Exception e){e.printStackTrace();}}}
```

别忘记了在配置文件中设置访问网络权限。

2.3 案例

2.3.1 SharedPreferences 存储个人信息

在Android应用程序中，通常会将软件常用的一些参数存储到本地，以便下次打开程序时不再去设置这些参数，所以会为每个应用程序添加一些配置文件。本案例采用SharedPreferences数据存储技术保持个人登录信息，支持保存、恢复、重置、清空和退出等功能。程序界面布局activity_main.xml代码如下：

```
<LinearLayout xmlns:android="http://schemas.android.com/apk/res/android"
    android:layout_width="fill_parent"
    android:layout_height="fill_parent"
    android:orientation="vertical">
    <LinearLayout
        android:layout_width="match_parent"
        android:layout_height="wrap_content">
        <TextView
            android:layout_width="match_parent"
            android:layout_height="wrap_content"
            android:background="#0000FF"
            android:gravity="center_horizontal"
            android:text="个人登录信息"
            android:textColor="#FFFFFF"
            android:textSize="30sp"/>
    </LinearLayout>
    <LinearLayout
        android:layout_width="match_parent"
        android:layout_height="wrap_content">
        <TextView
```

```
            android:id="@+id/textView1"
            android:layout_width="wrap_content"
            android:layout_height="wrap_content"
            android:text="用户名"/>
        <EditText
            android:id="@+id/etname"
            android:layout_width="match_parent"
            android:layout_height="wrap_content"
            android:layout_weight="1"
            android:ems="10"/>
    </LinearLayout>
    <LinearLayout
        android:layout_width="match_parent"
        android:layout_height="wrap_content">
        <TextView
            android:id="@+id/textView2"
            android:layout_width="wrap_content"
            android:layout_height="wrap_content"
            android:text="年龄"/>
        <EditText
            android:id="@+id/etage"
            android:layout_width="wrap_content"
            android:layout_height="wrap_content"
            android:layout_weight="1"
            android:ems="10"
            android:inputType="number">
            <requestFocus />
        </EditText>
    </LinearLayout>
    <LinearLayout
        android:layout_width="match_parent"
        android:layout_height="wrap_content">
        <TextView
            android:id="@+id/textView3"
            android:layout_width="wrap_content"
            android:layout_height="wrap_content"
            android:text="城市"/>
        <EditText
            android:id="@+id/etcity"
            android:layout_width="wrap_content"
            android:layout_height="wrap_content"
            android:layout_weight="1"
            android:ems="10">
```

```
            <requestFocus />
        </EditText>
    </LinearLayout>
    <LinearLayout
        android:layout_width="match_parent"
        android:layout_height="wrap_content"
        android:orientation="horizontal">
        <TextView
            android:id="@+id/textView4"
            android:layout_width="wrap_content"
            android:layout_height="wrap_content"
            android:text="性别"/>
        <RadioGroup
            android:id="@+id/rgsex"
            android:layout_width="wrap_content"
            android:layout_height="wrap_content"
            android:orientation="horizontal">
            <RadioButton
                android:id="@+id/radio0"
                android:layout_width="wrap_content"
                android:layout_height="wrap_content"
                android:checked="true"
                android:text="男"/>
            <RadioButton
                android:id="@+id/radio1"
                android:layout_width="wrap_content"
                android:layout_height="wrap_content"
                android:text="女"/>
        </RadioGroup>
    </LinearLayout>
    <LinearLayout
        android:layout_width="match_parent"
        android:layout_height="wrap_content">

        <Button
            android:id="@+id/btnSave"
            android:layout_width="wrap_content"
            android:layout_height="wrap_content"
            android:layout_weight="1"
            android:text="保存"/>
        <Button
            android:id="@+id/btnRestore"
            android:layout_width="wrap_content"
```

```
            android:layout_height="wrap_content"
            android:layout_weight="1"
            android:text="恢复"/>
        <Button
            android:id="@+id/btnReset"
            android:layout_width="wrap_content"
            android:layout_height="wrap_content"
            android:layout_weight="1"
            android:text="重置"/>
        <Button
            android:id="@+id/btnClear"
            android:layout_width="wrap_content"
            android:layout_height="wrap_content"
            android:layout_weight="1"
            android:text="清空"/>
        <Button
            android:id="@+id/btnExit"
            android:layout_width="wrap_content"
            android:layout_height="wrap_content"
            android:layout_weight="1"
            android:text="退出"/>
    </LinearLayout>
</LinearLayout>
```

MainActivity类程序Java代码如下：

```
public class MainActivity extends Activity implements OnClickListener{
  private static final String TAG="MainActivity";
  private final String sp_name="person";
  private static final String sNAME="name";
  private static final String sAGE="age";
  private static final String sCITY="city";
  private static final String sSEX="sex";
  private static final String sMALE="男";
  private static final String sFEMALE="女";
  private EditText etname,etage,etcity;
  private RadioGroup rgsex;
  private RadioButton radio0,radio1;
  private String sexString;
  private int  intSex;
  private Button btnSave,btnRestore,btnReset,btnClear,btnExit;
  public void onCreate(Bundle savedinstancestate){
      super.onCreate(savedinstancestate);
      setContentView(R.layout.activity_main);
```

```
        initView();                  //绑定组件
        setOnClickListener();    //设置按钮单击监听器
      doRestore();}
    private void initView(){      //绑定组件
        etname=(EditText)findViewById(R.id.etname);
        etage=(EditText)findViewById(R.id.etage);
        etcity=(EditText)findViewById(R.id.etcity);
        rgsex=(RadioGroup)findViewById(R.id.rgsex);
        radio0=(RadioButton)findViewById(R.id.radio0);
        radio1=(RadioButton)findViewById(R.id.radio1);
        btnSave=(Button)findViewById(R.id.btnSave);
        btnRestore=(Button)findViewById(R.id.btnRestore);
        btnReset=(Button)findViewById(R.id.btnReset);
        btnClear=(Button)findViewById(R.id.btnClear);
        btnExit=(Button)findViewById(R.id.btnExit);}
    private void setOnClickListener(){      //设置按钮单击监听器
        btnSave.setOnClickListener(this);
        btnRestore.setOnClickListener(this);
        btnReset.setOnClickListener(this);
        btnClear.setOnClickListener(this);
        btnExit.setOnClickListener(this);}
    private void doReset(){
        etname.setText("");etage.setText("0");etcity.setText("");
        rgsex.check(-1);sexString="";}
    private void doExit(){
        finish();}
    private void doRestore(){
        SharedPreferences sp=getSharedPreferences(sp_name,
          MainActivity.MODE_PRIVATE);
        etname.setText(sp.getString(sNAME, ""));
        etage.setText(String.valueOf(sp.getInt(sAGE, 0)));
        etcity.setText(sp.getString(sCITY, ""));
        sexString=sp.getString(sSEX, "");
        if(sexString.equalsIgnoreCase(sMALE)){
           radio0.setChecked(true);radio1.setChecked(false);}
        else if(sexString.equalsIgnoreCase(sFEMALE)){
           radio0.setChecked(false);radio1.setChecked(true);
    }else{
        rgsex.check(-1);}}
    private void doSave(){
        //获得sharedpreferences对象
        SharedPreferences sp=getSharedPreferences(
            sp_name, MainActivity.MODE_PRIVATE);
```

```
        Editor editor=sp.edit();  //获得sharedpreferences.editor
        //保存组件中的值
        editor.putString(sNAME, etname.getText().toString());
        intSex=Integer.valueOf(etage.getText().toString().trim());
        editor.putInt(sAGE, intSex);
        editor.putString(sCITY, etcity.getText().toString());
        switch(rgsex.getCheckedRadioButtonId()){
        case R.id.radio0: sexString=sMALE;break;
        case R.id.radio1: sexString=sFEMALE;break;
        default:  sexString="";break;}
        editor.putString(sSEX, sexString);
        editor.commit();          //提交保存的结果}
    protected void onStop(){
        super.onStop();doSave();}
    public void onClick(View v){  //做按钮事件回调处理
        switch(v.getId()){
        case R.id.btnSave:doSave();break;
        case R.id.btnRestore:doRestore();break;
        case R.id.btnReset:doReset();break;
        case R.id.btnClear:doReset();doSave();break;
        case R.id.btnExit:doExit();break;}}}
```

该程序的运行结果如图2-11所示。采用DDMS调试工具将person.xml从模拟器中导出（见图2-12），用记事本打开person.xml可验证所存储数据的正确性，如图2-13所示。

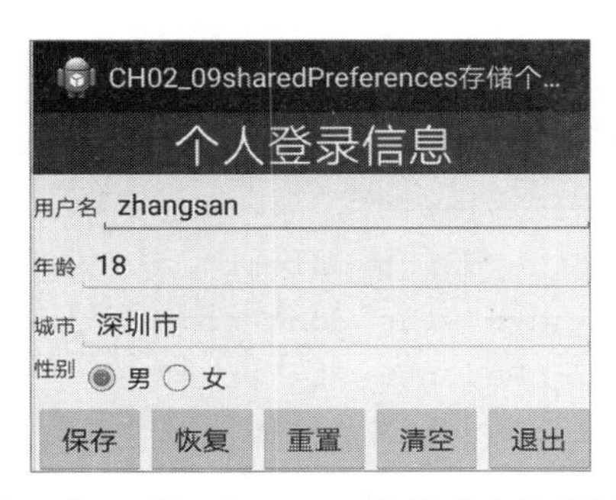

图 2-11　sharedPreferences 存储个人信息运行结果

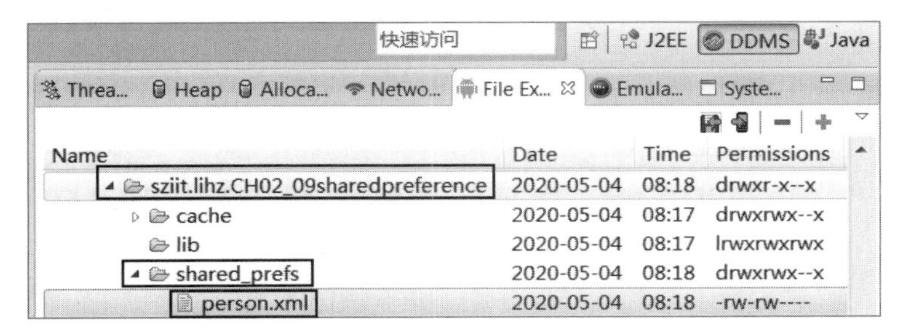

图 2-12　采用 DDMS 调试工具将 person.xml 从模拟器中导出

```
<?xml version=' 1.0' encoding=' utf-8' standalone=' yes' ?>
<map>
<string name="sex"> 男 </string>
<string name="name">zhangsan</string>
<int name="age"value="18"/>
<string name="city">shenzhen</string>
</map>
```

图 2-13　用记事本打开 person.xml

2.3.2　基于 SQLite 的设备状态信息显示

跟Web一些程序类似，在Android程序中通常会遇到一些数据需要存储、查询和显示，这时就需要使用到嵌入式平台上最流行的数据库SQLite存储数据，本案例利用SQLite数据库技术，显示物联网智能家居实训系统中的设备（如风扇、窗帘、壁灯和窗灯等）状态。该案例利用SQLite数据库存储设备信息，在程序启动时读取数据库中的数据呈现在界面上。

程序的界面布局activity_main.xml代码如下：

```
<RelativeLayout xmlns:android="http://schemas.android.com/apk/res/android"
    xmlns:tools="http://schemas.android.com/tools"
    android:layout_width="match_parent"
    android:layout_height="match_parent"
    android:paddingBottom="@dimen/activity_vertical_margin"
    android:paddingLeft="@dimen/activity_horizontal_margin"
    android:paddingRight="@dimen/activity_horizontal_margin"
    android:paddingTop="@dimen/activity_vertical_margin"
    tools:context=".MainActivity">
    <GridView
        android:id="@+id/gridView1"
        android:layout_width="match_parent"
        android:layout_height="wrap_content"
        android:layout_alignParentLeft="true"
        android:layout_alignParentTop="true"
        android:numColumns="1">
    </GridView>
</RelativeLayou>
```

equipment_list_item.xml程序清单如下：

```
<?xml version="1.0"encoding="utf-8"?>
<RelativeLayout xmlns:android="http://schemas.android.com/apk/res/android"
    android:layout_width="match_parent"
    android:layout_height="wrap_content">
    <ImageView
        android:id="@+id/img_image"
```

```
        android:layout_width="wrap_content"
        android:layout_height="wrap_content"
        android:layout_alignParentLeft="true"
        android:layout_alignParentTop="true"
        android:src="@drawable/ic_launcher"/>
    <TextView
        android:id="@+id/tv_datavalue"
        android:layout_width="wrap_content"
        android:layout_height="wrap_content"
        android:layout_alignBottom="@+id/img_image"
        android:layout_alignLeft="@+id/tv_name"
        android:text="TextView"/>
    <TextView
        android:id="@+id/tv_name"
        android:layout_width="wrap_content"
        android:layout_height="wrap_content"
        android:layout_alignParentTop="true"
        android:layout_toRightOf="@+id/img_image"
        android:layout_marginLeft="15dp"
        android:text="TextView"/>
    <TextView
        android:id="@+id/tv_area"
        android:layout_width="wrap_content"
        android:layout_height="wrap_content"
        android:layout_above="@+id/tv_datavalue"
        android:layout_alignParentRight="true"
        android:layout_alignParentTop="true"
        android:layout_marginRight="36dp"
        android:text="家用电器"/>
</RelativeLayout>
```

MainActivity类程序的Java代码如下：

```
public class MainActivity extends Activity{/***主页面**/
    private EquipmentDAO mEquipmentDAO=null; //设备业务类
    private GridView mGridView=null;  //显示设备的GridView
    private SimpleAdapter mSimpleAdapter=null;  //设备列表适配器
    //设备信息数据源
    private List<HashMap<String, Object>>mHashMapList=null;
    protected void onCreate(Bundle savedInstanceState){
    super.onCreate(savedInstanceState);
    setContentView(R.layout.activity_main);
    initial();    //初始化
    loadData();  //加载数据}
```

```
        private void initial(){/***初始化**/
        mEquipmentDAO=new EquipmentDAO(this);  //初始化设备业务类
        //初始化设备信息数据源
        mHashMapList=new ArrayList<HashMap<String,Object>>();
        mGridView=(GridView)findViewById(R.id.gridView1);  //初始化GridView
        //初始化适配器
        mSimpleAdapter=new SimpleAdapter(this,mHashMapList,R.layout. activity_
equipment_list_item,new String[]{"ItemImage","ItemNodeName","ItemDataValue"},
new int[]{R.id.img_image,R.id.tv_name,R.id.tv_datavalue});
        mGridView.setAdapter(mSimpleAdapter); }
         private void loadData(){/***加载SQLite数据库中的数据**/
            List<EquipmentEntity>equipmentList=mEquipmentDAO.getAll();
            if(equipmentList!=null){
                EquipmentEntity equipmentEntity=null;
                HashMap<String, Object>item=null;
                for(int i=0;i<equipmentList.size();i++){
                   equipmentEntity=equipmentList.get(i);
                   if(equipmentEntity!=null){
                      item=new HashMap<String, Object>();
                      item.put("ItemImage", R.drawable.ic_launcher);
                      item.put("ItemNodeName", equipmentEntity.getName());
                      item.put("ItemDataValue",equipmentEntity.getLastUpdateTime());
                      item.put("ItemArea", "");
                      mHashMapList.add(item); }}
           mSimpleAdapter.notifyDataSetChanged();}}}
```

EquipmentDAO类程序清单如下：

```
   public class EquipmentDAO{/***设备业务类**/
       private DBOpenHelper helper;
       public EquipmentDAO(Context context){
           helper=new DBOpenHelper(context);}
       public List<EquipmentEntity>getAll(){  /***获取所有设备**/
           ArrayList<EquipmentEntity>equpmentList=new ArrayList<EquipmentEntity>();
           SQLiteDatabase db=helper.getReadableDatabase();
           String sql="select [id],[name],[nodeID],[nodeTypeID],[lastUpdateTime] from
[equipment]";
           Cursor c=db.rawQuery(sql, null);
           while(c.moveToNext()){
               equpmentList.add(new EquipmentEntity(c.getInt(0), c.getString(1),
c.getString(2), c.getString(3), c.getString(4))); }
       c.close();db.close();return equpmentList; }}
```

DBOpenHelper类程序清单如下（数据库名为data.db，表名为equipment）：

```
    public class DBOpenHelper extends SQLiteOpenHelper{/***数据库助手类**/
        private static final String DATABASENAME="data.db"; /***Database name**/
        private static final int DATABASEVERSION=2; /***Database Version**/
        public DBOpenHelper(Context context){
            super(context, DATABASENAME, null, DATABASEVERSION);}
       public void onCreate(SQLiteDatabase db){
            db.execSQL("CREATE TABLE [equipment]("
                +"[id] integer not null primary key autoincrement,"
                +"[name] varchar(255),"
                +"[nodeID] varchar(255),"
                +"[nodeTypeID] varchar(255),"
                +"[lastUpdateTime] varchar(255))");
                db.execSQL("insert into [equipment]([name],[nodeID],[nodeTypeID],
[lastUpdateTime])values('风扇01','01','0001','2020/5/4 12:00:20')");
            db.execSQL("insert into [equipment]([name],[nodeID],[nodeTypeID],
[lastUpdateTime])values(' 窗帘02' ,' 02' ,' 0002' ,' 2020/5,/4 12:12:06' )");
            db.execSQL("insert into [equipment]([name],[nodeID],[nodeTypeID],
[lastUpdateTime])values(' 壁灯03' ,' 03' ,' 0003' ,' 2020/5/4 12:02:08' )");
            db.execSQL("insert into [equipment]([name],[nodeID],[nodeTypeID],
[lastUpdateTime])values(' 窗灯04' ,' 04' ,' 0004' ,' 2020/5/4 12:08:50' )");}
        public void onUpgrade(SQLiteDatabase db,int oldVersion,int newVersion){
            db.execSQL("DROP TABLE IF EXISTS [equipment]");
            onCreate(db); }}
```

EquipmentEntity类程序清单如下：

```
    package com.example.entity;
    public class EquipmentEntity{/***设备实体类**/
        private int id; /***编号**/
        private String name; /***节点名称**/
        private String nodeID; /***节点编号**/
        private String nodeTypeID; /***节点类型**/
        private String lastUpdateTime; /***最后更新时间**/
        public EquipmentEntity(){}
        public EquipmentEntity(int id,String name,String nodeID,String nodeTypeID,
String lastUpdateTime){
           this.id=id;  this.name=name;this.nodeID=nodeID;
           this.nodeTypeID=nodeTypeID;this.lastUpdateTime=lastUpdateTime;}
        public int getId(){return id;}
        public void setId(int id){this.id=id;}
        public String getName(){return name;}
        public void setName(String name){this.name=name;}
        public String getNodeID(){return nodeID;}
        public void setNodeID(String nodeID){this.nodeID=nodeID;}
```

```
public String getNodeTypeID(){return nodeTypeID;}
public void setNodeTypeID(String nodeTypeID){
   this.nodeTypeID=nodeTypeID;}
public String getLastUpdateTime(){return lastUpdateTime;}
public void setLastUpdateTime(String lastUpdate Time){
  this.lastUpdateTime=lastUpdateTime;}}
```

该程序的运行结果如图 2-14 所示。用 DDMS 导出 SQLite 数据库 data.db，如图 2-15 所示。用 Navicat 观察 SQLite 数据库 data.db 中 equipment 数据，如图 2-16 所示。

CH02_10SQLite设备状态信息...
风扇01
家用电器
2020/5/4 12:00:20
窗帘02
家用电器
2020/5/4 12:12:06
壁灯03
家用电器
2020/5/4 12:02:08
窗灯04
家用电器
2020/5/4 12:08:50

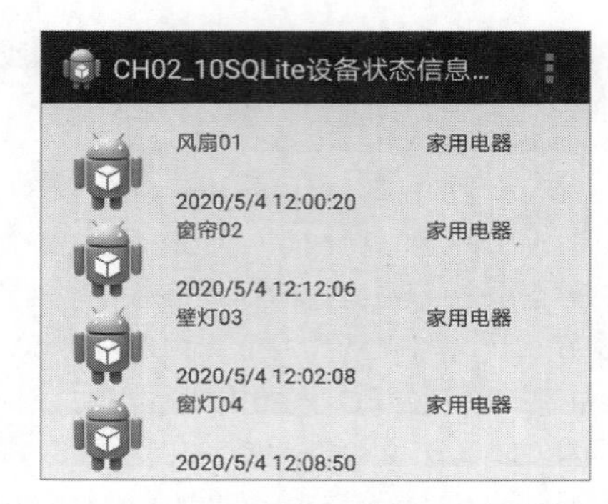

图 2-14　基于 SQLite 的设备状态信息显示

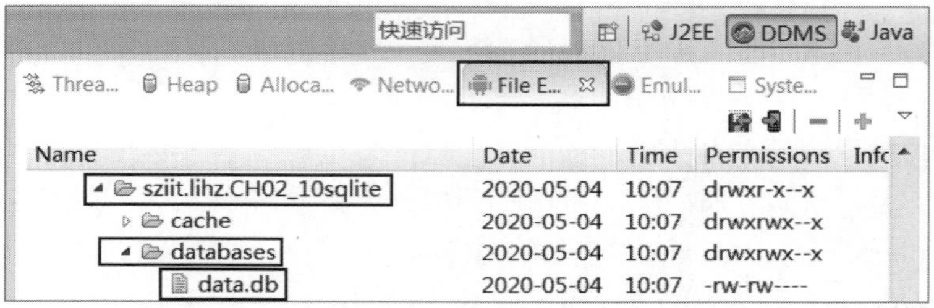

快速访问
J2EE
DDMS
Java
Threa...
Heap
Alloca...
Netwo...
File E...
Emul...
Syste...
Name
Date
Time
Permissions
Info
sziit.lihz.CH02_10sqlite 2020-05-04 10:07 drwxr-x--x
cache 2020-05-04 10:07 drwxrwx--x
databases 2020-05-04 10:07 drwxrwx--x
data.db 2020-05-04 10:07 -rw-rw----

图 2-15　用 DDMS 导出 SQLite 数据库 data.db

equipment @main (data) - 表 - Navicat Premium
文件 编辑 查看 表 收藏夹 工具 窗口 帮助
连接
新建查询
表
视图
A-Z
索引
触发器
用户
data
main
表
视图
索引
触发器
查询
报表
备份
对象
equipment @main (data) - 表
开始事务
文本
筛选
排序
导入
导出
id name nodeID nodeTypeID lastUpdateTime
1 风扇01 01 0001 2020/5/4 12:00:20
2 窗帘02 02 0002 2020/5/4 12:12:06
3 壁灯03 03 0003 2020/5/4 12:02:08
4 窗灯04 04 0004 2020/5/4 12:08:50

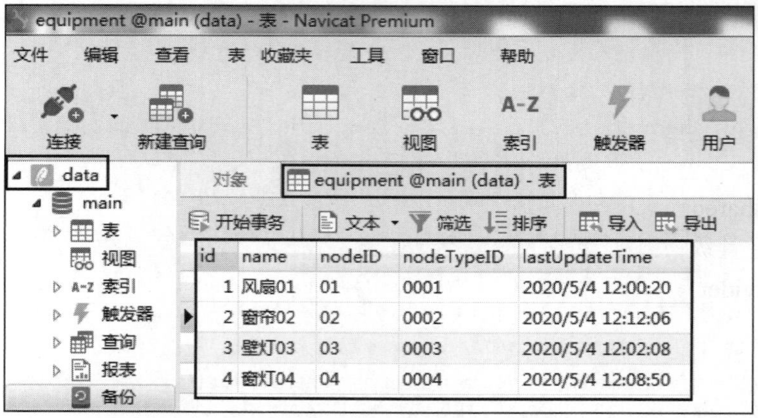

图 2-16　用 Navicat 观察 SQLite 数据库 data.db 中 equipment 数据

2.4 知识扩展

数据共享（ContentProvider）

（1）ContentProvider的基本概念。

（2）ConentProvider是一种特殊的存储数据的类型，它提供了一套标准的接口来获取、操作数据。

（3）ContentReslover。

（4）URI与URL。

（5）ContentReslover主要接口。

（6）ContentProvider的使用方法。

（7）ContentProvider典型应用案例。

本章小结

本章主要介绍了Android的5种不同的数据存储方式：SharedPreferences存储数据、文件存储数据、SQLite数据库存储数据、ContentProvider存储数据、网络存储数据的特点及其用法。

强化练习

一、填空题

1. Android的5种不同的数据存储方式为（　　）、（　　）、（　　）、（　　）、（　　）。
2. SharedPreferences提供了Android平台常规的（　　）、（　　）、（　　）、（　　）、（　　）数据数类的保存。
3. SharedPreferences是基于xml文件存储（　　）数据。
4. Activity提供了（　　）方法可以用于把数据输出到文件中。
5. SQLite基本上符合（　　）标准，和其他的主要SQL数据库没什么区别。
6. 可以调用（　　）返回的数据或者解析（　　）实现网络数据交互。

二、简答题

1. 简述实现SharedPreferences存储的步骤。
2. SQLite是轻量级嵌入式数据库引擎，其特点有哪些？
3. ContentProvider最大的特点是什么？

第3章 Android网络应用

学习目标

视频

Android网络应用

- 理解Android常用网络的几种基本概念。
- 掌握HTTP通信的实现。
- 掌握Socket通信的实现。
- 熟悉蓝牙技术；熟悉蓝牙的开发应用。
- 熟悉Wi-Fi技术；熟悉Wi-Fi的开发应用。
- 熟悉HttpClient技术简介和应用。
- 掌握GET和POST请求方式。
- 掌握响应数据的解析。

3.1 学习导入

Android的网络通信是数据交流的一个重要途径，也是打通Android应用程序与外界数据交流的主要途径。下面就介绍一下Android网络通信的基础知识。

3.1.1 网络协议

定义1：网络上各计算机间的一种通信语言，各计算机都必须遵守。

定义2：网络上所有设备（网络服务器、计算机及交换机、路由器和防火墙等）之间的通信规则集合，定义通信信息格式、含义，使得网络上各种设备能够相互交换信息。

TCP/IP是一种目前因特网上使用最广泛的网络协议。多数计算机网络一般都采用分层体系结构(见图3-1)，即每一层都建立在它的下一层之上，向它的上一层提供一定的服务，而把如何实现这一服务的细节对上一层加以屏蔽，一台设备上的一层与另一台设备上的同一层进行通信的规则就是这一层的协议。TCP网络协议分层结构：应用层、传输层、网际层、网络接口层。面向应用的协议：简单邮件传输协议（SMTP）、超文本传输协议（Hypertext Transfer Protocol，HTTP）、文件传输协议（FTP）和域名系统（DNS）等。传输层位于应用层之下，包括面向连接的传输控制协议（Transmission Control Protocol，TCP）和无连接的用户数据报协议（User Datagram Protocol，UDP）。TCP 是一种面向连接的保证可靠传输的协议，UDP 是一种无连接的协议，每个数据报都是一个独立的信息单位。

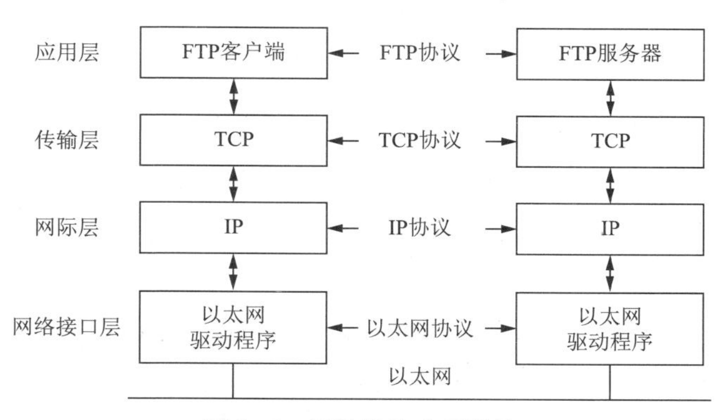

图 3-1　网络协议分层结构

3.1.2　HTTP 通信

HTTP（Hyper Text Transfer Protocol，超文本传输协议）通信是网络通信的重要方式，主要指在获得网络操作权限的前提下进行网络连接、数据发送和接收的通信方式。HTTP 是利用 TCP 在 Web 服务器和客户端间传输信息的协议。客户端使用 Web 浏览器发起 HTTP 请求给 Web 服务器，Web 服务器发送被请求的信息给客户端。HTTP 采用了请求/响应模型，用于传送 WWW 方式的数据。客户端向服务器发送一个请求，请求头包含了请求的方法、URI、协议版本，以及包含请求修饰符、客户信息和内容的类似于 MIME 的消息结构。服务器以一个状态行作为响应，响应的内容包括消息协议的版本、成功或者错误编码，还包括服务器信息、实体元信息，以及可能的实体内容。HTTP 通信模型示意图如图 3-2 所示。

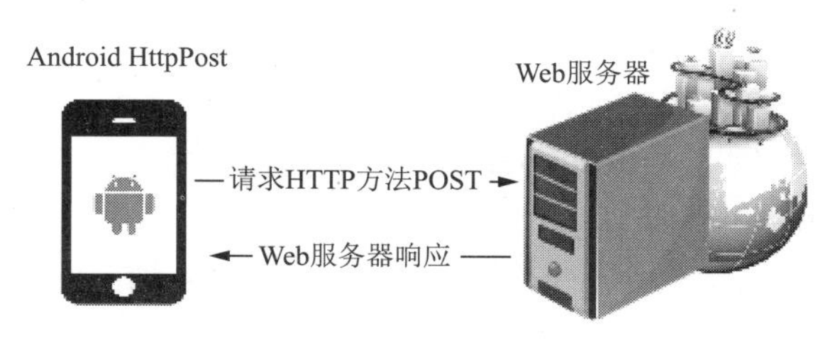

图 3-2　HTTP 通信模型示意图

HTTP 是一个应用层协议，由请求和响应构成，是一个标准的客户端服务器模型。HTTP 是一个无状态的协议，通常承载于 TCP 协议之上，一次 HTTP 请求操作，工作过程如下：

（1）客户端与服务器需建立连接。只要单击某个超链接，HTTP 的工作就开始。

（2）建立连接后，客户端发送一个请求给服务器，请求方式格式为：统一资源标识符（URL）、协议版本号，后面是 MIME 信息（包括请求修饰符、客户机信息和可能的内容）。

（3）服务器接到请求后，给予相应响应信息，其格式为一个状态行，包括信息的协议版本号、一个成功或错误的代码，后边是MIME信息（包括服务器信息、实体信息和可能的内容）。

（4）客户端接收服务器所返回信息通过浏览器显示在用户屏幕，然后客户端与服务器断开连接。

HTTP报文由从客户端到服务器的请求和从服务器到客户机的响应构成。

请求报文格式如下：

```
请求行 - 通用信息头 - 请求头 - 实体头 - 报文主体
```

应答报文格式如下：

```
状态行 - 通用信息头 - 响应头 - 实体头 - 报文主体
```

Google以网络搜索引擎著称，自然而然也会使Android SDK拥有强大的HTTP访问能力。在Android SDK中，Google集成了Apache的HttpClient模块。

3.1.3　Socket通信

Socket（套接字）用于描述IP地址和端口，是一个通信连接的句柄。Android Socket与Java Socket类同，Socket 提供了程序内部与外界通信的端口并为通信双方提供了数据传输通道。Socket通信在双方建立起连接后就可直接进行数据的传输，在连接时可实现信息的主动推送，不需要客户端每次向服务器发送请求。Socket通信模型示意图如图3-3所示。

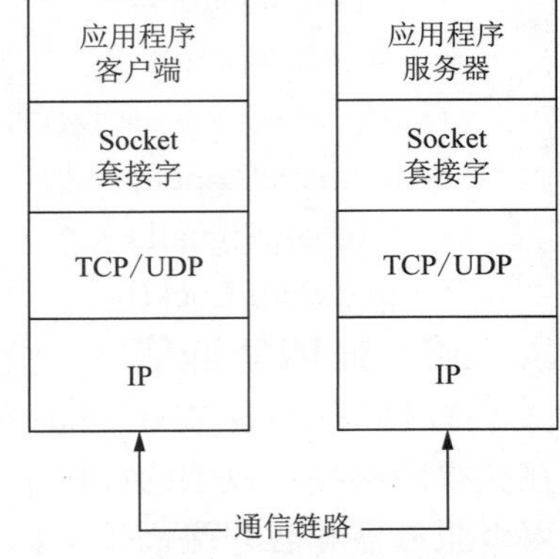

图 3-3　Socket 通信模型示意图

在使用Android执行网络操作时，需要请求网络通信权限，为此首先需在程序的配置文件AndroidManifest.xml中添加如下权限请求：

```
<uses-permission android:name="android.permission.INTERNET"/>
<uses-permission android:name="android.permission.ACCESS_NETWORK_STATE"/>
```

3.1.4　Wi-Fi

Wi-Fi（Wireless Fidelity，无线保真）是一种无线联网技术，常见的是使用无线路由器。在无线路由器信号覆盖的范围内都可采用Wi-Fi连接的方式进行联网。如果无线路由器连接了一个ADSL线路或其他的联网线路，则又被称为“热点”。

在Android系统中，提供了android.net.wifi包供操作，主要包括4个主要的类：ScanResult、wifiConfiguration、WifiInfo、WifiManager。

（1）ScanResult：主要用来描述已经检测出的接入点，包括接入点的地址、接入点的名称、身份认证、频率、信号强度等信息。

（2）WifiConfiguration：Wi-fi网络的配置，包括安全设置等。主要包括6个子类，其中WifiConfiguration.AuthAlgorthm 用来判断加密方法，WifiConfiguration.GroupCipher 获取使用GroupCipher 的方法进行加密，WifiConfiguration.KeyMgmt 获取使用KeyMgmt，WifiConfiguration.PairwiseCipher 获取使用WPA 方式的加密，WifiConfiguration.Protocol 获取使用协议的加密，WifiConfiguration.Status 获取当前网络的状态。

（3）WifiInfo：Wi-Fi无线连接的描述，包括接入点、网络连接状态、隐藏的接入点、IP地址、连接速度、MAC地址、网络ID、信号强度等信息。主要包括以下方法：

- getBSSID()：获取BSSID。
- getDetailedStateOf()：获取客户端的连通性。
- getHiddenSSID()：获得SSID 是否被隐藏。
- getIpAddress()：获取IP 地址。
- getLinkSpeed()：获得连接的速度。
- getMacAddress()：获得MAC地址。
- getRssi()：获得802.11n网络的信号。
- getSSID()：获得SSID。
- getSupplicanState()：返回具体客户端状态的信息。

（4）WifiManager：用来管理的Wi-Fi连接，主要包括以下方法：

- addNetwork(WifiConfiguration config)：通过获取到的网络的连接状态信息来添加网络。
- calculateSignalLevel(int rssi, int numLevels)：计算信号的等级。
- compareSignalLevel(int rssiA, int rssiB)：对比连接A和连接B。
- createWifiLock(int lockType, String tag)：创建一个Wi-Fi锁，锁定当前的Wi-Fi连接。
- disableNetwork(int netId)：让一个网络连接失效。
- disconnect()：断开连接。
- enableNetwork(int netId, Boolean disableOthers)：连接一个连接。
- getConfiguredNetworks()：获取网络连接的状态。
- getConnectionInfo()：获取当前连接的信息。
- getDhcpInfo()：获取DHCP 的信息。
- getScanResulats()：获取扫描测试的结果。
- getWifiState()：获取一个Wi-Fi接入点是否有效。
- isWifiEnabled()：判断一个Wi-Fi连接是否有效。
- pingSupplicant()：ping 一个连接，判断是否能连通。
- ressociate()：即使连接没有准备好，也要连通。
- reconnect()：如果连接准备好了，连通。
- removeNetwork()：移除某一个网络。
- saveConfiguration()：保留一个配置信息。
- setWifiEnabled()：让一个连接有效。
- startScan()：开始扫描。
- updateNetwork (WifiConfiguration config)：更新一个网络连接的信息。

Wi-Fi典型应用案例：附近Wi-Fi热点搜索

该案例通过Wi-Fi来搜索附近的Wi-Fi可用热点。由于模拟器中不支持Wi-Fi，所以程序的运行需要通过真机进行测试。首先设计程序界面布局大纲（见图3-4）；其次请求网络通信权限和访问Wi-Fi网络相关权限，通过ConnectivityManager对

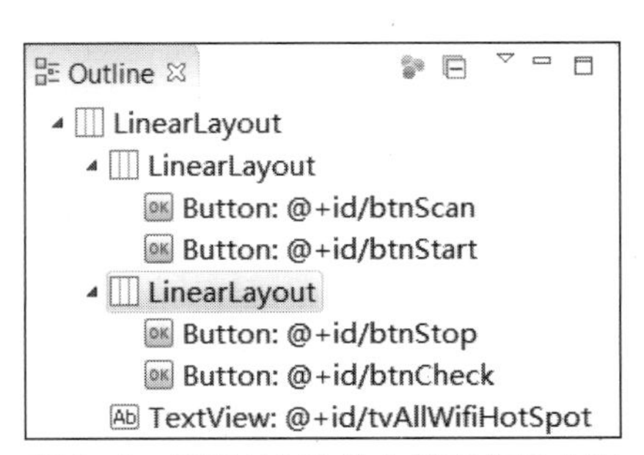

图 3-4　附近 Wi-Fi 热点搜索布局大纲

象检查网络连接；然后设计 WifiAdmin 类封装 Wi-Fi 相关调用方法，最后在 Activity 实现附近 Wi-Fi 热点搜索功能调用。主布局 activity_main.xml 源代码清单如下：

```
<LinearLayout xmlns:android="http://schemas.android.com/apk/res/android"
    xmlns:tools="http://schemas.android.com/tools"
    android:layout_width="match_parent"
    android:layout_height="match_parent"
    android:orientation="vertical"
    android:paddingBottom="@dimen/activity_vertical_margin"
    android:paddingLeft="@dimen/activity_horizontal_margin"
    android:paddingRight="@dimen/activity_horizontal_margin"
    android:paddingTop="@dimen/activity_vertical_margin"
    tools:context=".MainActivity">
    <LinearLayout
        android:layout_width="match_parent"
        android:layout_height="wrap_content"
        android:orientation="horizontal">
        <Button
            android:id="@+id/btnScan"
            android:layout_width="wrap_content"
            android:layout_height="wrap_content"
            android:layout_weight="1"
            android:text="扫描网络"/>
        <Button
            android:id="@+id/btnStart"
            android:layout_width="wrap_content"
            android:layout_height="wrap_content"
            android:layout_weight="1"
            android:text="打开Wi-Fi"/>
    </LinearLayout>
    <LinearLayout
        android:layout_width="match_parent"
        android:layout_height="wrap_content"
        android:orientation="horizontal">
        <Button
            android:id="@+id/btnStop"
            android:layout_width="wrap_content"
            android:layout_height="wrap_content"
            android:layout_weight="1"
            android:text="关闭Wi-Fi"/>
        <Button
            android:id="@+id/btnCheck"
            android:layout_width="wrap_content"
```

```
            android:layout_height="wrap_content"
            android:layout_weight="1"
            android:text="Wi-Fi 状态"/>
    </LinearLayout>
    <TextView
        android:id="@+id/tvAllWifiHotSpot"
        android:layout_width="wrap_content"
        android:layout_height="wrap_content"
        android:textSize="20sp"/>
</LinearLayout>
```

WifiAdmin类源代码清单如下：

```
public class WifiAdmin{
  private WifiManager manager;  //定义一个WifiManager对象
  private WifiInfo info;  //定义一个WifiInfo对象
  private List<ScanResult>list;  //扫描出的网络连接列表
  private List<WifiConfiguration>config;  //网络连接列表
  WifiLock lock;
  public WifiAdmin(Context context){  //取得WifiManager对象
      manager=(WifiManager)context
        .getSystemService(Context.WIFI_SERVICE);
      info=manager.getConnectionInfo();  //取得WifiInfo对象}
  public void openWifi(){  //打开Wi-Fi
      if(!manager.isWifiEnabled()){manager.setWifiEnabled(true);}}
  public void closeWifi(){  //关闭Wi-Fi
      if(manager.isWifiEnabled()){manager.setWifiEnabled(false);}}
  public int checkState(){  //检查当前Wi-Fi状态
      return manager.getWifiState();}
  public void acquireWifiLock(){lock.acquire();}  //锁定Wi-Fi
  public void releaseWifiLock(){  //解锁wifiLock
      if(lock.isHeld()){  //判断是否锁定
          lock.acquire();}}
  public void createWifiLock(){  //创建一个lock
      lock=manager.createWifiLock("hotspot");}
  //得到配置好的网络
  public List<WifiConfiguration>getConfiguration(){return config;}
  //指定配置好的网络进行连接
  public void connetionConfiguration(int index){
      if(index >config.size()){return;}
      //连接配置好指定ID的网络
    manager.enableNetwork(config.get(index).networkId,true);}
  public void startScan(){
      manager.startScan();
```

```
        list=manager.getScanResults();  //得到扫描结果
        //得到配置好的网络连接
        config=manager.getConfiguredNetworks();}
        public List<ScanResult>getWifiList(){return list;}  //得到网络列表
    public StringBuffer lookUpScan(){  //查看扫描结果
        StringBuffer sb=new StringBuffer();
        for(int i=0; i<list.size(); i++){
            sb.append("Index_"+new Integer(i+1).toString()+":");
            sb.append((list.get(i)).toString()).append("\n");}
        return sb;}
    public String getMacAddress(){
        return(info==null)? "NULL": info.getMacAddress();}
    public String getBSSID(){
        return(info==null)? "NULL": info.getBSSID();}
    public int getIpAddress(){
        return(info==null)? 0 : info.getIpAddress();}
    public int getNetWordId(){  //得到连接的ID
        return(info==null)? 0 : info.getNetworkId();}
    public String getWifiInfo(){  //得到wifiInfo的所有信息
        return(info==null)? "NULL": info.toString();}
    //添加一个网络并连接
    public void addNetWork(WifiConfiguration configuration){
        int wcgId=manager.addNetwork(configuration);
        manager.enableNetwork(wcgId, true);}
    public void disConnectionWifi(int netId){  //断开指定ID的网络
        manager.disableNetwork(netId);manager.disconnect();}}
```

主程序代码通过getSystemService(WIFI_SERVICE)获Wi-Fi服务，通过Wi-FiManager.getScanResults()获取有效可用的Wi-Fi热点。MainActivity源代码清单如下：

```
public class MainActivity extends Activity implements OnClickListener{
    private TextView tvAllWifiHotSpot;
    private Button btnScan,btnStart,btnStop,btnCheck;
    private WifiAdmin mWifiAdmin;
    private List<ScanResult>list;  //扫描结果列表
    private ScanResult mScanResult;
    private StringBuffer sb=new StringBuffer();
    @Override
    public void onCreate(Bundle savedInstanceState){
        super.onCreate(savedInstanceState);
        setContentView(R.layout.activity_main);
ConnectivityManager cm=(ConnectivityManager)
getSystemService(Context.CONNECTIVITY_SERVICE);
         NetworkInfo ni=cm.getActiveNetworkInfo();
```

```
        if(ni!=null && ni.isConnected()){
            Toast.makeText(getApplicationContext(),"网络连接正常!",
        Toast.LENGTH_LONG).show();
      }else{  //错误处理
            Toast.makeText(getApplicationContext(),"网络连接异常!",
            Toast.LENGTH_LONG).show();return;}
      mWifiAdmin=new WifiAdmin(MainActivity.this);
      init();}
  public void init(){findViews();setOnClickListener();}
  private void findViews(){
      tvAllWifiHotSpot=(TextView)findViewById(R.id.tvAllWifiHotSpot);
      btnScan=(Button)findViewById(R.id.btnScan);
      btnStart=(Button)findViewById(R.id.btnStart);
      btnStop=(Button)findViewById(R.id.btnStop);
      btnCheck=(Button)findViewById(R.id.btnCheck);}
  private void setOnClickListener(){
      btnScan.setOnClickListener(this);
      btnStart.setOnClickListener(this);
      btnStop.setOnClickListener(this);
      btnCheck.setOnClickListener(this);}
  public void getAllNetWorkList(){  //每次单击扫描之前清空上一次的扫描结果
      if(sb!=null){sb=new StringBuffer();}
      mWifiAdmin.startScan();  //开始扫描网络
      list=mWifiAdmin.getWifiList();
      if(list!=null){
          for(int i=0; i < list.size(); i++){
          mScanResult=list.get(i); //得到扫描结果
          sb=sb.append(mScanResult.BSSID+" ")
              .append(mScanResult.SSID+"  ")
              .append(mScanResult.capabilities+"  ")
              .append(mScanResult.frequency+"  ")
              .append(mScanResult.level+"\n\n");}
  tvAllWifiHotSpot.setText("扫描到的wifi网络: \n"+sb.toString());}}
  public void onClick(View v){
    switch(v.getId()){
    case R.id.btnScan:getAllNetWorkList();break;  //扫描网络
    case R.id.btnStart:mWifiAdmin.openWifi();  //打开Wi-Fi
  Toast.makeText(MainActivity.this,"当前Wi-Fi状态为: "+
  mWifiAdmin.checkState(), 1).show();break;
    case R.id.btnStop:mWifiAdmin.closeWifi();  //关闭Wi-Fi
  Toast.makeText(MainActivity.this,"当前Wi-Fi状态为: "+
  mWifiAdmin.checkState(), 1).show();break;
  case R.id.btnCheck:  //Wi-Fi状态
```

```
        Toast.makeText(MainActivity.this,"当前wifi状态为: "+
  mWifiAdmin.checkState(), 1).show();break;}}}
```

记得要在程序的AndroidManifest.xml中增加请求网络通信权限和Wi-Fi访问权限：

```
<uses-permission android:name="android.permission.CHANGE_NETWORK_STATE"/>
<uses-permission android:name="android.permission.CHANGE_WIFI_STATE"/>
<uses-permission android:name="android.permission.INTERNET"/>
```

该程序运行结果如图3–5所示。

3.1.5 蓝牙通信

蓝牙通信符合蓝牙协议(BlueTooth)V1.x，使用2.4 GHz的ISM（工业、科学、医学）频段。频道共有23个或79个，频道间隔均为1 MHz，采用时分双工方式，调制方式为BT= 0.5的GFSK。蓝牙的数据传输速率可达1 Mbit/s，与红外一样，蓝牙的传输距离也较短。

图3–5　附近Wi-Fi热点搜索运行结果

蓝牙技术的系统结构分为三大部分：底层硬件模块、中间协议层和高层应用。

底层硬件部分包括无线跳频（RF）、基带（BB）和链路管理(LM)。无线跳频层通过2.4 GHz无须授权的ISM频段的微波，实现数据位流的过滤和传输，本层协议主要定义了蓝牙收发器在此频带正常工作所需要满足的条件。基带负责跳频以及蓝牙数据和信息帧的传输。链路管理负责连接、建立和拆除链路并进行安全控制。

中间协议层包括逻辑链路控制和适应协议、服务发现协议、串口仿真协议和电话通信协议。逻辑链路控制和适应协议具有完成数据拆装、控制服务质量和复用协议的功能，该层协议是其他各层协议实现的基础。服务发现协议层为上层应用程序提供一种机制来发现网络中可用的服务及其特性。串口仿真协议层具有仿真9针RS–232串口的功能。电话通信协议层则提供蓝牙设备间语音和数据的呼叫控制指令。

主机控制接口层（HCI）是蓝牙协议中软硬件之间的接口，它提供了一个调用基带、链路管理、状态和控制寄存器等硬件的统一命令接口。蓝牙设备之间进行通信时，HCI以上的协议软件实体在主机上运行，而HCI以下的功能由蓝牙设备完成，二者之间通过一个对两端透明的传输层进行交互。

在蓝牙协议栈的最上部是各种高层应用框架。其中较典型的有拨号网络、耳机、局域网访问、文件传输等，它们分别对应一种应用模式。各种应用程序可以通过各自对应的应用模式实现无线通信。

蓝牙通信过程包括两步：搜索周围蓝牙设备；连接某一蓝牙设备。在Android系统中，提供了android.bluetooth包，里面提供了开发蓝牙应用所需要的主要接口。蓝牙功能包如表3–1所示。BluetoothAdapter中动作常量如表3–2所示。BluetoothAdapter常用方法如表3–3所示。

表 3-1　蓝牙功能包

功能包	功　能
BluetoothAdapter	本地蓝牙设备的适配类，所有的蓝牙操作都需要通过该类完成
BluetoothDevice	蓝牙设备类，代表了蓝牙通信过程中的远端设备
BluetoothSocket	蓝牙通信套接字，代表了与远端设备的连接点。使用 Socket 本地程序可以通过 inputstream 和 outputstream 与远端程序通信
BluetoothServerSocket	服务器通信套接字，与 TCP ServerSocket 类似
BluetoothClass	用于描述远端设备的类型、特点等信息，通过 getBluethoothClass() 方法获取代表远端设备属性的 BluetoothClass 对象

表 3-2　BluetoothAdapter 中动作常量

动作常量	说　明
ACTION_DISCOVERY_FINISHED	完成蓝牙搜索
ACTION_DISCOVERY_STARTED	开始搜索
ACTION_LOCAL_NAME_CHANGED	更改蓝牙的名字
ACTION_REQUEST_DISCOVERABLE	请求能够被搜索
ACTION_REQUEST_ENABLE	请求启动蓝牙
ACTION_SCAN_MODE_CHANGED	扫描模式已改变
ACTION_STATE_CHANGED	状态已改变

表 3-3　BluetoothAdapter 常用方法

常用方法	返回值
public boolean cancelDiscovery()	取消当前设备的发现查找进程
public static boolean checkBluetoothAddress(String address)	验证皆如 "00:43:A8:23:10:F0" 之类的蓝牙地址，字母必须为大写才有效。参数 address 为字符串形式的蓝牙模块地址返回值地址
public boolean disable()	关闭本地蓝牙适配器——不能在没有明确关闭蓝牙的用户动作中使用
public boolean enable()	打开本地蓝牙适配器——不能在没有明确打开蓝牙的用户动作中使用
public String getAddress()	返回本地蓝牙适配器的硬件地址。例如 "00:11:22:AA:BB:CC"
public Set<BluetoothDevice>getBondedDevices()	返回已经匹配到本地适配器的 BluetoothDevice 类的对象集合
public static BluetoothAdapter getDefaultAdapter()	获取对默认本地蓝牙适配器的操作权限
public String getName()	获取本地蓝牙适配器的蓝牙名称
public BluetoothDevice getRemoteDevice(String address)	为给予的蓝牙硬件地址获取一个 BluetoothDevice 对象
public int getScanMode()	获取本地蓝牙适配器的当前蓝牙扫描模式
public int getState()	获取本地蓝牙适配器的当前状态
public boolean isDiscovering()	如果当前蓝牙适配器正处于设备发现查找进程中，则返回真值
public boolean isEnabled()	如果蓝牙正处于打开状态并可用，则返回真值
public BluetoothServerSocket listenUsingRfcomm With Service Record (String name, UUID uuid)	创建一个正在监听的安全的带有服务记录的无线射频通信蓝牙端口
public boolean setName(String name)	设置蓝牙或者本地蓝牙适配器的昵称
public boolean startDiscovery()	开始对远程设备进行查找的进程

在程序中需要获取相应的权限才能够对蓝牙进行操作。对蓝牙的操作权限如下：

```
<uses-permission android:name="android.permission.BLUETOOTH"/>
<user-permission android:name="android.permission.BLUETOOTH_ADMIN"/>
<uses-permission android:name="android.permission.READ_CONTACTS"/>
```

在设置足够的权限后对蓝牙进行操作，首先就是要取得蓝牙适配器，通过BluetoothAdapter.getDefaultAdapter()方法获取本地的蓝牙适配器。若要获得远端的蓝牙适配器就要使用BluetoothDevice进行。在获取了本地蓝牙适配器后开启蓝牙，开启方法是通过一个Intent来进行，相关代码如下：

```
Intent startBT=new Intent(BluetoothAdapter.ACTION_REQUEST_ENABLE);
startActivityForResult(startBT,REQUEST_ENABLE);
```

通过BluetoothAdapter来请求系统允许蓝牙设备被搜索，只是Action的类型不一样而已：

```
Intent startSerach=new Intent(BluetoothAdapter.ACTION_REQUEST_DISCOVERABLE);
startActivityForResult(startSearch, REQUEST_DISCOVERABLE);
```

蓝牙典型应用案例：搜索周围的蓝牙设备

该案例演示蓝牙的基本应用，搜索周围的蓝牙设备。由于模拟器不支持蓝牙模块，所以程序需要在真机上进行演示，程序运行后的主界面如图3-6所示。DiscoveryBluetooth蓝牙设备搜索类的代码清单如下：

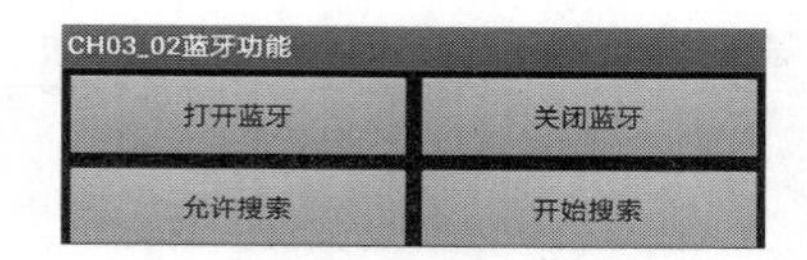

图3-6　程序主界面

```
public class DiscoveryBluetooth extends ListActivity{
    private Handler mHandler=new Handler();
    private BluetoothAdapter mBluetooth=BluetoothAdapter.getDefaultAdapter();
    private List<BluetoothDevice>devices=new ArrayList<BluetoothDevice>();
    //用来存储搜索到的蓝牙设备
    private volatile boolean isdiscoveryFinished;
    private Runnable workder=new Runnable(){
        public void run(){
        mBluetooth.startDiscovery();  /*开始搜索*/
        for(;;){if(isdiscoveryFinished){break;}
            try{Thread.sleep(100);}catch(InterruptedException e){}}}};
      /***接收器：当搜索蓝牙设备完成时调用*/
      private BroadcastReceiver mfoundReceiver=new BroadcastReceiver(){
          public void onReceive(Context context, Intent intent){
              /*从intent中取得搜索结果数据 */
              BluetoothDevice device=intent
              .getParcelableExtra(BluetoothDevice.EXTRA_DEVICE);
```

```
        devices.add(device); /*将结果添加到列表中*/
        showDevices();}};/*显示列表*/
   private BroadcastReceiver isdiscoveryReceiver=new BroadcastReceiver(){
      public void onReceive(Context context, Intent intent){
          unregisterReceiver(mfoundReceiver); /*卸载注册的接收器*/
          unregisterReceiver(this);isdiscoveryFinished=true;}};
      protected void onCreate(Bundle savedInstanceState){
          super.onCreate(savedInstanceState);
          getWindow().setFlags(WindowManager.LayoutParams.FLAG_BLUR_BEHIND,
WindowManager.LayoutParams.FLAG_BLUR_BEHIND);
          setContentView(R.layout.discovery);
      /*如果蓝牙适配器没有打开，则结果 */
          if(!mBluetooth.isEnabled()){finish();return;}
          /*注册接收器 */
          IntentFilter discoveryFilter=new IntentFilter(BluetoothAdapter.
ACTION_DISCOVERY_FINISHED);
          registerReceiver(isdiscoveryReceiver, discoveryFilter);
          IntentFilter foundFilter=new
   IntentFilter(BluetoothDevice.ACTION_FOUND);
          registerReceiver(mfoundReceiver, foundFilter);
          /*显示一个对话框,正在搜索蓝牙设备 */
          SamplesUtils.indeterminate(DiscoveryBluetooth.this, mHandler, "扫描...",
 workder, new OnDismissListener(){
     public void onDismiss(DialogInterface dialog){
         for(; mBluetooth.isDiscovering();){
               mBluetooth.cancelDiscovery();}
         isdiscoveryFinished=true;}}, true);}
     protected void showDevices(){  /*显示列表*/
         List<String>list=new ArrayList<String>();
         for(int i=0, size=devices.size(); i < size; ++i){
             StringBuilder b=new StringBuilder();
             BluetoothDevice d=devices.get(i);
             b.append(d.getAddress());b.append( '\n' );  b.append(d.getName());
             String s=b.toString();list.add(s);}
         final ArrayAdapter<String>adapter=new ArrayAdapter<String>(this, android.
R.layout.simple_list_item_1, list);
         mHandler.post(new Runnable(){
          public void run(){setListAdapter(adapter);}});}
     protected void onListItemClick(ListView l, View v, int position, long id)
{Intent result=new Intent();
         result.putExtra(BluetoothDevice.EXTRA_DEVICE, devices.get(position));
         setResult(RESULT_OK, result);finish();}}
```

CH03_2类代码清单如下：

```
public class CH03_2 extends Activity implements OnClickListener{
  private BluetoothAdapter Bluetooth=
      BluetoothAdapter.getDefaultAdapter();  //取得蓝牙适配器
  private static final int REQUEST_ENABLE=0x1;  //请求打开蓝牙
  private static final int REQUEST_DISCOVERABLE=0x2;  //请求能够被搜索
  private Button btnEnable,btnDisable,btnDiscove,btnSearch;
      public void onCreate(Bundle savedInstanceState){  //首次创建活动时调用
          super.onCreate(savedInstanceState);
          setContentView(R.layout.activity_main);
          findViews();setOnClickListener();}
  private void findViews(){
      btnEnable=(Button)findViewById(R.id.btnEnable);
          btnDisable=(Button)findViewById(R.id.btnDisable);
          btnDiscove=(Button)findViewById(R.id.btnDiscove);
          btnSearch=(Button)findViewById(R.id.btnSearch);}
  private void setOnClickListener(){
      btnEnable.setOnClickListener(this);
          btnDisable.setOnClickListener(this);
          btnDiscove.setOnClickListener(this);
          btnSearch.setOnClickListener(this);}
  public void onClick(View v){
      switch(v.getId()){
      case R.id.btnEnable:mBluetooth.enable();  break;  //打开蓝牙
      case R.id.btnDisable:mBluetooth.disable();break; //关闭蓝牙
      case R.id.btnDiscove:doDiscove();break;  //使设备能够被搜索
      case R.id.btnSearch:doSearch();break;  //开始搜索
  }}
  private void doDiscove(){  /*使设备能够被搜索*/
      Intent enabler=new Intent(BluetoothAdapter.ACTION_REQUEST_DISCOVERABLE);
      startActivityForResult(enabler, REQUEST_DISCOVERABLE);}
  private void doSearch(){  /*开始搜索 */
      Intent enabler=new Intent(this, DiscoveryBluetooth.class);
      startActivity(enabler);}}
```

首先用getDefaultAdapter()方法取得默认蓝牙适配器，创建存储搜索到蓝牙设备的List，在程序开始时注册搜索已完成和发现设备的两个BroadcastReceiver。然后，通过线程来控制蓝牙设备的搜索，当搜索中触发两个接收器事件时，就直接传递给接收器进行保存。最后，将保存在List中的蓝牙设备显示在ListView中。记得在AndroidManifest.xml添加访问蓝牙权限：

```
<uses-permission android:name="android.permission.BLUETOOTH"/>
<uses-permission android:name="android.permission.BLUETOOTH_ADMIN"/>
<uses-permission android:name="android.permission.READ_CONTACTS"/>
```

3.2 技术准备

3.2.1 Android 网络基础

Android目前有3种网络接口：标准Java接口、Apache接口、Android网络接口，其对应的功能包分别为java.net.*、org.apache和android.net.*。

1. 标准Java接口

Java.net.*提供与网络连接相关的类，java.net.*的包分为两部分：低级API和高级API。

低级API主要用于处理各项：

（1）地址：也就是网络标识符，如IP地址。

（2）套接字：也就是基本双向数据通信机制。

（3）接口：用于描述网络接口。

高级API主要用于处理各项：

（1）URI：表示统一资源标识符（Uniform Resource Identifier）。Web上可用的每种资源（HTML文档、图像、视频片段、程序等）可通过URI来定位。

（2）URL：表示统一资源定位符（Uniform Resource Locator），也被称为网页地址，是因特网上标准的资源地址。

（3）连接：表示到URL所指向资源的连接。

（4）地址：在整个java.net API中，地址或者用作主机标识符或者用作套接字端点标识符。

使用java.net.*创建连接的示意代码如下：

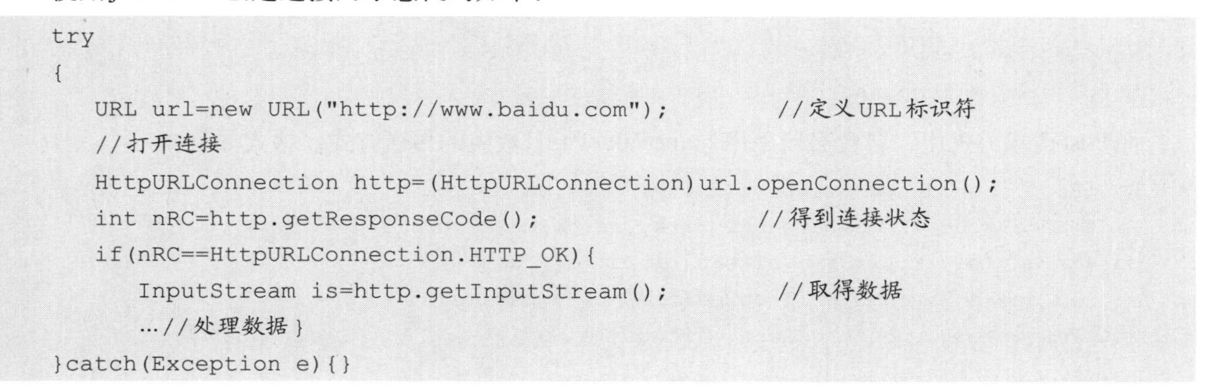

```
try
{
    URL url=new URL("http://www.baidu.com");        //定义URL标识符
    //打开连接
    HttpURLConnection http=(HttpURLConnection)url.openConnection();
    int nRC=http.getResponseCode();                 //得到连接状态
    if(nRC==HttpURLConnection.HTTP_OK){
        InputStream is=http.getInputStream();       //取得数据
        …//处理数据}
}catch(Exception e){}
```

2. Apache接口

HTTP是目前在Internet上使用最多、最重要的通信协议，越来越多的Java应用程序需要通过HTTP访问网络资源。Android系统引入了Apache HttpClient并对其进行封装和扩展，如设置默认的HTTP超时和缓存大小等，通过创建HttpClient、Get/Post、HttpRequest等对象，设置连接参数，执行HTTP操作，处理服务器返回结果等功能。使用android.net.http.*连接网络的示意代码如下：

```
try{
    HttpClient hc=new DefaultHttpClient();          //创建HttpClient使用默认属性
    HttpGet get=new HttpGet("http://www.baidu.com");//创建HttpGet实例
```

```
    HttpResponse rp=hc.execute(get);//连接
    if(rp.getStatusLine().getStatusCode()==HttpStatus.SC_OK){
       InputStream is=rp.getEntity().getContent();
       ...//处理数据}
}catch(IOException e){}
```

3. Android网络接口

Android.net.*包实际上是通过对Apache的HttpClient进行封装，实现一个HTTP接口，同时也提供了HTTP请求队列管理以及HTTP连接池管理，以提高并发情况下的处理效率。除此之外，还有网络状态监视等接口、网络访问的Socket，常用的Uri类以及有关Wi-Fi相关的类等。使用Android.net.*包中的Socket连接网络的示意代码如下：

```
try{
    InetAddress inetAddress=InetAddress.getByName("192.168.1.7");
    Socket client=new Socket(inetAddress,888888,true);
    InputStream in=client.getInputStream();
    OutputStream out=client.getOutputStream();    //处理数据
    out.close();in.close();client.close();
}catch(UnknownHostException e){}catch(IOException e){}
```

3.2.2 HTTP通信

1. HTTPGet与HTTPPost

HTTP通信中使用最多的就是Get和Post。Get请求方式中，参数直接放在URL字串后面，传递给服务器。格式如下：

```
HttpGet method=new HttpGet("http://www.baidu.com?admin=Get");
HttpResponse response=client.execute(method);
```

而Post请求方式中，参数必须采用NameValuePair[]数组的传送方式。格式如下：

```
HttpPost method=new HttpPost("http://www.baidu.com");
List<NameValuePair>params=new ArrayList<NameValuePair>();
params.add(new BasicNameValuePair("admin","Get"));
method.setEntity(new UrlEncodedFormEntity(params));
HttpResponse response=client.execute(method);
```

用代码获得服务器返回的状态码，具体代码如下：

```
HttpResponse httpResponse=new DefaultHttpClient().execute(method);
If(httpResponse.getStatusLine().getStatusCode()==200)
{//从URL获取数据}
else{//显示连接错误}
```

注意：Android请求连接网络前需要在AndroidManifest.xml中添加如下网络权限。

```
<uses-permission android:name="android.permission.INTERNET"/>
```

2. HttpURLConnection接口

在Android的SDK中，继承了网络连接中标准的Java接口，使得最基本的一些连接方式得以继续沿用。

注意：URLConnection与HttpURLConnection都是抽象类，无法直接实例化对象，其对象主要通过URL的openConnection()方法获得。

标准的Java接口格式如下：

```
URL url=new URL("http://www.baidu.com");
HttpURLConnection http=(HttpURLConnection)url.openConnection();
int response=http.getResponseCode();
if(200==response){  //从URL获取数据}
else{  //显示连接错误}
```

3. 网络接口

Android中的网络接口实际上是通过对Apache中HttpClient的封装来实现的一个HTTP编程接口。

4. 权限验证

首先需要明确的是，权限验证不同于参数传递。权限验证的方法如下：

```
DefaultHttpClient client=new DefaultHttpClient();
client.getCredentialsProvider().setCredentials(AuthScope.ANY,
new UsernamePasswordCredentials("admin", "password"));
```

5. 通过Http协议访问网络权限验证

在Android中，较为常见的HTTP编程有两种方式：HttpURLConnection和HttpClient。HttpClient可分为HttpGet和HttpPost两种方式。HttpURLConnection调用setRequestMethod()函数，通过传递GET和POST参数设置请求服务器的方式。

HTTP协议即超文本传输协议，是Web联网的基础，也是手机联网常用的协议之一，HTTP协议是建立在TCP协议之上的一种应用。HTTP连接最显著的特点是客户端发送的每次请求都需要服务器回送响应，在请求结束后，会主动释放连接。从建立连接到关闭连接的过程称为“一次连接”。其特点如下：

（1）在HTTP 1.0中，客户端的每次请求都要求建立一次单独的连接，在处理完本次请求后，就自动释放连接。

（2）在HTTP 1.1中则可以在一次连接中处理多个请求，并且多个请求可以重叠进行，不需要等待一个请求结束后再发送下一个请求。

HttpClinet访问网络的通用步骤如下：

（1）创建HttpGet或HttpPost客户端对象client，将要请求的URL通过构造方法传入HttpGet或HttpPost对象。

（2）使用DefaultHttpClient类实例化HttpClient对象。

（3）设置请求方法（输入网址）：Get方式为HttpGet；POST方式为HttpPost。

（4）设置请求的参数/请求头信息/连接超时时间/读取数据超时时间等参数。

（5）执行请求，调用execute()方法，发送HTTP GET或HTTP POST请求，并返回HttpResponse

对象 response。

（6）获取状态码：response.getStatusLine().getStatusCode(); 若状态码是 200，获取服务器返回的数据，存放在输入流中 InputStream is=response.getEntity().getContent()。或通过 HttpResponse 接口的 getEntity() 方法返回响应信息，并进行相应的处理。

（7）操作结束后断开连接：client.getConnectionManager().shutdown()。

AndroidManifest.xml 文件添加如下代码以获得网络访问权限，否则无法运行。

```
<uses-permission android:name="android.permission.INTERNET"/>
```

使用 HttpURLConnection 的基本步骤如下：

（1）调用 URL 对象的 openConnection() 方法获得 HttpURLConnection 实例对象。例如：

```
URL url=new URL("https://www.baidu.com/");
HttpURLConnection con=(HttpURLConnection)url.openConnection();
```

（2）设置 HTTP 请求方法。常用的 HTTP 请求方法主要有 GET 和 POST 两种（注意大写）。GET 方法通常仅希望从服务器返回数据，POST 则可向服务器提交数据。例如：

```
con.setRequestMethod("POST"); 或  con.setRequestMethod("GET");
```

（3）设置请求相关参数。例如，可设置连接和请求超时（单位为毫秒）。

```
con.setConnectTimeout(2000);con.setReadTimeout(2000);
```

如果采用 POST 方式，则需要使用 DataOutputStream 添加向服务器提交的数据。向服务器提交的数据采用键值对的方式表示，键值对之间用 & 符号分隔。例如：

```
con.setRequestMethod("POST");con.setDoOutput(true);
DataOutputStream out=new DataOutputStream(con.getOutputStream());
out.writeBytes("user=admin&pwd=12345678");
```

（4）处理返回结果。调用 HttpURLConnection 对象的 getInputStream() 方法，获得服务器返回结果的 InputStream，从中可获取服务器的返回结果。例如：

```
InputStream inputStream=con.getInputStream();
reader=new BufferedReader(new InputStreamReader(inputStream));
StringBuilder sb=new StringBuilder();String s;s=reader.readLine();
while(s!=null){sb.append(s);s=reader.readLine();}
```

HTTP 协议支持 GET 和 POST 两种不同请求方式向服务器请求和提交数据。

GET 方式请求数据时，会把数据附加到 URL 中传递给服务器，比如常见的路径 http: //192.168.1.7: 8080/HelloWorld02/index.jsp?user=admin，包含的数据为 user=admin。

POST 方式则是将请求的数据放到 HTTP 请求头中，作为请求头的一部分传入服务器。

6. Tomcat 环境搭建和 JavaWeb 服务器编程

在 Android HTTP 编程学习前，要搭建 Tomcat 环境并采用 MyEclipse10 编写相应的 JavaWeb 服务器程序。编程测试时，关闭 PC 防火墙设置，Android 手机（或模拟器）和 PC 通过相同 Wi-Fi 登录网络，并保证 Android 手机（或模拟器）和 PC 在相同网段内。搭建 Tomcat 服务器的主要步骤如下：

（1）下载最新版本的tomcat，并解压到目标位置。

（2）启动tomcat，即运行bin文件夹下的startup.bat文件。

（3）测试tomcat是否运行成功。在浏览器输入网址http://localhost:8080/。

7. GET和POST乱码解决方式

编写Android网络程序从服务器获取数据时，经常遇到内容乱码问题。原因在于Android默认编码是UTF-8编码，Tomcat服务器默认使用ISO-8859-1编码。而ISO-8859-1是不支持中文的，因而出现中文显示乱码。具体的解决方法如下：

（1）对GET方式，在响应服务器端的doGet()方法中加入：

```
String name=new String(request.getParameter("name").getBytes("ISO-8859-1"),"UTF-8");
```

（2）对POST方式，在响应服务器端的doPost()方法中加入：

```
request.setCharacterEncoding("UTF-8");
```

下面通过实例学习HttpClinet和HttpURLConnection编程方法。主程序布局文件activity_main.xml代码如下：

```
<LinearLayout xmlns:android="http://schemas.android.com/apk/res/android"
    xmlns:tools="http://schemas.android.com/tools"
    android:layout_width="match_parent"
    android:layout_height="match_parent"
    tools:context="sziit.lihz.httpdemo.MainActivity">
    <LinearLayout
        android:layout_width="match_parent"
        android:layout_height="match_parent"
        android:orientation="vertical">
        <TextView
            android:id="@+id/tvTitle"
            android:layout_width="match_parent"
            android:layout_height="wrap_content"
            android:text="使用HTTP访问网络"/>
        <Button
            android:id="@+id/btnHTTPURLCon"
            android:layout_width="match_parent"
            android:layout_height="wrap_content"
            android:text="使用HTTPURLConnection方法"/>
        <Button
            android:id="@+id/btnHttpclient"
            android:layout_width="match_parent"
            android:layout_height="wrap_content"
            android:text="使用Httpclient方法"/>
    </LinearLayout>
</LinearLayout>
```

学习HttpClinet的GET和POST方式的页面布局activity_httpclient.xml代码如下：

```
<?xml version="1.0"encoding="utf-8"?>
<LinearLayout xmlns:android="http://schemas.android.com/apk/res/android"
    android:layout_width="match_parent"
    android:layout_height="match_parent"
    android:orientation="vertical">
    <TextView
        android:id="@+id/tvWebTitle"
        android:layout_width="match_parent"
        android:layout_height="wrap_content"
        android:gravity="left"
        android:text="使用Httpclient!"/>
    <LinearLayout
        android:layout_width="match_parent"
        android:layout_height="wrap_content"
        android:orientation="horizontal">
        <Button
            android:id="@+id/btnHttpClientGet"
            android:layout_width="wrap_content"
            android:layout_height="wrap_content"
            android:layout_weight="1"
            android:text="Get方式"/>
        <Button
            android:id="@+id/btnHTTPClientPost"
            android:layout_width="wrap_content"
            android:layout_height="wrap_content"
            android:layout_weight="1"
            android:text="Post方式"/>
    </LinearLayout>
    <ScrollView
        android:layout_width="match_parent"
        android:layout_height="wrap_content">
        <TextView
            android:id="@+id/tvHttpClient"
            android:layout_width="match_parent"
            android:layout_height="match_parent"
            android:text="HTTPClient案例"/>
    </ScrollView>
</LinearLayout>
```

学习HttpURLConnection编程技巧的页面布局activity_httpurlcon.xml代码如下：

```
<?xml version="1.0"encoding="utf-8"?>
<LinearLayout xmlns:android="http://schemas.android.com/apk/res/android"
```

```
    android:layout_width="match_parent"
    android:layout_height="match_parent"
    android:orientation="vertical">
    <TextView
        android:id="@+id/tvURLTitle"
        android:layout_width="match_parent"
        android:layout_height="wrap_content"
        android:gravity="left"
        android:text="使用HTTPURLConnection!"/>
    <LinearLayout
        android:layout_width="match_parent"
        android:layout_height="wrap_content"
        android:orientation="horizontal">
        <Button
            android:id="@+id/btnGetPic"
            android:layout_width="wrap_content"
            android:layout_height="wrap_content"
            android:layout_weight="1"
            android:text="获取图片"/>
        <Button
            android:id="@+id/btnGet"
            android:layout_width="wrap_content"
            android:layout_height="wrap_content"
            android:layout_weight="1"
            android:text="Get方式"/>
        <Button
            android:id="@+id/btnPost"
            android:layout_width="wrap_content"
            android:layout_height="wrap_content"
            android:layout_weight="1"
            android:text="Post方式"/>
    </LinearLayout>
    <ImageView
        android:id="@+id/imageView"
        android:layout_width="match_parent"
        android:layout_height="wrap_content"
        android:src="@drawable/ic_launcher"/>
    <ScrollView
         android:layout_width="match_parent"
        android:layout_height="wrap_content">
        <LinearLayout
            android:layout_width="match_parent"
            android:layout_height="match_parent"
```

```
            android:orientation="vertical">
            <TextView
                android:id="@+id/tvInfo"
                android:layout_width="wrap_content"
                android:layout_height="wrap_content"
                android:text="HTTPURLConnection案例"/>
        </LinearLayout>
    </ScrollView>
</LinearLayout>
```

主活动MainActivity的源代码如下：

```
public class MainActivity extends Activity implements OnClickListener{
    public static String BASE="http://192.168.1.7:8080/HelloWorld02/ch2/";
    private Button btnHTTPURLCon,btnHttpclient;
    protected void onCreate(Bundle savedInstanceState){
        super.onCreate(savedInstanceState);
        setContentView(R.layout.activity_main);
        setTitle("CH03_03使用HTTP访问网络案例");
        findViews();setOnClickListener();}
    private void findViews(){
        btnHTTPURLCon=(Button)findViewById(R.id.btnHTTPURLCon);
        btnHttpclient=(Button)findViewById(R.id.btnHttpclient);}
    private void setOnClickListener(){
        btnHTTPURLCon.setOnClickListener(this);
        btnHttpclient.setOnClickListener(this);}
    public void onClick(View v){
        switch(v.getId()){
        case R.id.btnHTTPURLCon:doHTTPURLCon();break;
        case R.id.btnHttpclient:doHttpclient();break;}}
    private void doHttpclient(){
startActivity(new Intent(MainActivity.this,HttpclientActivity.class));}
    private void doHTTPURLCon(){
startActivity(new Intent(MainActivity.this,HTTPURLConActivity.class));}}
```

学习HttpClinet编程的HttpclientActivity的源代码如下：

```
public class HttpclientActivity extends Activity implements OnClickListener{
    private final String getWebSite=MainActivity.BASE+"2.10/index.html";
    private final String postWebSite=MainActivity.BASE+"2.11/index.html";
    private TextView tvInfo,tvWebTitle=null;
    private Button btnGet=null, btnPost=null;
    private String resultGet="",resultPost="";
    protected void onCreate(Bundle savedInstanceState){
         super.onCreate(savedInstanceState);
```

```
        setContentView(R.layout.activity_httpclient);
        setTitle("CH03_03使用Httpclient访问网络案例");
        findView();setClickListener();}
    private void findView(){
        tvInfo=(TextView)findViewById(R.id.tvHttpClient);
        tvWebTitle=(TextView)findViewById(R.id.tvWebTitle);
        btnGet=(Button)findViewById(R.id.btnHttpClientGet);
        btnPost=(Button)findViewById(R.id.btnHTTPClientPost);}
    private void setClickListener(){
        btnGet.setOnClickListener(this);
        btnPost.setOnClickListener(this);}
    private Runnable HttpClientGet=new Runnable(){
    public void run(){
        try{HttpGet httpRequest=new HttpGet(getWebSite);
            HttpClient httpclient=new DefaultHttpClient();
            HttpResponse httpResponse=httpclient.execute(httpRequest);
    if(httpResponse.getStatusLine().getStatusCode()==HttpStatus.SC_OK){
        resultGet=EntityUtils.toString(httpResponse.getEntity(),"utf-8");
             tvInfo.post(new Runnable(){public void run(){
                     tvInfo.setText(resultGet);}});} else{
                     Log.d(TAG, "Get请求失败");}}catch(MalformedURLException e){
                     e.printStackTrace();
          } catch(IOException e){e.printStackTrace();}}};
    private Runnable HttpClientPost=new Runnable(){
        public void run(){
        try{HttpPost httpRequest=new HttpPost(postWebSite);
            List<NameValuePair>params=new ArrayList<NameValuePair>();
            params.add(new BasicNameValuePair("name","HTTPClientPost"));
            HttpEntity httpEntity=new UrlEncodedFormEntity(params,"GBK");
            httpRequest.setEntity(httpEntity);
            HttpClient httpclient=new DefaultHttpClient();
            HttpResponse httpResponse=httpclient.execute(httpRequest);
  if(httpResponse.getStatusLine().getStatusCode()==HttpStatus.SC_OK){
      resultPost=EntityUtils.toString(httpResponse.getEntity(),"utf-8");
      tvInfo.post(new Runnable(){public void run(){
      tvInfo.setText(resultPost);}});}
  else{Log.i(TAG,"Post请求失败");}}catch(MalformedURLException e){
         e.printStackTrace();}catch(IOException e){e.printStackTrace();}}};
         public void onClick(View v){
              switch(v.getId()){
              case R.id.btnHttpClientGet:
                  tvWebTitle.setText("使用Httpclient:Get方式获得网站资源!");
                  new Thread(HttpClientGet).start();break;
```

```
                case R.id.btnHTTPClientPost:
                tvWebTitle.setText("使用Httpclient:Post方式获得网站资源!");
                new Thread(HttpClientPost).start();break;}}}
```

学习HttpURLConnection编程的HttpclientActivity的源代码如下：

```
public class HTTPURLConActivity extends Activity implements OnClickListener
{  private ImageView imageView=null;
   private Bitmap bitmap=null;
   private URL getPicUrl=null,getUrl=null,postUrl=null;
   private TextView tvInfo ,tvURLTitle;
   private Button btnGet=null,btnPost=null,btnGetPic=null;
   private final String picWeb=MainActivity.BASE+"/sj1/bike.jpg";
   private final String getWeb=MainActivity.BASE+"2.7/index.html";
   private final String postWeb=MainActivity.BASE+"2.4/index.html";
   private String resultGet="",resultPost="";
   protected void onCreate(Bundle savedInstanceState){
       super.onCreate(savedInstanceState);
       setContentView(R.layout.activity_httpurlcon);
       setTitle("CH03_03使用HttpURLConnection!");
       findView();setClickListener();}
   private void findView(){
       imageView=(ImageView)findViewById(R.id.imageView);
       tvURLTitle=(TextView)findViewById(R.id.tvURLTitle);
       tvInfo=(TextView)findViewById(R.id.tvInfo);
       btnGet=(Button)findViewById(R.id.btnGet);
       btnPost=(Button)findViewById(R.id.btnPost);
       btnGetPic=(Button)findViewById(R.id.btnGetPic);}
   private void setClickListener(){
       btnGet.setOnClickListener(this);
       btnPost.setOnClickListener(this);
       btnGetPic.setOnClickListener(this);}
   private Runnable HttpUrlGetPic=new Runnable(){
   public void run(){
   try{getPicUrl=new URL(picWeb);
        HttpURLConnection urlConnGetPic=(HttpURLConnection)
    getPicUrl.openConnection();
        urlConnGetPic.setDoInput(true);urlConnGetPic.connect();
        InputStream is=urlConnGetPic.getInputStream();
        bitmap=BitmapFactory.decodeStream(is);
        if(bitmap!=null){imageView.post(new Runnable(){
             public void run(){imageView.setImageBitmap(bitmap);
                 tvInfo.setText("Get图片成功! ");}});
        }else{tvInfo.setText("获得图片失败");}
```

```
                is.close();urlConnGetPic.disconnect();
            }catch(MalformedURLException e){e.printStackTrace();}
catch(IOException e){e.printStackTrace();}}};
    private Runnable HTTPUrlConGet=new Runnable(){
        public void run(){
            try{getUrl=new URL(getWeb);
HttpURLConnection urlConnGet=(HttpURLConnection)getUrl.openConnection();
        urlConnGet.setDoInput(true);urlConnGet.connect();
        InputStreamReader in=new InputStreamReader(
            urlConnGet.getInputStream());
            BufferedReader bufReaderGet=new BufferedReader(in);
            String inputLine=null;
            while(((inputLine=bufReaderGet.readLine())!=null)){
resultGet+=inputLine+"\n";}
            tvInfo.post(new Runnable(){public void run(){
            tvInfo.setText(resultGet);}});
            in.close();urlConnGet.disconnect();
        }catch(MalformedURLException e){e.printStackTrace();}
catch(IOException e){e.printStackTrace();}}};
    private Runnable HTTPUrlConPost=new Runnable(){
        public void run(){
        try{postUrl=new URL(postWeb);
        HttpURLConnection urlConnPost=(HttpURLConnection)postUrl
        .openConnection();
        urlConnPost.setDoOutput(true);urlConnPost.setDoInput(true);
        urlConnPost.setRequestMethod("POST");
urlConnPost.setUseCaches(false);
        urlConnPost.setInstanceFollowRedirects(true);
        urlConnPost.connect();
        DataOutputStream out=new DataOutputStream(
        urlConnPost.getOutputStream());
        String content="title="
           +URLEncoder.encode("深圳信息职业技术学院", "utf-8");
        out.writeBytes(content);out.flush();out.close();
        int responseCode=urlConnPost.getResponseCode();
        if(responseCode!=200){
           Log.d("HTTPUrlConPost","错误==="+responseCode);}else{
           Log.d("HTTPUrlConPost","Post成功!");}
        BufferedReader bufReaderPost=new BufferedReader(
        new InputStreamReader(urlConnPost.getInputStream(),"utf-8"));
        String linePost="";
        while((linePost=bufReaderPost.readLine())!=null){
           resultPost+=linePost+"\n";}
```

```
                tvInfo.post(new Runnable(){public void run(){
                 tvInfo.setText(resultPost);}});
        urlConnPost.disconnect();}catch(MalformedURLException e){
        e.printStackTrace();}catch(IOException e){e.printStackTrace();}}};
        public void onClick(View v){
        switch(v.getId()){
        case R.id.btnGetPic:
             tvURLTitle.setText("使用HTTPURLConnection的Get方式获得图片！");
             new Thread(HttpUrlGetPic).start();break;
        case R.id.btnGet:
             tvURLTitle.setText("使用HTTPURLConnection的Get方式获得文本数据！");
             new Thread(HTTPUrlConGet).start();break;
        case R.id.btnPost:
             tvURLTitle.setText("使用HTTPURLConnection的Post方式获得文本数据！");
             new Thread(HTTPUrlConPost).start();break;}}}
```

为运行该程序先通过ipconfig命令获得本机IP地址192.168.1.7，如图3-7所示。

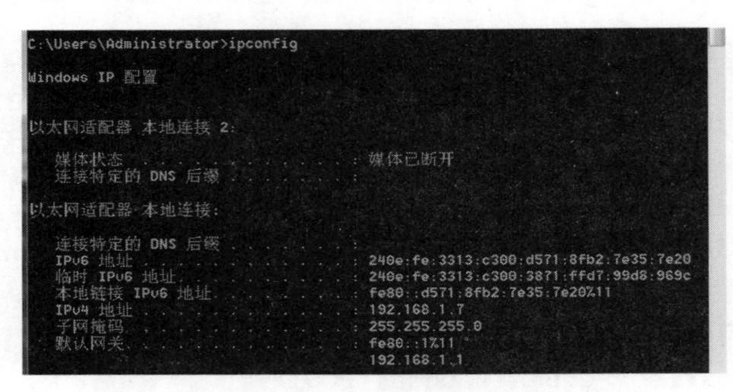

图 3-7　通过 ipconfig 命令获得本机 IP 地址

然后在MyEclipse10中启动Web服务（见图3-8），其HttpServletTest.java源代码清单如下，访问图片Web服务地址为http://192.168.1.7:8080/HelloWorld02/ch2/sj1/bike.jpg（源代码详见CH03_03WebServer）：

```
public class HttpServletTest extends HttpServlet{
   protected void doGet(HttpServletRequest req, HttpServletResponse resp)
           throws ServletException, IOException{
      //resp.setCharacterEncoding("GBK");    //设置响应的编码类型为GBK
      resp.setCharacterEncoding("utf-8");    //设置响应的编码类型为utf-8
      PrintWriter out=resp.getWriter();      //获取输出对象
      out.println("<html>");
      out.println("<head>");
      out.println("<title>Android高级应用编程实战(第二版)</title>");
      out.println("</head>");
      out.println("<body>");
      String name=req.getParameter("name");  //获取请求的参数
```

```
        if(name==null||name.equals("")){
            name="lihuazhong";}
        out.println("<h2>您好，"+name+"<br>这是Android高级应用编程实战(第二版)使用
HttpServlet测试HTTP编程的服务端案例</h2>");
        out.println("</body>");
        out.println("</html>");
        out.close();}  //关闭输出对象
    protected void doPost(HttpServletRequest req, HttpServletResponse resp)
            throws ServletException, IOException{
        this.doGet(req, resp);}}
```

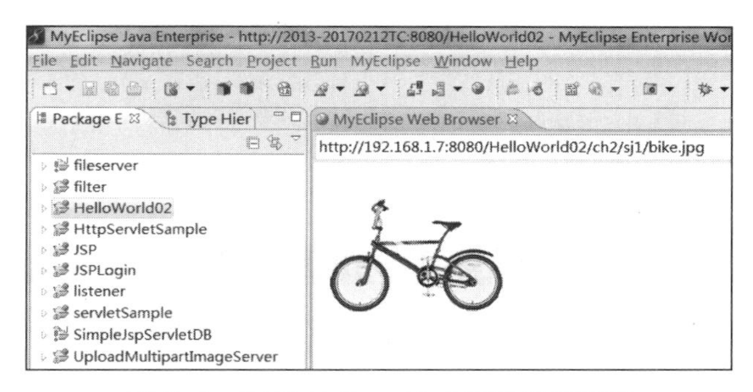

图 3-8　在 MyEclipse10 中启动服务程序

本程序在Android手机或模拟器上的运行结果如图3-9~图3-11所示。

图 3-9　程序主界面运行效果

图 3-10　用 HttpURLConnection 从 Web 服务器获取图片、Get 方式和 Post 方式运行效果

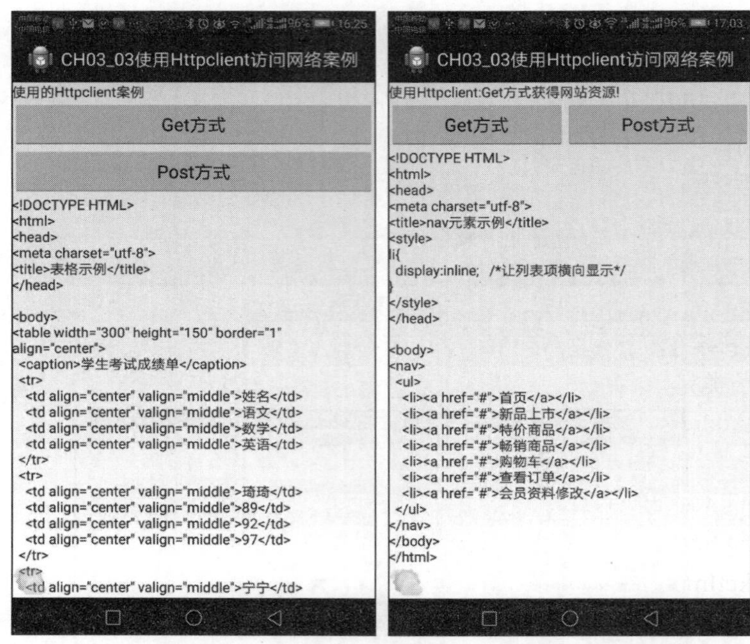

图 3-11　用 HttpClient 从 Web 服务器 Get 方式和 Post 方式运行效果

视频

调用系统内置浏览器应用实战

3.2.3　Socket 通信

所谓Socket通常也称作“套接字"，应用程序通常通过“套接字”向网络发出请求或者应答网络请求，用于描述IP地址和端口，是一个通信链的句柄。

常用的Socket有两种类型，分别是：流式Socket（SOCK_STREAM）和数据报式Socket（SOCK_DGRAM）。流式是一种面向连接的Socket，针对于面向连接的TCP（Transmission Control Protocol，传输控制协议）服务应用；数据报式Socket是一种无连接的Socket，对应于无连接的UDP（User Datagram Protocol，用户数据报协议）服务应用。基于TCP的Socket通信模型如图3-12所示。

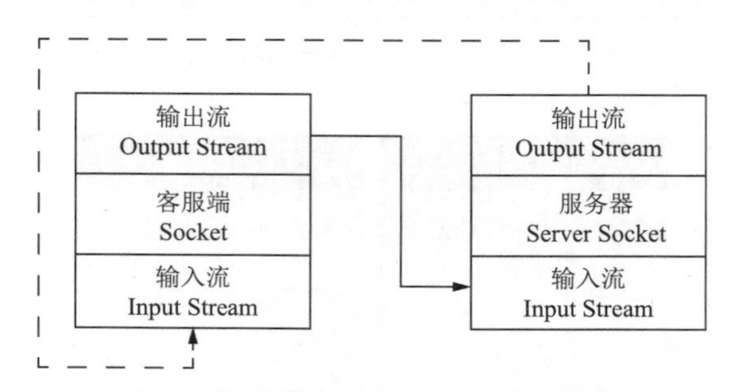

图 3-12　基于 TCP 的 Socket 通信模型

Socket通信的原理比较简单，大致分为以下几个步骤：

服务器端步骤：

（1）建立服务器端的Socket，开始侦听整个网络的连接请求。

（2）当检测到来自客户端的请求时，向客户端发送收到连接请求的信息，并与之建立连接。

（3）当通信完成后，服务器关闭与客户端的Socket连接。

客户端步骤：

（1）建立客户端的Socket，确定要连接的服务器的主机名和端口。

（2）发送连接请求到服务器，并等待服务器的回馈信息。

（3）建立连接后，与服务器开始进行数据传输与交互。

（4）数据处理完毕后，关闭自身的Socket连接。

在Java中，在java.net包中提供了两个类Socket和ServerSocket。ServerSocket用于服务器端，Socket是建立网络连接时使用的。在连接成功时，应用程序两端都会产生一个Socket实例，操作这个实例，完成所需的会话。对于一个网络连接来说，套接字是平等的，并没有差别，不因为在服务器端或在客户端而产生不同级别。不管是Socket还是ServerSocket，它们的工作都是通过SocketImpl类及其子类完成的。下面了解一下Socket的构造方法：

- Socket(InetAddress address,int port)。
- Socket(InetAddress address,int port,boolean stream)。
- Socket(String host,int port)。
- Socket(String host,int port,boolean stream)。
- Socket(SocketImpl impl)。
- Socket(String host,int port,InetAddress localAddr,int localPort)。
- Socket(InetAddress address,int port,InetAddress localAddr, int localport)。
- ServerSocket(int port)。
- ServerSocket(int port,int backlog)。
- ServerSocket(int port,int backlog,InetAddress bindAddr)。

其中，address、host和port分别是双向连接中另一方的IP地址、主机号和端口号；stream指明Socket是流式Socket还是数据报式Socket；localPort表示本地主机的端口号；loalAddr和bindAddr是本机器的地址；impl是Socket的父类，既可以用来创建ServerSocket，又可以用来创建Socket。

下面介绍几个常用的方法：

- accept()方法：用于产生"阻塞"，直到接收到一个连接，并且返回一个客户端的Socket对象实例。"阻塞"是一个术语，它使程序运行暂时"停留"在这个地方，直到一个会话产生，然后程序继续；通常"阻塞"是由循环产生的。
- getInputStream()方法获得网络连接输入，同时返回一个InputStream对象实例。
- getOutputStream()方法连接的另一端将得到输入，同时返回一个OutputStream对象实例。

注意：其中getInputStream()和getOutputStream()方法均会产生一个IOException，它必须被捕获，因为它们返回的流对象，通常都会被另一个流对象使用。

最后，在每一个Socket对象使用完毕时，要使其关闭，使用Socket对象的close()方法，在关闭Socket之前，应将与Socket相关的所有输入流和输出流关闭，以释放所有资源。先关闭输入流和输出流，再关闭Socket。

1. PC服务器与Android客户端的简易通信

本案例实现一个PC服务器与Android客户端的简易通信，服务器用Java工程实现，客户端用Android工程实现。PC服务器端代码如下：

```
public class PCServerSocket implements Runnable{
```

```
    public void run(){
    try{//创建ServerSocket,IP=192.168.1.7,端口号=6666
    ServerSocket serverSocket=new ServerSocket(6666);
    while(true){  //接受客户端请求
        System.out.println("服务器正等待客服端连接请求！");
        Socket client=serverSocket.accept();
        System.out.println("已经接受客服端连接请求！");
        try{  //接收客户端消息
        BufferedReader in=new BufferedReader(
        new InputStreamReader(client.getInputStream(),"UTF-8"));
        String str=in.readLine();
        System.out.println("读取:"+str);
        //向客户端发送消息
        PrintWriter out=new PrintWriter(new BufferedWriter(
  new OutputStreamWriter(client.getOutputStream(),"UTF-8")),true);
        out.println("服务器发送的消息!");out.close();  //关闭Socket输出流
        in.close();  //关闭Socket输入流
        }catch(Exception e){
        System.out.println(e.getMessage());e.printStackTrace();}
  finally{client.close();  //关闭
        System.out.println("关闭Socket连接！");}
        serverSocket.close();}
      }catch(Exception e){
        System.out.println(e.getMessage());}}
    /***main()函数，开启服务器*/
    public static void main(String[] args){
        Thread thread=new Thread(new PCServerSocket());
        thread.start();}}
```

Android客户端实现源码代码如下：

```
private void connectServer(){  //连接服务器
    Socket socket=null;  //定义Socket对象
  //获得要发送的消息
    String clientMsg=etClientMsg.getText().toString()+"\r\n";
    try{//创建Socket，ipconfig查看服务器IP和端口号
        socket=new Socket("192.168.1.7",6666);  //构建套节字Socket对象
        //封装Socket输出流OutputStream对象，向网络发送消息
        PrintWriter out=new PrintWriter(new BufferedWriter(
      new OutputStreamWriter(socket.getOutputStream(),"UTF-8")),true);
        out.println(clientMsg);  //向服务器发送消息
        //接收来自服务器的消息：构造BufferedReader对象，字符输入流中读取文本
        //并缓冲字符，以便有效地读取字符，数组和行
        //封装Socket的输入流InputStream对象，读取消息
```

```
BufferedReader reader=new BufferedReader(
new InputStreamReader(socket.getInputStream(),"UTF-8"));
String serverMsg=reader.readLine();  //读一行数据
if(serverMsg!=null){tvServerMsg.setText(serverMsg);}
else{tvServerMsg.setText("数据错误!");}
out.close();      //关闭PrintWriter对象流
reader.close();  //关闭BufferedReader对象
socket.close();  //关闭Socket}catch(Exception e){
Log.e("", e.toString());}}
```

启动服务器之后，运行Android Socket客户端，运行结果如图3-13和图3-14所示。

```
Problems @ Javadoc Declaration Console
<terminated> PCServerSocket [Java Application] C:\MyEclipse10
服务器正等待客服端连接请求!
已经接受客服端连接请求!
读取:来自客服端消息!
关闭Socket连接!
服务器正等待客服端连接请求!
Socket is closed
```

图 3-13 PC 服务器运行效果

图 3-14 Android 客户端运行效果

2. Android服务器与PC客户端的简易通信

本案例实现一个Android服务器与PC客户端的简易通信，服务器用Android工程实现，PC客户端用Java工程实现。PC客户端代码如下：

```
public class PCSocketClient{
    public static void main(String[] args){
        try{
          Socket socket=new Socket("192.168.1.88",8888);
          //向客户端发送消息
          PrintWriter out=new PrintWriter(new BufferedWriter(
              new OutputStreamWriter(socket.getOutputStream(),
                  "UTF-8")),true);
          out.println("这是PC和Android间Socket通信案例");
          //使用InputStream获取服务器端的数据
          BufferedReader bufferedReader=new BufferedReader(
               new InputStreamReader(socket.getInputStream()));
          String line=bufferedReader.readLine();  //读取数据
          System.out.println(line); //在控制台输出数据
          bufferedReader.close(); //闭bufferedReader
          out.close();  //关闭Socket输出流
          socket.close();}  //关闭socket
        //捕获异常(无法找到远程的服务器主机)
        catch(UnknownHostException ee){ee.printStackTrace();}
        //捕获异常(输入输出异常)
        catch(IOException e){e.printStackTrace();}}}
```

Android服务器代码如下：

```
private void startThread(){
    new Thread(new Runnable(){  //构建线程
      public void run(){
      try{//创建服务器ServerSocket对象,服务器端口号PORT=8888
          ServerSocket serverSocket=new ServerSocket(PORT);
          while(true){
          Socket socket=serverSocket.accept();  //等到接受客户端连接
          //获得网络输出流对象
          OutputStream outputStream=socket.getOutputStream();
          if(strInput.equals("")){
              strInput=getResources().getString(R.string.etInput);}
              //向客户端发送消息
              outputStream.write(strInput.getBytes("utf-8"));
              outputStream.close();socket.close();}
    }catch(Exception e){e.printStackTrace();}}}).start();}
```

Android服务器运行结果如图3-15所示。

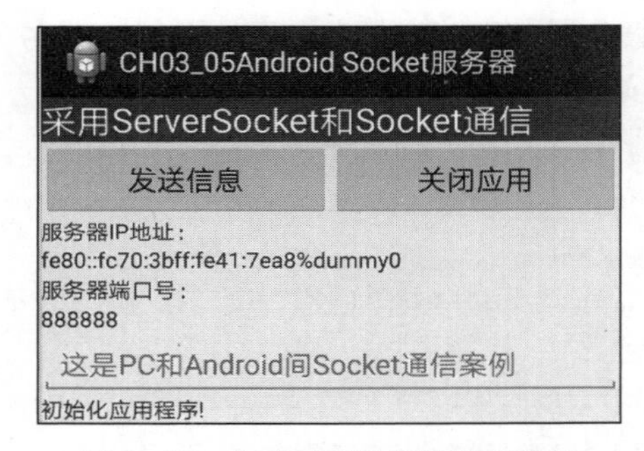

图 3-15　Android 服务器运行效果

3. Socket高级通信

本案例采用线程实现Socket高级通信，可以使服务器能同时与多个客户端相互连接。服务器实现代码如下：

```
public class AdvancedPCServer{
    public static void main(String[] args)throws IOException{
        try{//创建ServerSocket服务器对象，端口号设置为1000
            ServerSocket server=new ServerSocket(1000);
            while(true){
                Socket client;  //创建Socket对象，负责与客户端通信
                System.out.println("等待客户端连接请求！");
                client=server.accept();
                ReceServer rs=new ReceServer(client);
                rs.start();
```

```
                Thread.currentThread().sleep(500);}
        } catch(InterruptedException e){e.printStackTrace();
        } catch(IOException e){e.printStackTrace();}}}
class ReceServer extends Thread{
    private Socket client;
    ReceServer(Socket client){this.client=client;}
    public void run(){
        try{//创建输入流对象
            BufferedReader in=new BufferedReader(new InputStreamReader(
                client.getInputStream(),"UTF-8"));
            //向客户端发送消息
            PrintWriter out=new PrintWriter(new BufferedWriter(
                new OutputStreamWriter(client.getOutputStream(),
                    "UTF-8")), true);
            //创建输出流对象
            while(true){String str=in.readLine();
                if(str==null)  break;
                System.out.println(str);
                out.println("服务器发送消息: "+str);
                out.flush();
                if(str.equals("over"))break;}
            out.close();in.close();client.close();
        } catch(IOException e){e.printStackTrace();}}}
```

客户端实现的代码如下：

```
public class MainActivity extends Activity{
    private static final String TAG="MainActivity";
    private static final String IP="192.168.1.7";
    private static final int PORT=1000;
    private Button btnSend;
    private EditText etSend, etReciveServer;
    private Thread thread=null;
    protected void onCreate(Bundle savedInstanceState){
    super.onCreate(savedInstanceState);
    setContentView(R.layout.activity_main);
    etSend=(EditText)findViewById(R.id.etSend);
    etReciveServer=(EditText)findViewById(R.id.etReciveServer);
    btnSend=(Button)findViewById(R.id.btnSend);
    btnSend.setOnClickListener(new OnClickListener(){
        public void onClick(View arg0){
            thread=new Thread(new Runnable(){//创建线程
                public void run(){Socket socket=null;
                try{socket=new Socket(IP, PORT);
```

```
                    String msg=etSend.getText().toString();
                    //将客户端的消息写入PrintWriter流中，通过out.println(msg)
                    //方法客户端将消息放入输出流中，以便服务器读取
            PrintWriter out=new PrintWriter(
                new BufferedWriter(new OutputStreamWriter(
                    socket.getOutputStream())),true);
            out.println(msg);  out.flush();
            BufferedReader in=new BufferedReader(
                new InputStreamReader(socket.getInputStream()));
            String str=in.readLine();
            Message message=new Message();
            Bundle bundle=new Bundle();
            bundle.putString("msg", str);message.setData(bundle);
            handler.sendMessage(message);
        }catch(Exception e){e.printStackTrace();}}});
        thread.start();  //启动线程
    }});}
  Handler handler=new Handler(){
      public void handleMessage(Message msg){
          Bundle bundle=msg.getData();
          String returnMsg=bundle.get("msg").toString();
          etReciveServer.setText(returnMsg);};};}
```

高级通信实现了多个客户端同时可以与服务器进行交互，可以开两个模拟器，同时运行客户端应用，发现两个客户端通信互不影响。高级通信程序运行结果如图3-16和图3-17所示。

Problems | @ Javadoc | Declaration | Console
AdvancedPCServer [Java Application] C:\MyEclipse10\
等待客户端连接请求!
春暖花开
等待客户端连接请求!
over
等待客户端连接请求!
团结就是力量，脱贫攻坚再加把劲。
等待客户端连接请求!

图 3-16　服务器运行效果

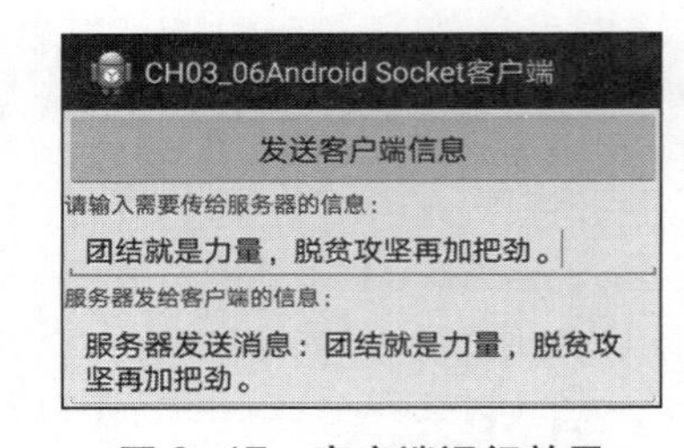

图 3-17　客户端运行效果

3.3 案例

视频

使用WebView浏览网页应用实战

3.3.1 WebView 迷你浏览器

该WebView迷你浏览器实现嵌入式浏览器功能，它包括5个组件，一个文本框用于接收用户输入URL；一个WebView用于加载并显示该URL页面，3个按钮分别支持Web搜索、前进和后退功能。该程序布局大纲如图3-18所示，程序布局activity_webwiew.xml代码如下：

```
<LinearLayout xmlns:android="http://schemas.
android.com/apk/res/android"
    android:layout_width="fill_parent"
    android:layout_height="fill_parent"
    android:orientation="vertical">
    <LinearLayout
        android:layout_width="fill_parent"
        android:layout_height="wrap_content"
        android:orientation="horizontal">
        <Button
            android:id="@+id/btnForward"
            android:layout_width="wrap_content"
            android:layout_height="wrap_content"
            android:text="@string/forward_str"/>
        <Button
            android:id="@+id/btnBack"
            android:layout_width="wrap_content"
            android:layout_height="wrap_content"
            android:text="@string/back_str"/>
        <EditText
            android:id="@+id/etURL"
            android:layout_width="wrap_content"
            android:layout_height="wrap_content"
            android:layout_weight="1"
            android:hint="@string/url_str"
            android:lines="1"
            android:text="https://www.sziit.edu.cn/"/>
        <Button
            android:id="@+id/btnSearch"
            android:layout_width="wrap_content"
            android:layout_height="wrap_content"
            android:text="@string/search_str"/>
    </LinearLayout>
    <WebView
        android:id="@+id/webView"
        android:layout_width="fill_parent"
        android:layout_height="0dip"
        android:layout_weight="1.0"
        android:focusable="false"/>
</LinearLayout>
```

Outline
LinearLayout
LinearLayout
btnForward (Button) - "前进"
btnBack (Button) - "后退"
etURL (EditText) - "https://www.sziit.edu.cn/"
btnSearch (Button) - "搜索"
webView

图 3-18　WebView 迷你浏览布局大纲

WebViewActivity 类程序的 Java 代码如下：

```
public class WebViewActivity extends Activity implements OnClickListener,
```

```
OnKeyListener{
      private WebView webView;    //声明WebView组件的对象
      private EditText etURL;     //声明作为地址栏的EditText对象
      private Button btnSearch;   //声明Search搜索按钮对象
      private Button btnForward; //声明前进按钮对象
      private Button btnBack;     //声明后退按钮对象
      protected void onCreate(Bundle savedInstanceState){  //重写onCreate
          super.onCreate(savedInstanceState);
          setContentView(R.layout.activity_webwiew);
          initView();       //初始化组件
          initWebView();  //初始化WebView组件
          setListener();  //设置事件监听器，处理事件回调函数
      }
      private void initView(){  //初始化组件
          etURL=(EditText)findViewById(R.id.etURL);  //获取布局中添加地址栏
          btnSearch=(Button)findViewById(R.id.btnSearch);  //获取布局中GO按钮
          btnForward=(Button)findViewById(R.id.btnForward);  //获取前进按钮
          btnBack=(Button)findViewById(R.id.btnBack);}  //获取布局后退按钮
      private void initWebView(){  //初始化WebView组件
          webView=(WebView)findViewById(R.id.webView);  //获取WebView组件
          //设置JavaScript可用
          webView.getSettings().setJavaScriptEnabled(true);
          //处理JavaScript对话框
          webView.setWebChromeClient(new WebChromeClient());
          //处理各种通知和请求事件，如果不使用该句代码，将使用内置浏览器访问网页
          webView.setWebViewClient(new WebViewClient());}
      private void setListener(){  //设置事件监听器，处理事件回调函数
      btnForward.setOnClickListener(this);  //“前进”按钮添加单击事件监听器
      btnBack.setOnClickListener(this);  //为“后退”按钮添加单击事件监听器
      btnSearch.setOnClickListener(this); //为Search按钮添加单击事件监听器
      etURL.setOnKeyListener(this);  //为地址栏添加键被按下事件监听器
      }
      private void openURL(){  //打开网页方法
          webView.loadUrl(etURL.getText().toString());}
      private void showDialog(){  //显示对话框
          new AlertDialog.Builder(WebViewActivity.this).setTitle("网页浏览器")
              .setMessage("请输入要访问的网址")
              .setPositiveButton("确定", new DialogInterface.OnClickListener(){
                public void onClick(DialogInterface dialog,int which){
                  Log.d("WebWiew", "单击确定按钮");}
          }).show();}
      public void onClick(View v){
          switch(v.getId()){
```

```
        case R.id.btnSearch:
        if(!"".equals(etURL.getText().toString())){
            openURL();}else{showDialog();  //弹出提示对话框}break;
      case R.id.btnForward:webView.goForward();break;  //前进
      case R.id.btnBack:webView.goBack();break;  //后退
  }}
  public boolean onKey(View v, int keyCode, KeyEvent event){
      if(keyCode==KeyEvent.KEYCODE_ENTER){  //如果为回车键
         if(!"".equals(etURL.getText().toString())){openURL();  //打开浏览器
              return true;}else{showDialog();}}  //弹出提示对话框
      return false;}}
```

该程序的运行结果如图 3-19 所示。

图 3-19　WebView 迷你浏览器运行结果

3.3.2　获取 Web 服务器数据

在开发一些基于 Web 系统的 Android 客户端程序开发过程中，通常要获取服务器数据是需要访问一个 Web 网址来获取的，再将内容进行解析、处理和显示。本案例主要实现获取 Web 服务器数据，该程序的界面布局大纲如图 3-20 所示。

图 3-20　获取 Web 服务器数据

程序的界面布局 activity_main.xm 代码如下：

```
<LinearLayout xmlns:android="http://schemas.android.com/apk/res/android"
    xmlns:tools="http://schemas.android.com/tools"
    android:layout_width="match_parent"
    android:layout_height="match_parent"
    tools:context="sziit.lihz.GetWebData.MainActivity">
    <LinearLayout
        android:layout_width="match_parent"
        android:layout_height="match_parent"
        android:orientation="vertical">
        <TextView
            android:id="@+id/tvWebTitle"
```

```
            android:layout_width="match_parent"
            android:layout_height="wrap_content"
            android:gravity="left"
            android:text="使用WebView访问网络"/>
        <Button
            android:id="@+id/btnWebBrowser"
            android:layout_width="match_parent"
            android:layout_height="wrap_content"
            android:gravity="left"
            android:text="使用WebView浏览网页"/>
        <Button
            android:id="@+id/btnWebHtml"
            android:layout_width="match_parent"
            android:layout_height="wrap_content"
            android:gravity="left"
            android:text="使用Webview加载HTML代码"/>
    </LinearLayout>
</LinearLayout>
```

WebBrowserActivity程序的Java代码如下：

```
public class WebBrowserActivity extends Activity implements OnClickListener{
    private WebView webViewShow=null;
    private Button btnUrlOk;
    private EditText etUrl=null;
    private String strUrl="http://192.168.1.7:8080/HelloWorld02/ch3/index.jsp";
    protected void onCreate(Bundle savedInstanceState){
         super.onCreate(savedInstanceState);
           setContentView(R.layout.activity_webview);
           etUrl=(EditText)findViewById(R.id.etUrl);
           btnUrlOk=(Button)findViewById(R.id.btnUrlOk);
           btnUrlOk.setOnClickListener(this);
           webViewShow=(WebView)findViewById(R.id.webViewShow);
           webViewShow.loadUrl(strUrl);}
      public void onClick(View v)
           switch(v.getId()){
           case R.id.btnUrlOk:browseWeb();break;}}
      private void browseWeb(){
           strUrl=etUrl.getText().toString();
           if(!strUrl.equals(""))
           {webViewShow.loadUrl(strUrl);}
           else{Toast.makeText(getApplicationContext(), "网址不可以为空",
Toast.LENGTH_SHORT).show();}}}
```

程序清单 CH03\ CH03_8\ com\example\demohttpclient\WebHtmlActivity.java。

```
public class WebHtmlActivity extends Activity implements OnClickListener{
    private TextView tvWebHtml=null;
    private Button btnLoadData,Button btnLoadURLData;
    private WebView webViewHtml=null;
    private StringBuilder sbHtml,sbLoadData,sbLoadURLData;
    protected void onCreate(Bundle savedInstanceState){
        super.onCreate(savedInstanceState);
        setContentView(R.layout.activity_webhttp);
        findView();  setClickListener();initWebView();}
    private void findView(){
        tvWebHtml=(TextView)findViewById(R.id.tvWebHttp);
        btnLoadData=(Button)findViewById(R.id.btnloadData);
        btnLoadURLData=(Button)findViewById(R.id.btnLoadURLData);}
    private void setClickListener(){
        btnLoadData.setOnClickListener(this);
        btnLoadURLData.setOnClickListener(this);}
    private void initWebView(){
        webViewHtml=(WebView)findViewById(R.id.webViewHttp);
        sbHtml=new StringBuilder();
        sbHtml.append("<div>");
        sbHtml.append("<h1>使用Webview加载HTML代码案例实现方法</h1>");
        sbHtml.append("<p><b>步骤1</b>: 创建StringBuilder对象并向其添加HTML
代码; </p>");
        sbHtml.append("<p><b>步骤2</b>: 从StringBuilder导出HTML代码显示在文本
视图TextView上; </p>");
        sbHtml.append("<p><b>步骤3</b>: 将对应的HTML代码加载到WebView上显示
出来! </p>");
        tvWebHtml.setText(sbHtml.toString());
        webViewHtml.loadDataWithBaseURL(null,sbHtml.toString(),
          "text/html", "utf-8", null);}
    private void doLoadDataClick(){
        sbLoadData=new StringBuilder();
        sbLoadData.append("<div>");
        sbLoadData.append("<h1>1级标题—HTML5标记语言</h1>");
        sbLoadData.append("<h2>2级标题—文字排版标记</h2>");
        sbLoadData.append("<h3>3级标题—标题标记</h3>");
        sbLoadData.append("<h4>4级标题—居左对齐</h4>");
        sbLoadData.append("<h5>5级标题—居中对齐</h5>");
        sbLoadData.append("<h6>6级标题—居右对齐</h6>");
        sbLoadData.append("</div>");
        tvWebHtml.setText(sbLoadData.toString());
```

```
            webViewHtml.getSettings().setDefaultTextEncodingName("utf-8");
            webViewHtml.loadData(sbLoadData.toString(), "text/html;
charset=UTF-8",null);}
        private void doLoadURLDataClick(){
            sbLoadURLData=new StringBuilder();
            sbLoadURLData.append("<div>");
            sbLoadURLData.append("<img src=\"bike.png\">");
            sbLoadURLData.append("</div>");
            sbLoadURLData.append("<div>");
            sbLoadURLData.append("<ul>");
            sbLoadURLData.append("<li>Android高级应用编程实战课程</li>");
            sbLoadURLData.append("<li>意图与服务</li>");
            sbLoadURLData.append("<li>Android数据永久存储应用</li>");
            sbLoadURLData.append("<li>Android网络应用</li>");
            sbLoadURLData.append("</ul>");
            sbLoadURLData.append("</div>");
            tvWebHtml.setText(sbLoadURLData.toString());
            webViewHtml.getSettings().setDefaultTextEncodingName("utf-8");
            webViewHtml.loadDataWithBaseURL("file:///android_asset/",
               sbLoadURLData.toString(),"text/html", "utf-8",null);}
        public void onClick(View v){
            switch(v.getId()){
            case R.id.btnloadData:doLoadDataClick();  break;
            case R.id.btnLoadURLData:doLoadURLDataClick();break;}}}
```

该程序的运行结果如图 3-21 所示。

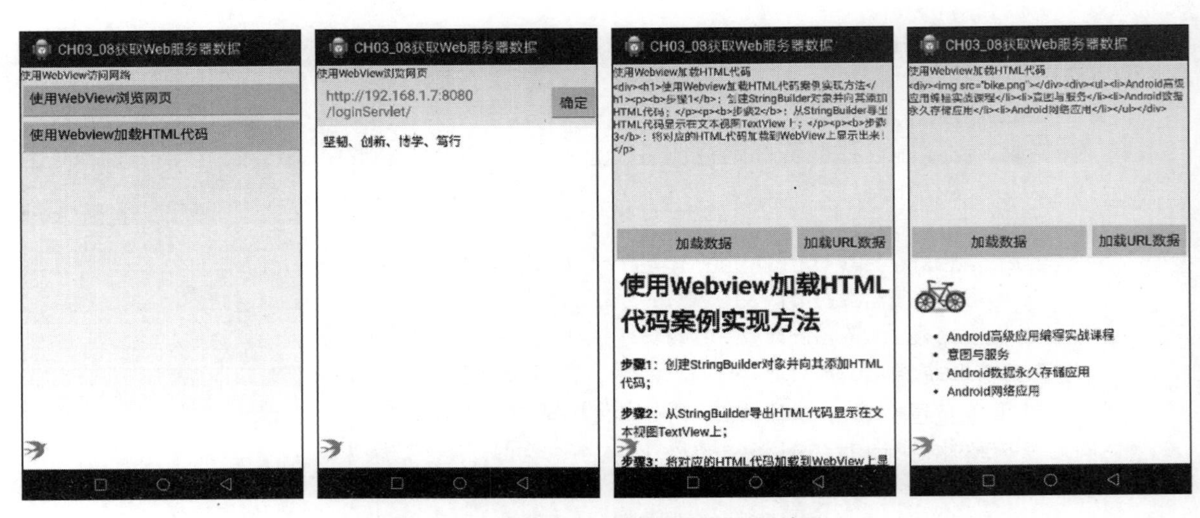

图 3-21　获取 Web 服务器数据运行效果

3.4 知识扩展

3.4.1 使用 WebView 浏览网页

WebView组件支持浏览网页功能，它的内核基于开源的WebKit引擎。

WebView的用法与普通ImageView组件的用法很相似，其常用的函数如下：

```
void  goBack();                    //后退
void  goForward();                 //前进
void loadUrl(String url);          //加载指定URL对应的网页
boolean zoomln();                  //放大网页
boolean zoomOut();                 //缩小网页
```

3.4.2 使用 WebView 中 JavaScript 脚本调用 Android 方法

WebView加载的页面上常带有JavaScript脚本，比如用户单击按钮时弹出提示框或列表框等。在HTML页面上的按钮只能激发一段JavaScript脚本，这就需要JavaScript脚本来调用Adnroid的相应方法。在WebView的JavaScript中调用Android方法的步骤如下：

（1）调用WebView关联的WebSettings的JavaScript启用JavaSrcipt调用功能。

（2）调用WebView的addJavascriptInterface方法将Object对象暴露给JavaScript。

（3）在JavaScript脚本中通过刚才暴露的name对象调用Android的方法。

本章小结

本章主要讲解了Android平台上网络与通信的开发，主要包括对无线网络技术的简单介绍，以及网络通信中常见的HTTP和Socket通信，最后又介绍了Wi-Fi以及蓝牙技术在Android中的应用。

强化练习

一、填空题

1. 计算机网络是现代通信技术与计算机技术相结合的产物，它是利用（　　）、（　　）和（　　）等网络设备，把分散在各地的计算机连接起来，并通过特定的软件（网络协议）实现计算机之间的相互通信和资源共享。

2. 网络按覆盖范围分类（　　）、（　　）和（　　）。

3. 网络按网络拓扑结构分类（　　）、（　　）和（　　）。

4. Android目前有3种网络接口可以使用，分别为（　　）、（　　）和（　　）。

5. HTTP通信中使用最多的两种方式是（　　）和（　　）。

6. HttpClient中get是将参数在（　　）中传递，post是将参数用（　　）传递。

7. HttpURLConnection编程时，con.setRequestMethod("POST")设置（　　）方式提交请求，con.setDoInput(true)设置该URLConnection是可（　　）的，con.setDoOutput(true);设置该URLConnection是可（　　）的。

二、单选题

1. HttpClient请求访问Web服务器时设置读取超时属性的方法为（　　）。

   A. setTimeout()　　B.setRequestMethod()　　C.setConnectTimeout()　　D.setReadTimeout()

2. 访问Wi-Fi的权限名称为（　　）。

   A. android.permission.INTERNET

   B. android.permission.ACCESS_CHECKIN_PROPERTIES

   C. android.permission.ACCESS_WIFI_STATE

   D. android.permission.CALL_PHONE

3. HttpClient请求访问Web服务器数据成功的状态码为（　　）。

   A. 200　　B. 100　　C. 0　　D. 1

4. 连接Wi-Fi后获取当前连接速度应使用WifiInfo的方法为（　　）。

   A. getSpeed()　　B. getSSID()

   C. getLinkSpeed()　　D. getConnectionInfo()

5. 采用HTTP编程时，获取HttpURLConnection类对象的方法为（　　）。

   A. con=new HttpURLConnection();

   B. con=(HttpURLConnection)url.getConnection();

   C. con=HttpURLConnection.newInstance();

   D. con=(HttpURLConnection)url.openConnection();

6. HttpURLConnection采用Get方式请求数据时,必须设置设置的属性为（　　）。

   A. connection.connect()　　B. connection.setDoInput(true)

   C. connection.setDoOutput(true)　　D. connection.setRequestMethod("POST")

7. WebView用loadUrl()方法加载assets目录下index.jsp文件需传入的参数为（　　）。

   A. file:///androidasset/index.jsp　　B. file:///assets/html/index. jsp

   C. file:///asset/index. jsp　　D. file:///android_asset/index.jsp

8. HttpClient采用Get方式请求数据时,可用（　　）类来构建Http请求。

   A. HttpGet　　B. HttpPost

   C. Get　　D. URLConnection

9. 支持打开和关闭Wi-Fi功能的类为（　　）。

   A. WifiManager　　B. WifiService　　C. WifiState　　D. Wifi

10. 表示Wi-Fi为关闭状态的为（　　）。

    A. WifiManager.WIFI_STATE_DISABLED　　B. WifiManager.WIFI_STATE_DISABLE

    C. WifiManager.WIFI_STATE_CLOSED　　D. WifiManager.WIFI_STATE_CLOSE

11. 已知HttpURLConnection con=(HttpURLConnection)url.openConnection();则返回数据的输入流的方法为（　　）。

    A. con.getReader();　　B. con.getEntity().getStream();

C. con.getEntity().getInputStream(); D. con.getInputStream();

12. HttpClient通过Post方式访问url时要传递pwd参数，下列正确方式为（ ）。
    A. httpGet(url, pwd);
    B. httpGet.addParams("pwd",123456);
    C. url=url+"?pwd="+123456;
    D. list.add(new BasicNameValuePair("pwd", 123456));
13. 打开和关闭Wi-Fi都要使用到（ ）类的方法。
    A. WifiManager B. Wifi C. WifiState D. WifiService
14. 表示Wi-Fi为打开状态的为（ ）。
    A. WifiManager.WIFI_STATE_OPENING B. WifiManager.WIFI_STATE_ENABLING
    C. WifiManager.WIFI_STATE_OK D. WifiManager.WIFI_STATE_USING

三、简答题、操作题或编程题

1. HTTP Get与HTTP Post两种方式的格式是什么？
2. 如何获得服务器返回的状态？
3. 在访问网页时，权限验证的方法有哪些？
4. 简述Socket通信的步骤。
5. 用Socket实现简单的通信。

第4章

Android 调用外部数据

学习目标

- 理解XML的概念。
- 掌握SAX解析XML文件的方法。
- 掌握DOM解析XML文件的方法。
- 掌握PULL解析XML文件的方法。
- 掌握JSON（JavaScript Object Notation）数据交换方法。
- 掌握基于位置的服务的实现。

4.1 学习导入

XML（Extensible Markup Language，可扩展标记语言）是标准通用标记语言的子集，用于标记电子文件使其具有结构性的标记语言，可用来标记数据、定义数据类型，是一种允许用户对自己的标记语言进行定义的源语言。它非常适合Web传输，提供统一的方法来描述和交换独立于应用程序或供应商的结构化数据。XML是一种类似于HTML的标记语言，其标记不是在XML中预定义的，而是由用户自定义的；XML使用文档类型定义（DTD）或者模式（Schema）来描述数据；XML主要用来描述数据和存放数据。

4.2 技术准备

4.2.1 SAX解析器

SAX（Simple API for XML, XML简单API）是一个公共的基于事件的XML文档解析标准，能够通过一个简单的、快速的方法来对XML文档进行处理。SAX解析器框架如图4–1所示。

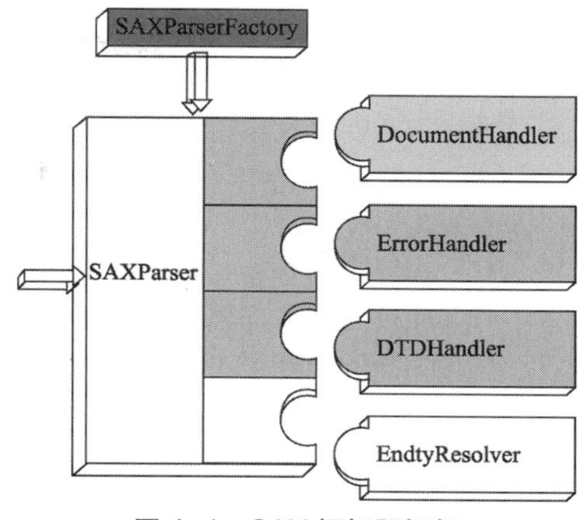

图 4–1　SAX 解析器框架

SAX既是一个接口，也是一个软件包。作为接口，SAX是事件驱动型XML解析的一个标准接口，对文档进行顺序扫描，当扫描到文档开始与结束、元素开始与结束、文档开始与结束等地方时通知事件处理函数，由事件处理函数做相应动作，然后继续同样的扫描，直至文档结束。

图4–1中最上方的SAXParserFactory用来生成一个分析器实例。XML文档是从左侧箭头所示处读入，当分析器对文档进行分析时，就会触发在DocumentHandler、ErrorHandler、DTDHandler以及EntityResolver接口中定义的回调方法。SAX解析器的工作方式是自动将事件推入事件处理器进行处理，用户不能控制事件的处理主动结束。

视频

SAX解析器工作机制

SAX会产生的主要事件类型包括：

（1）在文档的开始和结束时触发文档处理事件。

（2）在文档内每个XML元素接受解析的前后触发元素事件。

（3）任何元数据通常由单独的事件处理。

SAX解析XML文件的主要步骤包括：

（1）创建事件处理程序（即编写ContentHandler的实现类，一般继承自DefaultHandler类，采用adapter模式）。

（2）创建SAX解析器。

（3）将事件处理程序分配到解析器。

（4）对文档进行解析，将每个事件发送给事件处理程序。

下面就对SAX分析器中的几个主要API接口进行简单介绍。

1. ContentHandler接口

ContentHandler是Java类包中一个特殊的SAX接口，位于org.xml.sax包中。该接口封装了一些对事件处理的方法，当XML解析器开始解析XML输入文档时，它会遇到某些特殊的事件，如文档的开头和结束、元素开头和结束以及元素中的字符数据等事件。当遇到这些事件时，XML解析器会调用ContentHandler接口中相应的方法来响应该事件。

ContentHandler接口的方法有以下几种：

（1）void startDocument()：文档解析开始的处理。

（2）void endDocument()：文档解析结束的处理。

（3）void startElement(String uri, String localName, String qName, Attributes atts)：ElementNode开始的处理。其参数说明如下：

- uri：名称空间URI。如果元素没有任何名称空间URI，或者没有正在执行名称空间处理，则为空字符串。
- localName：本地名称。如果没有正在执行名称空间处理，则为空字符串。

- qName：限定名称（带有前缀）。如果限定的名称不可用，则为空字符串。
- Attributes：附加到元素的属性。如果没有属性，则它将是空的Attributes对象。

（4）void endElement(String uri, String localName, String qName)：ElementNode结束的处理。

（5）void characters(char[ ] ch, int start, int length)：具体在某一节点中的处理。其参数说明：

- ch：正在读取的文本节点。
- start：字符数组中的开始位置。
- length：从字符数组中使用的字符数。

2. DTDHandler接口

DTDHandler用于接收基本的DTD（Document Type Definition，文档类型定义）相关事件的通知。该接口位于org.xml.sax包中，仅包括DTD事件的注释和未解析的实体声明部分。SAX解析器可按任何顺序报告这些事件，而不管声明注释和未解析实体时所采用的顺序；但是，必须在文档处理程序的startDocument()事件之后，在第一个startElement()事件之前报告所有的DTD事件。DTDHandler接口包括以下两个方法：

（1）void notationDecl(String name,String publicId,String systemId)：接收注释声明事件的通知。

（2）void unparsedEntityDecl(String name,String publicId,String systemId,String notationName)：接收未分析实体声明事件的通知。

3. EntityResolver接口

EntityResolver接口是用于解析实体的基本接口，该接口位于org.xml.sax包中。该接口只有如下一个方法：

public InputSource resolveEntity(String publicId, String systemId)：应用程序解析外部实体。

解析器将在打开任何外部实体前调用此方法。此类实体包括在DTD内引用的外部DTD子集和外部参数实体和在文档元素内引用的外部通用实体等。如果SAX应用程序需要实现自定义处理外部实体，则必须实现此接口。

4. ErrorHandler接口

ErrorHandler接口是SAX错误处理程序的基本接口。如果SAX应用程序需要实现自定义的错误处理，则它必须实现此接口，然后解析器将通过此接口报告所有的错误和警告。

该接口的方法如下：

（1）void error(SAXParseException exception)：接收可恢复错误的通知。

视频

SAX解析器应用实战

（2）void fatalError(SAXParseException exception)：接收不可恢复错误的通知。

（3）void warning(SAXParseException exception)：接收警告通知。

DefaultHandler类已经实现了EntityResolver接口、DTDHandler接口、ContentHandler接口和ErrorHandler的默认实现。一个典型的SAX应用程序通常从DefaultHandler派生子类，然后重写ContentHandler接口startDocument、endDocument、startElement、endElement和characters等事件回调方法。

SAX解析器解析xml字符串的基本解析步骤如下：

（1）创建解析器工厂：SAXParserFactory factory=SAXParserFactory.newInstance()。

（2）获得解析器：SAXParser parser=factory.newSAXParser()。

（3）获得XML读取器：XMLReader xmlReader=parser.getXMLReader()。

（4）继承DefaultHandler生成SAX解析器SaxHandler。

（5）设置内容处理程序：xmlReader.setContentHandler(baseHandler)。

（6）创建字符串读取器：StringReader read=new StringReader(xmlString)。

（7）创建输入源：InputSource source=new InputSource(read)。

（8）解析文件：xmlReader.parse(source)。

（9）关闭字符串读取器：read.close()。

下面通过采用SAX方法解析assets中student.xml案例详细讲解SAX解析器的使用技巧。

（1）了解assets中student.xml文件格式，其内容如下：

```
<?xml version="1.0"encoding="UTF-8"?>
<students>
<student sn='0001' >
    <id>1</id>
    <name>倩倩</name>
    <age>18</age>
</student>
<student sn='0002' >
    <id>2</id>
    <name>莹莹</name>
    <age>20</age>
</student>
<student sn='0003' >
    <id>3</id>
    <name>莉莉</name>
    <age>18</age>
</student>
</studets>
```

（2）定义与student.xml文件对应的学生实体类Student，其源代码如下：

```
public class Student{/***学生实体类Student**/
    private String sn,id,name,age;  //定义私有数据成员
    public String getSn(){return sn;}
    public void setSn(String sn){this.sn=sn;}
    public String getId(){return id;}
    public void setId(String id){this.id=id;}
    public String getName(){return name;}
    public void setName(String name){this.name=name;}
    public String getAge(){return age;}
    public void setAge(String age){this.age=age;}
    public String toString(){return "学生信息[姓名="+name+",身份号="+
id+",序号="+sn+",年龄="+age+"]";}}
```

（3）根据DefaultHandler类定义抽象类BaseHandler，其源代码如下：

```
public abstract class BaseHandler extends DefaultHandler{
//缓存！公共数据的存放类，所有数据都从这里存取
public static Map sMap=new HashMap();  //公共数据的存放Map
public abstract boolean parse(String xmlString);  //解析方法
public static void parserXml(BaseHandler baseHandler, String xmlString)
throws Exception{if(xmlString==null||xmlString.length()==0)return;
    SAXParserFactory factory=SAXParserFactory.newInstance();  //建立工厂对象
    SAXParser parser=factory.newSAXParser();  //获得SAX解析器对象
    XMLReader xmlReader=parser.getXMLReader();  //获得XML读取器对象
    xmlReader.setContentHandler(baseHandler);  //设置内容处理程序
    StringReader read=new StringReader(xmlString);  //构造字符串读取器对象
        InputSource source=new InputSource(read);  //构造输入源对象
        xmlReader.parse(source);  //解析XML数据
        read.close();  //关闭字符串读取器对象
}
    //定义以下SAX解析器抽象事件回调方法
public abstract void characters(char[] ch, int start, int length)
    throws SAXException;
public abstract void endDocument()throws SAXException;
public abstract void endElement(String uri, String localName, String qName)
throws SAXException;
public abstract void startDocument()throws SAXException;
public abstract void startElement(String uri, String localName,
    String qName, Attributes attributes)throws SAXException;}
```

（4）设计SAX解析student.xml的实现类，其源代码如下：

```
public class StudentHandler extends BaseHandler{
    private Stack<String>tagStack=new Stack<String>();  //定义标签堆栈
    private Vector<Student>students=new Vector<Student>();
    public void startDocument()throws SAXException{
        Log.d(TAG, "startDocument()");}
    public void startElement(String uri, String localName, String qName,
        Attributes attributes)throws SAXException{
        if(qName.equals("student")){Student student=new Student();
        student.setSn(attributes.getValue("sn"));
        students.addElement(student);}
        tagStack.push(qName);}
    public void characters(char[] ch, int start, int length)
        throws SAXException{
        String chars=new String(ch, start, length).trim();
        if(chars!=null){
```

```
        String tagName=(String)tagStack.peek();  //查看栈顶对象而不移除它
      Student student=students.lastElement();
      if(tagName.equals("id")){student.setId(chars);}
      else if(tagName.equals("name")){student.setName(chars);}
      else if(tagName.equals("age")){student.setAge(chars);}}}
  public void endElement(String uri, String localName, String qName)
      throws SAXException{
    tagStack.pop();  //移除栈顶对象并作为此函数的值返回该对象}
  public boolean parse(String xmlString){
    try{super.parserXml(this, xmlString);return true;}
    catch(Exception e){e.printStackTrace();return false;}}
  public void endDocument()throws SAXException{
    sMap.put("students",students);  //保存入sMap，即Vector对象
    students=null;}}
```

（5）设计活动布局，其界面大纲如图 4-2 所示。

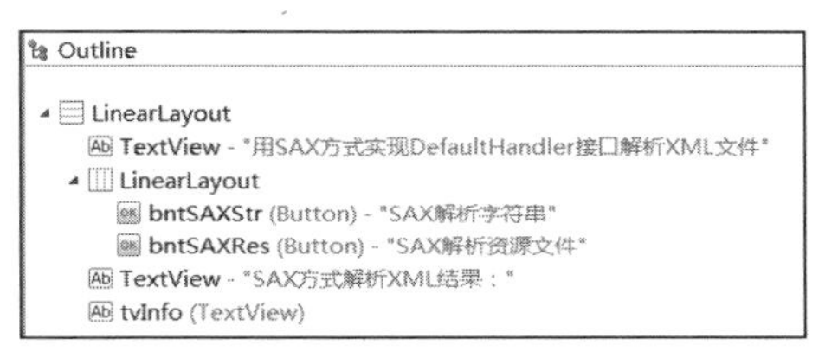

图 4-2　SAX 解析 XML 案例活动布局大纲

activity_sax.xml 源代码如下：

```
<LinearLayout xmlns:android="http://schemas.android.com/apk/res/android"
    xmlns:tools="http://schemas.android.com/tools"
    android:layout_width="match_parent"
    android:layout_height="match_parent"
    android:orientation="vertical"
    tools:context=".SAXActivity">
    <TextView
        android:layout_width="match_parent"
        android:layout_height="wrap_content"
        android:background="#0000FF"
        android:text="@string/note_msg"
        android:textColor="#FFFFFF"
        android:textSize="20sp"/>
    <LinearLayout
        android:layout_width="match_parent"
        android:layout_height="wrap_content"
        android:orientation="horizontal">
        <Button
```

```
            android:id="@+id/bntSAXStr"
            android:layout_width="wrap_content"
            android:layout_height="wrap_content"
            android:layout_weight="1"
            android:text="@string/bntsax_str"/>
        <Button
            android:id="@+id/bntSAXRes"
            android:layout_width="wrap_content"
            android:layout_height="wrap_content"
            android:layout_weight="1"
            android:text="@string/bntsax_res"/>
    </LinearLayout>
    <TextView
        android:layout_width="match_parent"
        android:layout_height="wrap_content"
        android:background="#0000FF"
        android:text="@string/prompt_str"
        android:textColor="#FFFFFF"
        android:textSize="24sp"/>
    <TextView
        android:id="@+id/tvInfo"
        android:layout_width="match_parent"
        android:layout_height="wrap_content"
        android:hint="@string/result_hint"/>
</LinearLayout>
```

（6）SAXActivity活动类实现代码如下：

```
public class SAXActivity extends Activity{
    private static final String data="<students><student sn='0001'><id>1</
id><name>倩倩</name><age>18</age></student><student sn='0002'><id>2</id><name>
莹莹</name><age>20</age></student></students>";
    private BaseHandler handler;  //定义处理程序
    private StringBuffer sb;  //定义字符串缓冲区
    private TextView tvInfo;  //定义显示解析结果的文本视图对象
    private Button bntSAXStr,bntSAXRes;  //定义按钮对象
    private AssetManager assetManager;  //定义Asset管理器对象
    private static String xmlData=new String();
    protected void onCreate(Bundle savedInstanceState){  //重写onCreate()方法
      super.onCreate(savedInstanceState);  //调用基类onCreate()方法
      setContentView(R.layout.activity_sax);  //设置活动布局UI
      tvInfo=(TextView)findViewById(R.id.tvInfo);  //绑定TextView对象
      bntSAXStr=(Button)findViewById(R.id.bntSAXStr);  //绑定Button对象
      bntSAXRes=(Button)findViewById(R.id.bntSAXRes);  //绑定Button对象
```

```
    bntSAXStr.setOnClickListener(new OnClickListener(){  //设置单击监听器
        public void onClick(View v){  //按钮单击事件监听器回调函数
      sb=new StringBuffer();  //构造字符串缓存
      handler=new StudentHandler();  //构建处理程序
      handler.parse(data);  //解析XML数据
      showParseResult();  //显示SAX解析XML文件结果}});
  bntSAXRes.setOnClickListener(new OnClickListener(){  //设置单击监听器
    public void onClick(View v){  //按钮单击事件监听器回调函数
      sb=new StringBuffer();  //构造字符串缓存
      handler=new StudentHandler();  //构建处理程序
      xmlData=readAssetXML("student.xml");  //读取asset中XML文件
      tvInfo.setText(xmlData);
      if(xmlData.length()>1){handler.parse(xmlData);  //解析XML数据
         showParseResult();  //显示SAX解析XML文件结果
      }
      else{tvInfo.setText("读文件student.xml失败!");}}});}
  private void showParseResult(){  //显示SAX解析XML文件结果
    //SAX解析XML结果均存放在sMap中
    Vector students=(Vector)BaseHandler.sMap.get("students");
    for(int i=0;i<students.size();i++){  //遍历sMap
      Student student=(Student)students.elementAt(i);
      if(student==null){sb.append("student 为空");}
      else{sb.append("序号="+student.getSn()+":"+
          "身份号="+student.getId()+":"+
          "姓名="+student.getName()+":"+
          "年龄="+student.getAge()+"\n");}}
    tvInfo.setText(sb.toString());}
  /***先将XML文件存到assets目录，用AssetMananage对象调用open()方法获得InputStream对象，
然后构造BufferedReader对象，将XML文件内容读到StringBuffer中*/
  private String readAssetXML(String name){
    //使用当前应用上下文得到AssetManager对象
    assetManager=getApplicationContext().getAssets();
    //定义StringBuffer对象,存放所有页面数据
    StringBuffer buffer=new StringBuffer();
    try{InputStream mis=assetManager.open(name);
      BufferedReader br=new BufferedReader(new InputStreamReader(mis));
      String str=null;
      while((str=br.readLine())!=ull){buffer.append(str.trim());}
      br.close();mis.close();
    }catch(IOException e){tvInfo.setText(e.toString());
      e.printStackTrace();}
    return buffer.toString();}}
```

（7）本SAX解析XML文件案例运行结果如图4-3所示。

图4-3　本SAX解析XML文件案例运行效果

4.2.2　DOM解析器

DOM（Document Object Model，文档对象模型）在本质上是一种文档平台。DOM解析器是W3C推荐处理XML的一种方式。DOM解析器将XML数据转换为一个对象模型的集合，用树这种数据结构对信息进行存储。通过DOM接口，应用程序可在任何时候访问XML文档中的任何一部分数据，因此利用DOM接口访问的方式也称为随机访问。DOM解析器框架结构如图4-4所示。javax.xml.parsers包中，定义了几个工厂类。可通过调用这些工厂类，得到对XML文档进行解析的DOM和SAX解析器对象。javax.xml.parsers包中的DocumentBuilderFactory用于创建DOM模式的解析器对象，DocumentBuilderFactory是一个抽象工厂类，它不能直接实例化，但该类提供了一个newInstance()方法，该方法会根据本地默认安装的解析器自动创建一个工厂对象并返回。DOM方式解析会读取完整个XML文档，在内存中构建代表整个DOM树的Document对象，从而再对XML文档进行操作：增、删、改、查。

视频

DOM解析器
工作机制

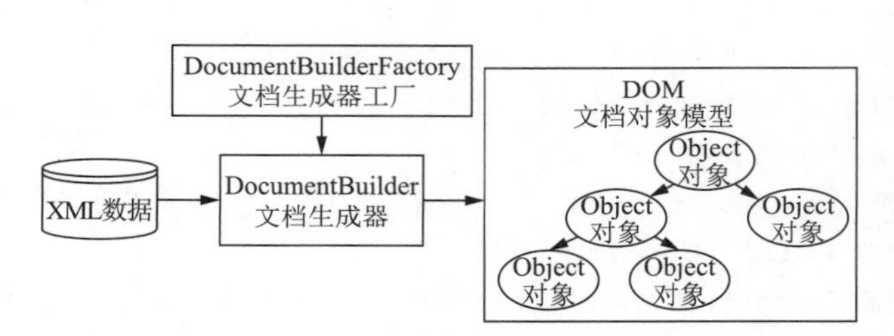

图4-4　DOM解析器框架结构

DOM解析中有以下4个核心操作接口：

（1）Document接口：该接口代表了整个XML文档，表示的是整棵DOM树的根，提供了对文档中的数据进行访问和操作的入口，通过Document节点可访问XML文件中所有的元素内容。Document接口常用方法包括：

- NodeList getElementsByTagName(String name)：根据节点的名称获得节点。
- Element getElementById(String elementId)：获得给定ID的元素值。
- Element getDocumentElement()：获取文档根节点。

（2）Node接口：该接口在整个DOM树中具有举足轻重的地位，DOM操作的核心接口中有很大一部分接口是从Node接口继承过来的。Node接口的常用方法包括：

- NodeList getChildNodes()：获得当前节点所有的子节点。
- NamedNodeMap getAttributes()：获得当前节点所有的属性和值。
- String getTextContent()：获得当前子节点的文本。

- String getNodeName()：获得当前节点的名称。
- short getNodeType()：获得基本对象类型。
- String getNodeValue()：获得节点值。

（3）NodeList 接口：该接口表示的是一个节点的集合，一般用于表示有顺序关系的一组节点。NodeList 接口的常用方法包括：

- Node item(int index)：返回集合中的第 index 项。
- int getLength()：列表中的节点数。

（4）NamedNodeMap 接口：该接口表示的是一组节点和其唯一名字对应的一一对应关系，该接口主要用于属性节点的表示上。NamedNodeMap 接口的常用方法为：

Node getNamedItem(String name)：检索按名称指定的节点。

DOM 解析器执行读取操作的步骤如下：

（1）创建 DocumentBuilderFactory:DocumentBuilderFactory factory = DocumentBuilderFactory.newInstance()。

（2）创建 DocumentBuilder:DocumentBuilder builder=factory.newDocumentBuilder()。

（3）创建 Document:Document doc=builder.parse("要解析的 XML 文件路径")。

（4）创建 NodeList:NodeList nl =doc.getElementsByTagName("读取节点")。

（5）进行 XML 信息读取。

视频

DOM解析器应用实战

下面通过采用 DOM 方法解析 assets 中 person.xml 案例详细讲解 DOM 解析器的使用技巧。

（1）了解 assets 中 person.xml 文件格式，其内容如下：

```
<?xml version="1.0"encoding="UTF-8"?>
<persons>
    <person id="1001">
        <name>倩倩</name>
        <age>18</age>
    </person>
    <person id="1002">
        <name>婷婷</name>
        <age>20</age>
     </person>
     <person id="1003">
        <name>磊磊</name>
        <age>21</age>
     </person>
</persons>
```

（2）定义与 person.xml 文件对应的实体类 Person，其源代码如下：

```
/***功能：设计实体类Person**/
public class Person{
    private int id;
    private String name;
    private int age;
```

```
    public int getId(){return id;}
    public void setId(int id){this.id=id;}
    public String getName(){return name;}
    public void setName(String name){this.name=name;}
    public int getAge(){return age;}
    public void setAge(int age){this.age=age;}
    public String toString(){
        return "个人信息[ID="+id+",姓名="+name+",年龄="+age+"]";}}
```

（3）设计采用DOM解析person.xml的实现类，其源代码如下：

```
public class DomParsePerson{
    public List<Person>getPersons(InputStream inputStream)throws Exception{
        List<Person>persons=new ArrayList<Person>();
        //创建DOM解析工厂(文档生成器工厂)实例
        DocumentBuilderFactory factory=DocumentBuilderFactory.newInstance();
        DocumentBuilder builder=factory.newDocumentBuilder(); //创建DOM解析器
        //开始解析XML文档并且得到整个文档的对象模型
        Document document=builder.parse(inputStream);
        Element root=document.getDocumentElement(); //得到根节点<persons>
        //得到根节点下所有标签为<person>的子节点
        NodeList personList=root.getElementsByTagName("person");
        for(int i=0;i<personList.getLength();i++){ //遍历person节点
            Person person=new Person(); //首先创建一个Person
            //得到本次Person元素节点
            Element personElement=(Element)personList.item(i);
            //得到Person节点中的ID
            person.setId(new Integer(personElement.getAttribute("id")));
            //得到Person节点下的所有子节点
            NodeList personChilds=personElement.getChildNodes();
            //遍历person节点下的所有子节点
            for(int j=0; j<personChilds.getLength();j++){
                  //如果是元素节点
                  if(personChilds.item(j).getNodeType()==Node.ELEMENT_NODE){
                        //得到该元素节点
                        Element childElement=(Element)personChilds.item(j);
                        //如果该元素节点是name节点
                        if("name".equals(childElement.getNodeName())){
                            //得到name节点下的第一个文本子节点的值
                            person.setName(childElement.getFirstChild().
                        getNodeValue());}
                            //如果该元素节点是age节点
                  else if("age".equals(childElement.getNodeName())){
                         //得到age节点下的第一个文本字节点的值
```

```
                person.setAge(new Short(childElement.getFirstChild()
                        .getNodeValue()));}}}
        //遍历完person下的所有子节点后将person元素加入到集合中
        persons.add(person);}return persons;}}
```

（4）设计活动布局，其界面大纲如图4-5所示。

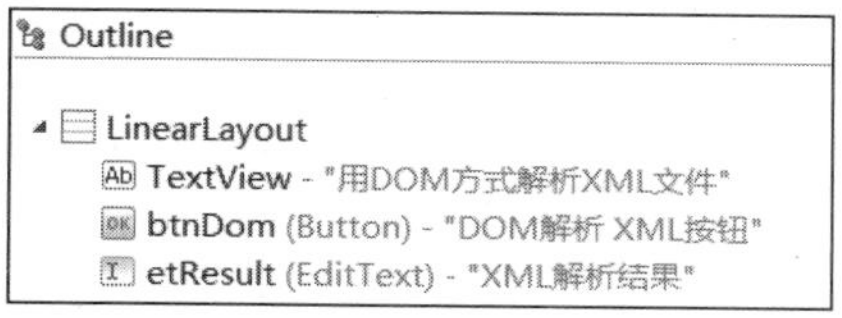

图 4-5　DOM 解析 XML 案例活动布局大纲

activity_dom.xml源代码如下：

```
<LinearLayout xmlns:android="http://schemas.android.com/apk/res/android"
    android:layout_width="fill_parent"
    android:layout_height="fill_parent"
    android:orientation="vertical">
    <TextView
        android:layout_width="fill_parent"
        android:layout_height="wrap_content"
        android:gravity="center_horizontal"
        android:text="用DOM方式解析XML文件"
        android:textColor="#FFFFFF"
        android:background="#0000FF"
        android:textSize="20sp"/>
    <Button
        android:id="@+id/btnDom"
        android:layout_width="fill_parent"
        android:layout_height="wrap_content"
        android:text="DOM解析 XML按钮"/>
    <EditText
        android:id="@+id/etResult"
        android:layout_width="fill_parent"
        android:layout_height="wrap_content"
        android:editable="false"
        android:scrollbars="vertical"
        android:text="XML解析结果"
        android:textSize="20sp"/>
</LinearLayout>
```

（5）DOMActivity活动类实现代码如下：

```
public class DOMActivity extends Activity{
    private static final String TAG="DOMActivity";
```

```
    private List<Person>listPerson;
    private InputStream inputStream;
    private EditText etResult;
    private Button btnDom;
    public void onCreate(Bundle savedInstanceState){
        super.onCreate(savedInstanceState);
        Log.d(TAG, "onCreate()");
        setContentView(R.layout.activity_dom);
        listPerson=new ArrayList<Person>();
        etResult=(EditText)findViewById(R.id.etResult);
        openAssetsXML();
        btnDom=(Button)findViewById(R.id.btnDom);
        btnDom.setOnClickListener(new OnClickListener(){
              public void onClick(View v){doClickDomXML();}});}
    private void openAssetsXML(){  //打开assets下XML文件
        try{inputStream=getAssets().open("person.xml");
        }catch(IOException e){etResult.setText(e.toString());
        e.printStackTrace();}}
    private void doClickDomXML(){
        String str="";
        try{
           listPerson=new DomParsePerson().getPersons(inputStream);
           str="DOM方式解析XML文件结果: \n";}catch(Exception e){
           etResult.setText(e.toString());e.printStackTrace();}
       for(int i=0; i<listPerson.size(); i++){
           Person person=listPerson.get(i);
           str=str+"ID号:"+person.getId()+"姓名:"+person.getName()
           +"  年龄:"+person.getAge()+"\n";}
        etResult.setText(str);}}
```

（6）本DOM解析XML文件案例运行结果如图4-6所示。

图4-6 本DOM解析XML文件案例运行结果

4.2.3 PULL解析器

1. PULL解析器简介

PULL是Android内置的XML解析器，其运行方式与SAX解析器相似。它提供了类似的事件，例如，开始元素和结束元素事件，使用parser.next()可进入下一个元素并触发相应事件。事件将作为数值代码被发送，因此可使用一个switch对感兴趣的事件进行处理。PULL是一个while循环，随时可跳出。当元素开始解析时，用parser.nextText()方法可获取下一个Text类型元素的值。PULL解析器的工作方式允许应用程序代码主动从解析器中获取事件，正因为是主动获取事件，因此可在满足了需要的条件后不再获取事件，结束解析。XmlSerializer可以将对象转化成xml。

2. PULL解析器特点

（1）小巧轻便，解析速度快，简单易用，非常适合在Android移动设备中使用。

（2）简单的结构，一个接口和工厂组成了PULL解析器。

视频

PULL解析器工作机制

（3）简单易用，PULL解析器只有一个重要的方法next()，用来检索下一个事件，而其事件也仅仅只有5个：START_DOCUMENT、START_TAG、TEXT、END_TAG、END_DOCUMENT。

（4）最小的内存消耗。

3. PULL解析器的运行机制

XmlPullParser接口是一个定义解析功能的接口，基于事件对XML文档进行分析。

常用的5个标签为：START_DOCUMENT、START_TAG、TEXT、END_DOCUMENT、END_TAG。其中，START_DOCUMENT指文档开始，END_DOCUMENT指文档结束，START_TAG指标签开始，END_TAG指标签结束，TEXT指文本。

- 读取到XML文档的开始标签将返回：START_DOCUMENT事件。
- 读取到XML文档的结束标签将返回：END_DOCUMENT事件。
- 读取到XML元素的开始标签将返回：START_TAG事件。
- 读取到XML元素的结束标签将返回：END_TAG事件。
- 读取到XML元素的文本内容将返回：TEXT事件。

Android内置PULL解析器用到的类为XmlPullParser，其解析XML数据主要步骤如下：

（1）创建PULL解析器：

```
XmlPullParser parser=Xml.newPullParser();
```

（2）将XML文档设置解析器要处理的输入流对象和编码方式：

```
parser.setInput(inputStream,"utf-8");
```

（3）返回当前事件的类型（START_TAG, END_TAG, TEXT等）：

```
int type=parser.getEventType();
```

（4）循环读取节点：

视频

PULL解析器应用实战

```
while(type!=XmlPullParser.END_DOCUMENT){
    switch(type){
        case XmlPullParser.START_DOCUMENT:break;  //处理文档开始事件
        case XmlPullParser.START_TAG:break;  //处理开始标签事件
        case XmlPullParser.TEXT:break;  //处理文本事件
        case XmlPullParser.END_TAG:break;}  //处理结束标签事件
    type=parser.next();  //获得下一个事件类型}
```

4. PULL解析器常用方法

（1）void setInput(InputStream inputstream, string inputEncoding)：设置解析器将要处理的输入

流。执行该方法，将会重置解析器的状态，并且设置事件类型为START_DOCUMENT。其中，参数inputStream表示读取XML文档的输入流；参数inputEncoding表示输入流的编码。

（2）int getEventType()：获得事件类型，例如START_DOCUMENT（文档开始）、END_DOCUMENT（文档结束）、START_TAG（标签开始）、END_TAG（标签结束）等。

（3）String getName()：如果是START_TAG或者END_TAG事件，获取当前元素的名字。

（4）String getAttributeName（int index）：根据节点属性的序号来获取属性名。

（5）String getAttributeValue（String namespace, String name）：根据给定的命名空间和属性名称，返回相应的属性值。参数namespace表示属性的命名空间，如果不存在，则参数设置为null；参数name表示属性名称。

（6）String getAttributeValue(int index)：根据索引index返回元素相应的属性值。参数index从0开始。如果当前事件类型不是START_TAG或者索引越界，将会抛出异常。

（7）int next()：获得下一个解析事件，会产生START_TAG、END_DOCUMENT、START_TAG、END_TAG、TEXT等事件。

（8）String nextText()：如果当前事件是START_TAG，则返回当前元素的文本内容；若下一个事件是END_TAG，则返回空字符串，否则会抛出异常。

（9）int getAttributeCount()：返回元素属性的数量，如果当前事件类型不是START_TAG，则返回－1。

（10）String getText()：返回当前事件的文本内容。返回的内容依赖于当前事件的类型，如对于TEXT事件，则返回元素的文本内容。

（11）boolean isWhitespace()：检查当前TEXT事件是否只包含空白字符。

（12）void setInput（Reader in）：设置解析器将要处理的输入流。执行该方法，将会重置解析器的状态，并且设置事件类型为START_DOCUMENT。参数in表示读取XML文档的输入流。

（13）int nextTag()：调用next()方法并返回事件，但要求返回事件的必须是START_TAG或者END_TAG，否则会抛出异常。

PULL解析过程就是一个文档遍历过程，每次调用next()、nextTag()、nextToken()和nextText()都会向前推进文档，产生Event，并使解析器停留在某些事件上面，然后再处理Event。

下面通过采用PULL方法解析文件book.xml案例详细讲解PULL解析器的使用技巧。

（1）首先了解book.xml文件格式，其内容如下：

```
<?xml version="1.0"encoding="UTF-8"?>
<books>
    <book id="1">
        <name>Android高级应用编程实战</name>
        <author>李华忠、梁永生</author>
        <year>2014</year>
        <price>32</price>
    </book>
     <book id="2">
        <name>Android应用程序设计基础</name>
        <author>李华忠、陈勖、但唐仁</author>
```

```
        <year>2016</year>
        <price>31</price>
    </book>
        <book id="3">
        <name>Android应用程序设计</name>
        <author>李华忠、梁永生、刘涛</author>
        <year>2013</year>
        <price>28</price>
    </book>
</books>
```

（2）定义与book.xml文件对应的实体类Book，其源代码如下：

```
public class Book{/***功能：实体类Book**/
    private int id;         //书号
    private String name;    //书名
    private String author;  //作者
    private String year;    //出版年份
    private String price;   //书价
    public int getId(){return id;}
        public void setId(int id){this.id=id;}
        public String getName(){return name;}
        public void setName(String name){this.name=name;}
        public String getAuthor(){return author;}
        public void setAuthor(String author){this.author=author;}
        public String getYear(){return year;}
        public void setYear(String year){this.year=year;}
        public String getPrice(){return price;}
        public void setPrice(String price){this.price=price;}
        public String toString(){
            return "书籍信息["+"书号="+id+", 书名="+name+", 作者="+author+
                    ", 出版年份="+year+", 书价="+price+"]";}}
```

（3）设计PULL解析器解析XML文件的XmlPullParseBook类，其源代码如下：

```
public class XmlPullParseBook{
    public static List<Book>getBooks(InputStream inputStream){
        List<Book>bookList=null;Book book=null;
        //创建xml Pull解析器(XmlPullParser)对象
        XmlPullParser parser=Xml.newPullParser();
        try{  //初始化Pull XML解析，设置解析器要处理的输入流和编码方式
            parser.setInput(inputStream,"utf-8");  //设置输入流和编码方式
            int type=parser.getEventType();  //获取当前事件类型
            //循环读取节点，循环各种事件类型
            while(type!=XmlPullParser.END_DOCUMENT){  //直到遇到文档结束事件
```

```
                switch( type ){  //分支处理不同事件类型
                    case XmlPullParser.START_TAG:  //标签开始事件
                        if("books".equals(parser.getName())){  //获得标签名
                            bookList=new ArrayList<Book>();  //创建列表对象
                        }else if("book".equals(parser.getName())){
                            book=new Book();  //创建Book对象
                            book.setId(Integer.parseInt(parser.getAttributeValue(0)));
                        }else if("name".equals(parser.getName())){
                            book.setName(parser.nextText());
                        }else if("author".equals(parser.getName())){
                            book.setAuthor(parser.nextText());
                        }else if("year".equals(parser.getName())){
                            book.setYear(parser.nextText());  //获得标签文本
                        }else if("price".equals(parser.getName())){
                            book.setPrice(parser.nextText());}break;
                    case XmlPullParser.END_TAG:  //标签结束事件
                        if("book".equals(parser.getName())){  //获得标签名称
                            bookList.add(book);  //将book添加到列表
                            book=null;}break;}
                type=parser.next();  //获得下一个事件类型}
        }catch(Exception e){e.printStackTrace();}
        return bookList;  //返回解析结果}}
```

（4）设计活动布局，其界面大纲如图4-7所示，activity_xmlpull.xml源代码如下：

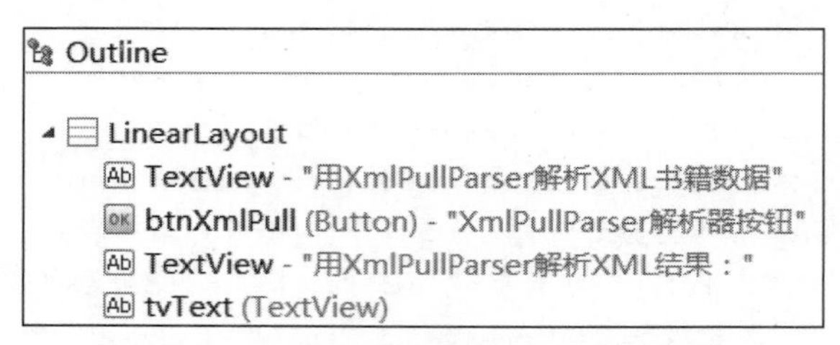

图4-7　PULL 解析 XML 案例活动布局大纲

（5）XmlPullActivity活动类实现代码如下：

```
public class DomParsePerson{
    public class XmlPullActivity extends Activity{
        private static final String TAG="XmlPullActivity";
        private TextView tvText;
        private Button btnXmlPull;
        private String bookStr="";
        /***首次创建活动时调用.*/
        @Override
        public void onCreate(Bundle savedInstanceState){
```

```
        super.onCreate(savedInstanceState);
        Log.d(TAG, "onCreate()");
        setContentView(R.layout.activity_xmlpull);
        tvText=(TextView)findViewById(R.id.tvText);
        btnXmlPull=(Button)findViewById(R.id.btnXmlPull);
        btnXmlPull.setOnClickListener(new OnClickListener(){
          @Override
          public void onClick(View v){
            doXmlPullParse();}});}
    private void doXmlPullParse(){
        //使用类加载器加载XML文件,XmlPullParser解析器解析XML文件
        List<Book>bookList=XmlPullParseBook.getBooks(XmlPullActivity.class.
    getClassLoader().getResourceAsStream("book.xml"));
        for(Book data: bookList){bookStr+=data.toString()+"\n";}
        tvText.setText(bookStr);}}
```

（6）本PULL解析XML文件案例运行结果如图4-8所示。

图 4-8　本 PULL 解析 XML 文件案例运行结果

4.2.4　解析 JSON 数据

1. JSON简介

JSON（JavaScript Object Notation，JS对象标记）是一种简单的数据格式，比XML更轻巧。JSON是一种轻量级的数据交换格式，非常适合于服务器与JavaScript的交互。JSON是JavaScript原生格式，这意味着在JavaScript中处理JSON数据不需要任何特殊的API或工具包。JSON采用完全独立于语言的文本格式，但是也使用了类似于C语言家族的习惯（包括C、C++、C#、Java、JavaScript、Perl、Python等）。这些特性使JSON成为理想的数据交换语言。

视频

解析JSON数据

2. JSON的两种结构

JSON简单地说就是JavaScript中的对象和数组，所以这两种结构就是对象和数组两种结构，通过这两种结构可以表示各种复杂的结构。

（1）对象：对象在JavaScript中表示为“{}”扩起来的内容，数据结构为{key：value,key：value,...}的键值对的结构，在面向对象的语言中，key为对象的属性，value为对应的属性值，取值方法为对象.key获取属性值，该属性值的类型可以是数字、字符串、数组、对象几种。

（2）数组：数组在JavaScript中是中括号“[]”扩起来的内容，数据结构为["C","C++","android",...]，取值方式和所有语言中一样，使用索引获取，字段值的类型可以是数字、字符串、数组、对象几种。经过对象、数组2种结构就可组合成复杂的数据结构。

JSON特点：对象表示方式是键值对；每个属性之间使用逗号隔开；大括号只能保存对象；中括号中保存集合。

JSON语法：

```
{
    属性1:值1,
    属性2:值2
}
```

（1）对象结构以“{”开始，以“}”结束。中间部分由0或多个以“,”分隔的“key(关键字)/value(值)”对构成，关键字和值之间以“：”分隔。

（2）其中关键字是字符串，而值可以是字符串、数值、true、false、null、对象或数组。数组结构以“[”开始，以“]”结束。中间由0或多个以“,”分隔的值列表组成。

JSON主要以键值对的方式表示数据。例如：

```
{
    "sziit":"深圳信息职业技术学院",
    "users":[{"id":"admin","pwd":"666"},{"id":"lihz","pwd":"888"}]
}
```

最外围的花括号表示这是一个JSON格式的对象数据，该对象有两个键：sziit和users。sziit的值是一个字符串，users的值是一个数组，数组有两个对象。使用org.json包提供的JSONArray、JSONObject等类可轻松完成JSON数据解析。下面的代码可用于解析前面的这个JSON字符串：

```
try{JSONObject json=new JSONObject(data);  //构造JSONObject对象
    String result="学院名="+json.getString("sziit")+"\n";  //获得指定键的值
    JSONArray users=json.getJSONArray("users");  //获得指定键的数组
    for(int i=0;i<users.length();i++){  //遍历数组
       JSONObject item=users.getJSONObject(i);  //获得一个数组元素
       result+="用户"+(i+1)+"id="+item.getString("id")+",";  //获取键值
       result+="密码="+item.getString("pwd")+"\n";}return result;
    } catch(Exception e){e.printStackTrace();}
```

3. 字符串转成JSON对象

JSON格式的字符串，是“名称/值”对的集合，{}之间包含数据、名称和值，必须以""号引起。名称和值之间用“：”格式。String json指字符串格式，JSONObject是JSON对象类，getString()方法根据键值获取数据。将字符串转成JSON对象的示例代码如下：

```
try{//{"name":"张三",age:20,"address","深圳市龙岗区"};
    String json="{\"name\":\"磊磊\",\"age\":18,\"address\":\"深圳市\"}";
    JSONObject jsonObject=new JSONObject(json);
    str="姓名:"+jsonObject.getString("name")+"\n"+"年龄:"
    +jsonObject.getString("age")+"\n"+"地址:"
```

```
        +jsonObject.getString("address")+"\n";
        tvJSON.setText(str);
    }catch(JSONException e){e.printStackTrace();}
```

4. 数组JSON对象

数组型JSON字符串格式，是用[]包含，它之间的数据是{}，用“,”号隔开。将数组型JSON字符串转化为JSON对象是JSONArray类。将数组型JSON字符串转成JSON对象的代码如下：

```
    try{String json="{\"name\":\"倩倩\",age:18,\"memo\":[{\"diploma\":\"计算机
等级证书\","+"\"date\":\"2020年\"},{\"diploma\":\"项目管理师\",\"date\":\"2020年
\"}],"+"\"address\":\"深圳\"}";
        JSONObject jsonObject1=new JSONObject(json);
        str="姓名:"+jsonObject1.getString("name")+"\n";
        str=str+"年龄:"+jsonObject1.getString("age")+"\n"+"证书:";
        JSONArray jsonArray=jsonObject1.getJSONArray("memo");
        for(int i=0; i<jsonArray.length(); i++){
            JSONObject jsonObjectSon=(JSONObject)jsonArray.opt(i);
            str=str+jsonObjectSon.getString("diploma")+","+"年份: "
            +jsonObjectSon.getString("date")+"\n";}
            str=str+"地址:"+jsonObject1.getString("address");tvJSON.setText(str);
    }catch(JSONException e){e.printStackTrace();}
```

5. 数组里面套数组

数组中套数组的JSON数据，在进行解析时，需要用到两个for循环：

```
    String json="["+"{\"poi\":\"0001\",\"gps\":[{\"time\":\"2020-05-12\","
    +"\"latitude\":\"22.3243\",\"longitude\":\"114.031\"},{\"time\":"+
    "\"2020-05-12\",\"latitude\":\"22.3244\",\"longitude\":\"114.032\"}],"
    +"\"addrname\":\"粤0001\"},"  +"{\"poi\":\"0002\",\"gps\":[{\"time\":\
"2020-05-12\","+"\"latitude\":\"22.3245\",\"longitude\":\"114.033\"},
{\"time\":"+"\"2020-05-12\",\"latitude\":\"22.3246\",\"longitude\":\"114.034\"}],
\"addrname\":\"粤002\"},"+"{\"poi\":\"0003\",\"gps\":[{\"time\":\"2020-05-12\",
"+"\"latitude\":\"22.3247\",\"longitude\":\"114.035\"},{\"time\":\"2020-05-12\
","+"\"latitude\":\"22.3248\",\"longitude\":\"114.036\"}],\"addrname\":\
"粤003\"}"+"]";
    try{JSONArray jsonArray=new JSONArray(json);     //构建JSON数组
        for(int i=0; i < jsonArray.length(); i++){ //遍历JSON数组
        String tpstr="";
        JSONObject jsonObject=(JSONObject)jsonArray.opt(i);  //第i个数组对象
        str=str+"第"+i+"个信息点:"+jsonObject.getString("poi");
        JSONArray jsonArray1=jsonObject.getJSONArray("gps");
        for(int j=0;j<jsonArray1.length(); j++){
        JSONObject jsonOb1Son=(JSONObject)jsonArray1.opt(j);
        tpstr ="采集时间:"+jsonOb2Son.getString("time")+"\n"+"纬度:"
        +jsonOb1Son.getString("latitude")+","+"经度:"
```

```
        +jsonOb1Son.getString("longitude");}
        str=str+tpstr+"\n";}tvJSON.setText(str);
    } catch(JSONException e){e.printStackTrace();}
```

6. 集合对象转成JSON数据

Gson是谷歌提供的在Java对象和JSON数据间进行映射的Java类库。可将一个JSON字符串转成一个Java对象，或者反过来。Android自带的JSON没有将对象转化成JSON数据，如果用到Gson，则需要加载Gson开发包。例如：

```
List<String>list=new ArrayList<String>();  //构造列表集合对象
list.add("id:0001");list.add("姓名：莹莹");list.add("性别：女");
list.add("年龄:18");list.add("籍贯：深圳");Gson gson=new Gson();
//将列表集合对象转化成JSON数据
String jsonString=gson.toJson(list);tvJSON.setText(jsonString);}
catch(JSONException e){e.printStackTrace();}
```

7. Map对象转成JSON数据

```
//构造Map对象集合对象
List<Map<String,Object>>allMenus=new ArrayList<Map<String,Object>>();
Map map1=new HashMap();map1.put("menuId",0001);
map1.put("menuName","豆腐炒韭菜");map1.put("menuPrice",12);
map1.put("menuNubmer",8);map1.put("menuSummary","土菜");
map1.put("menuCompany","洪湖鱼米香");
map1.put("menuCompanyAdress","五联");allMenus.add(map1);
Map map2=new HashMap();map2.put("menuId",0002);
map2.put("menuName","爆炒韭菜北极虾");map2.put("menuPrice",24);
map2.put("menuNubmer",6);map2.put("menuSummary","特色菜");
map2.put("menuCompany","洪湖鱼米香");
map2.put("menuCompanyAdress","五联");allMenus.add(map2);
//Map对象转化成JSON数据
Gson gson=new Gson();String jsonString=gson.toJson(allMenus);
tvJSON.setText(jsonString);
```

可通过案例学习解析JSON数据，运行结果如图4-9所示。

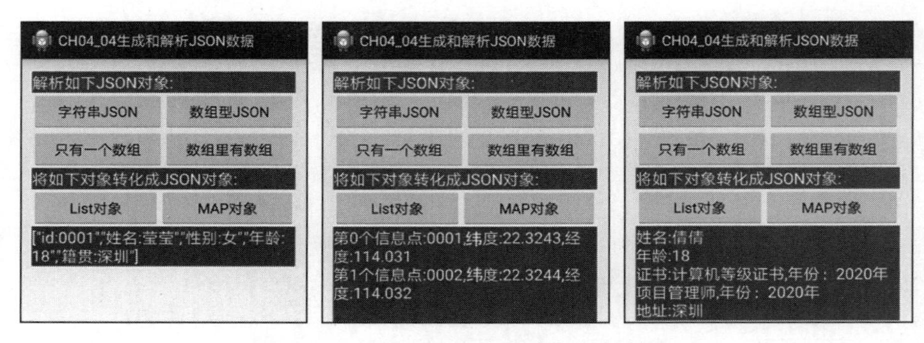

图4-9 解析JSON数据案例运行效果

4.2.5　基于位置的服务

所谓基于位置服务（Location-Based Services，LBS）又称定位服务或基于位置的服务，融合了GPS定位、移动通信、导航等多种技术，提供了与空间位置相关的综合应用服务。它包括两层含义：首先是确定移动设备或用户所在的地理位置；其次是提供与位置相关的各类信息服务。意指与定位相关的各类服务系统，简称“定位服务”。Android平台支持提供位置服务的API，在开发过程中主要用到LocationManager和LocationProviders对象。LocationManager可以用来获取当前的位置，追踪设备的移动路线，或设置敏感区域，在进入或离开敏感区域时设备会发出特定警报。LocationProvider是能够提供定位功能的组件集合，集合中的每种组件以不同的技术提供设备的当前位置，区别在于定位的精度、速度和成本等方面。在LocationManager中包含了两种不同的定位技术：GPS定位技术GPS_PROVIDER和网络定位技术NETWORK_PROVIDER。LocationManager是用于周期性获得当前设备位置信息的类。要获得LocationManager的实例，需要调用Context.getSystemService()方法，并传入服务名，这里要用到的服务就是LOCATION_SERVICE，这样就获得了LocationManager的一个实例。然后，通过LocationManager获得要使用的定位技术，若使用GPS定位技术则为LocationManager.GPS_PROVIDER；若使用网络定位则为LocationManager.NETWORK_PROVIDER。最后通过getLastKnownLocation()方法将上一次LocationManager获得的有效位置信息以Location的对象形式返回。具体使用方法如下面的代码段所示：

```
String serviceString=Context.LOCATION_SERVICE;  //获取服务
LocationManager locationManager=(LocationManager)getSystemService(serviceString);
String provider=LocationManager.GPS_PROVIDER;   //通过GPS获得位置信息
//获得location对象
Location location=locationManager.getLastKnownLocation(provider);
```

在上面的代码中获取到的服务是LOCATION_SERVICE，在Android系统中有不同的服务，表4-1中列出几个常用的系统级服务。通过上面可以获得所在的地理位置，但是由于是处于不断移动状态中，如果GPS信息变化了，它不能及时反映到界面上，除非重启程序才能获得新的GPS信息，显然这种方法是相当笨拙且不能满足需要的。要完成上面的需求，需要启动LocationManager的监听器，通过调用requestLocationUpdates()方法为其设置一个LocationListener（位置监听器）。同时requestLocationUpdates()方法还需要指定要使用的位置服务类型以及位置更新的时间和最小位移，以确保在满足用户需求的前提下最低的耗电量。requestLocationUpdates(provider, minTime, minDistance, listener)：其中provider为提供服务的类型；minTime为更新的最小间隔；minDistance为最小的位移变化；listener为监听方法，为LocationListener对象。

表 4-1　Android 系统常用系统级服务

Context 类的静态常量	值	返回对象	说　明
LOCATION_SERVICE	location	LocationManager	控制位置等设备的更新
WINDOW_SERVICE	window	WindowManager	最顶层的窗口管理器
LAYOUT_INFLATER_SERVICE	layout_inflater	LayoutInflater	将 XML 资源实例化为 View
POWER_SERVICE	power	PowerManager	电源管理
ALARM_SERVICE	alarm	AlarmManager	在指定时间接受 Intent

续表

Context 类的静态常量	值	返 回 对 象	说　　明
NOTIFICATION_SERVICE	notification	NotificationManager	后台事件通知
KEYGUARD_SERVICE	keyguard	KeyguardManager	锁定或解锁键盘
SEARCH_SERVICE	search	SearchManager	访问系统的搜索服务
VIBRATOR_SERVICE	vibrator	Vibrator	访问支持振动的硬件
CONNECTIVITY_SERVICE	connection	ConnectivityManager	网络连接管理
WIFI_SERVICE	wifi	WifiManager	Wi-Fi 连接管理
INPUT_METHOD_SERVICE	input_method	InputMethodManager	输入法管理

下面通过基于GPS位置的服务应用案例重点学习LocationManager和LocationProviders应用技巧(源代码详见光盘CH04_05GPSLocation)。

（1）设计GPSActivity活动界面，其布局大纲如图4-10所示。

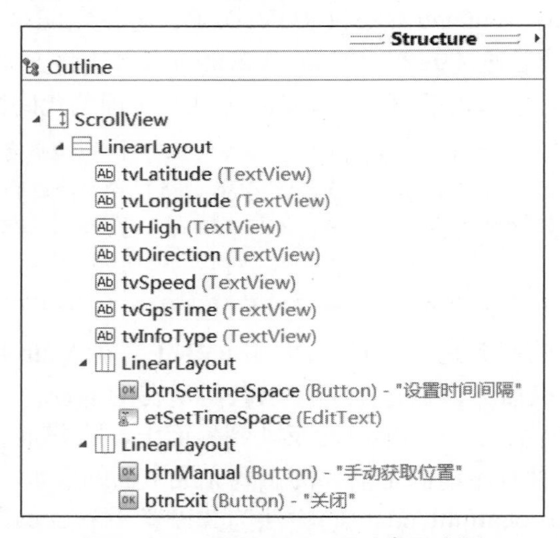

图 4-10　GPSActivity 活动布局大纲

（2）从SQLiteOpenHelper派生子类SQLiteDB实现保存gps数据功能。

```
public class GPSSQLiteDB{
    private static final String dbname="gps.db";  //定义保存GPS数据的数据名
    private final Context ct;
    private SQLiteDatabase db;
    private SQLiteDB sdb;
    private static class SQLiteDB extends SQLiteOpenHelper{
       public SQLiteDB(Context context){  //构造器，创建SQLite数据库"gps.db"
           super(context, dbname, null, 1);}
       public void onCreate(SQLiteDatabase sdb){  //建表
           sdb.execSQL("create table gps_table(infotype integer,
latitude integer,longitude integer,high double,direct double,
```

```
speed double,gpstime date);");}
        public void onUpgrade(SQLiteDatabase sdb, int oldVersion,int newVersion){
            sdb.execSQL("drop table if exists gps_table");  //删除表
            onCreate(sdb);}}
    //初始化数据库
    public GPSSQLiteDB(Context context){ct=context;sdb=new SQLiteDB(ct);}
    public void openDB(){db=sdb.getWritableDatabase();}  //打开数据库
    public void closeDB(){sdb.close();}  //关闭数据库
    public boolean addGpsData(gpsdata data){  //将GPS数据插入到数据库中
        boolean result=true;
        try{String StrSql=String.format("insert into gps_table
(infotype,latitude,longitude,high,direct,speed,gpstime)values
(%d,%d,%d,%.1f,%.1f,%.1f,'%s' )",data.InfoType, data.Latitude,
data.Longitude,data.High,data.Direct,data.Speed,data.GpsTime);
         db.execSQL(StrSql);result=true;
    }catch(Exception e){result=false;}
    return result;}}
```

（3）设计GPSActivity类，实现基于GPS定位功能，其核心代码如下：

```
private void initLocation(){  //获得LocationManager对象
    lm=(LocationManager)getSystemService(Context.LOCATION_SERVICE);
    if(lm.isProviderEnabled(LocationManager.GPS_PROVIDER)){
        criteria=new Criteria();
        criteria.setAccuracy(Criteria.ACCURACY_FINE);  //高精度
        criteria.setAltitudeRequired(true);  //显示海拔
        criteria.setBearingRequired(true);  //显示方向
        criteria.setSpeedRequired(true);  //显示速度
        criteria.setCostAllowed(false);  //不允许有花费
        criteria.setPowerRequirement(Criteria.POWER_LOW);  //低功耗
        provider=lm.getBestProvider(criteria,true);
        //位置变化监听，默认1s一次，距离10m以上
        lm.requestLocationUpdates(provider,1000,10,locListener);
    }else{showInfo(null, -1);}}
//构建位置监听器LocationListener对象
private final LocationListener locListener=new LocationListener(){
    public void onLocationChanged(Location arg0){  //位置已经改变
        showInfo(getLastPosition(), 2);}
    public void onProviderDisabled(String arg0){showInfo(null,-1);}
    public void onProviderEnabled(String arg0){}
    public void onStatusChanged(String arg0,int arg1,Bundle arg2){}};
//获得GPS数据
private gpsdata getLastPosition(){
    gpsdata result=new gpsdata();
```

```
    location=lm.getLastKnownLocation(provider);
    if(location!=null){
        result.Latitude=(int)(location.getLatitude()*1E6);
        result.Longitude=(int)(location.getLongitude()*1E6);
        result.High=location.getAltitude();
        result.Direct=location.getBearing();
        result.Speed=location.getSpeed();Date d=new Date();
        d.setTime(location.getTime()+28800000);  //UTC时间，转北京时间+8小时
        result.GpsTime=DateFormat.format("yyyy-MM-dd hh:mm:ss",d)
.toString();d=null;}return result;}
```

（4）在AndroidManifest.xml中配置允许GPS精确定位ACCESS_FINE_LOCATION权限。

```
<uses-permission android:name="android.permission.ACCESS_FINE_LOCATION"/>
```

（5）本案例运行结果如图4-11所示。

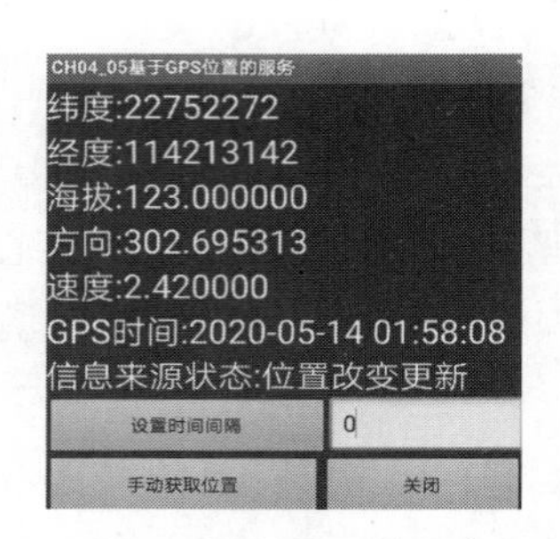

图4-11　基于GPS位置的服务运行结果

4.3 案例——Web服务中的XML数据解析

在多平台系统开发过程中，为了适应多个平台的数据交换，通常会提供Web服务接口供各个终端调用。Web服务接口一般会返回XML或JSON格式的字符串，各个终端需要对这些数据进行解析处理。下面的实例是将Web服务返回的JSON字符串数据反序列化成Java对象，方便在Android开发中使用这些数据。

案例1　查询天气信息：该程序从中国天气网查询城市天气信息。WebWeather类核心代码如下：

（1）重载onCreate()方法创建活动页面。

```
public void onCreate(Bundle savedInstanceState){
    super.onCreate(savedInstanceState);
    Log.d(TAG, "onCreate()");
    setContentView(R.layout.activity_main);
    setTitle("HttpURLConnection查询和PULL解析天气数据");
    findView();  //绑定UI组件
    setOnClickListener();    //设置按钮单击监听事件
}
```

（2）实现按钮单击事件启动线程查询天气数据。

```
private void doQueryWeather(){
    body.removeAllViews();  //移除当前的所有结果
    city=value.getText().toString();
    Toast.makeText(WebWeather.this,
      "正在查询天气信息...", Toast.LENGTH_LONG).show();
    Thread thread=new Thread(WebWeather.this);
    thread.start();
}
public void run(){
    removeAllElements();
    parseData();  //获取数据
    downImage();  //下载图片
    sendMessage();
}
```

（3）采用PULL方法解析从服务器获得的XML天气数据。

```
public void parseData(){
    int i=0;String sValue;
    //city变量表示城市名字的拼音
  String weatherUrl="http://flash.weather.com.cn/wmaps/xml/"+city+".xml";
    //表示天气情况图标的基础网址
    String weatherIcon="http://m.weather.com.cn/img/c";
    try{URL url=new URL(weatherUrl);
        conn=(HttpURLConnection)url.openConnection();  //建立天气预报查询连接
        conn.setRequestMethod("GET");  //采用GET请求方法
        ism=conn.getInputStream();  //打开数据输入流
        InputStreamReader in=new InputStreamReader(conn.getInputStream());
        //创建BufferedReader对象
        BufferedReader buffer=new BufferedReader(in);
        String inputLine="";String resultData="";
        while((inputLine=buffer.readLine())!=null)  //循环读取获得的数据
       {   resultData+=inputLine+"\n";  //在每一行后面加上一个"\n"来换行
       }
        XmlPullParser xmlParser=Xml.newPullParser();  //获得PULL解析器
        ByteArrayInputStream bais=null;
        bais =new ByteArrayInputStream(resultData.getBytes());
        xmlParser.setInput(bais,"UTF-8");  //设置输入流对象和编码方式
   //获得解析到的事件类别，包括开始文档START_DOCUMENT，结束文档END_DOCUMENT，
   //开始标签START_TAG，结束标签END_TAG，文本TEXT等事件
   int evtType=xmlParser.getEventType();
   while(evtType!=XmlPullParser.END_DOCUMENT){  //一直循环，直到文档结束
```

```
        switch(evtType){
            case XmlPullParser.START_TAG:  //标签开始
              String tag=xmlParser.getName();
                //如果是city标签开始，则开始实例化对象
            if(tag.equalsIgnoreCase("city")){
                //城市天气预报
                cityname.addElement(xmlParser.getAttributeValue(null,"cityname")+"天气: ");
                //天气情况概述
                summary.addElement(xmlParser.getAttributeValue(null,"stateDetailed"));
                //最低温度
                low.addElement("最低: "+xmlParser.getAttributeValue(null,"tem2"));
                //最高温度
                high.addElement("最高: "+xmlParser.getAttributeValue(null,"tem1"));
                //天气情况图标网址
                icon.addElement(weatherIcon+xmlParser.getAttributeValue(null,"state1")+".
gif");}break;
                case XmlPullParser.END_TAG:default:break;  //标签结束
            }
             evtType=xmlParser.next();  //如果xml没有结束，则导航到下一个节点
        }
      }catch(Exception ex){ex.printStackTrace();
      }finally{  //释放连接
          try{ism.close();conn.disconnect();
          }catch(Exception ex){ex.printStackTrace();}}}
```

（4）该程序的运行结果如图4-12所示。

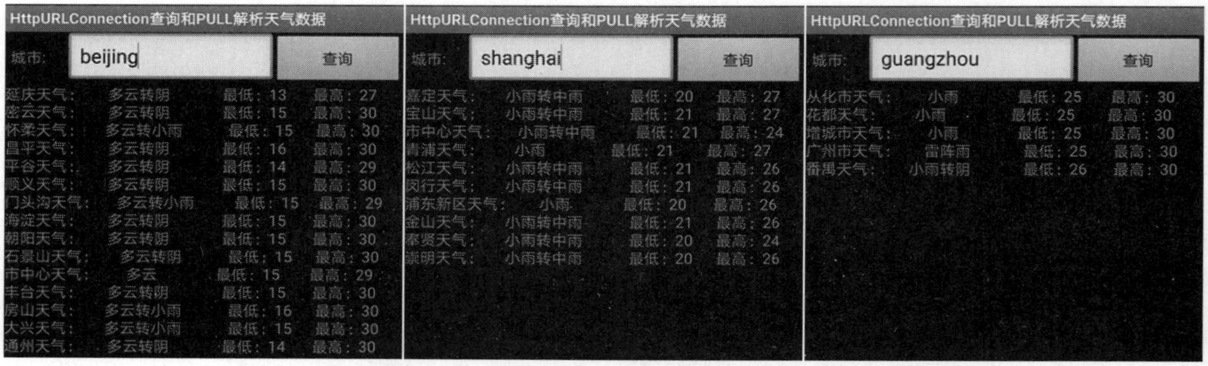

图4-12　Web服务中的XML数据解析案例运行主界面

案例2　采用SAX、DOM和PULL方式解析学生信息：该程序采用3种解析方式解析位于assets目录下的students.xml。主要实现步骤描述如下：

（1）设计案例activity_main.xml主界面，其布局大纲如图4-13所示。

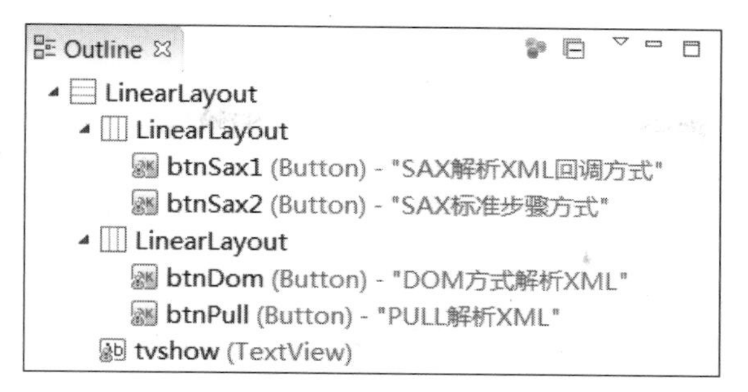

图 4-13　activity_main 布局大纲

（2）分析assets目录下的students.xml数据，其数据如下：

```
<?xml version="1.0"encoding="utf-8"?>
<students>
    <student group="g1"id="8001">
        <name>李茜</name>
        <sex>女</sex>
        <age>18</age>
        <email>lixi@sziit.edu.cn</email>
        <birthday>1992-08-24</birthday>
        <memo>三好学生</memo>
    </student>
    <student group="g2"id="8002">
        <name>李倩</name>
        <sex>女</sex>
        <age>18</age>
        <email>liqian@sziit.edu.cn</email>
        <birthday>1992-08-06</birthday>
        <memo>团支书</memo>
    </student>
    <student group="g3"id="8003">
        <name>张伟</name>
        <sex>男</sex>
        <age>18</age>
        <email>zhangwei@sziit.edu.cn</email>
        <birthday>1992-08-06</birthday>
        <memo>优秀干部</memo>
    </student>
    <student group="g4"id="8004">
        <name>亮亮</name>
        <sex>男</sex>
        <age>18</age>
        <email>liang@sziit.edu.cn</email>
        <birthday>1992-08-04</birthday>
```

```
        <memo>好学生</memo>
    </student>
    <student group="g5"id="8005">
        <name>熊熊</name>
        <sex>男</sex>
        <age>18</age>
        <email>xiong@sziit.edu.cn</email>
        <birthday>1992-06-08</birthday>
        <memo>优秀学生</memo>
    </student>
</students>
```

（3）设计能描述students.xml数据的实体类Student，其源代码如下：

```
public class Student{
    private String id,group,name,age,sex,email,birthday,memo;
        public String getId(){return id;}
        public void setId(String id){this.id=id;}
        public String getGroup(){return group;}
        public void setGroup(String group){
            this.group=group;}
        public String getName(){return name;}
        public void setName(String name){this.name=name;}
        public String getAge(){return age;}
        public void setAge(String age){this.age=age;}
        public String getSex(){return sex;}
        public void setSex(String sex){this.sex=sex;}
        public String getEmail(){return email;}
        public void setEmail(String email){this.email=email;}
        public String getBirthday(){return birthday;}
        public void setBirthday(String birthday){this.birthday=birthday;}
        public String getMemo(){return memo;}
        public void setMemo(String memo){this.memo=memo;}
        public String toString(){
            return "学生信息{"+"学号='"+id+'\''+",小组='"+group+\''+
                ",姓名='"+name +'\''+",年龄='"+age+'\''+
                ",性别='"+sex+'\''+",邮箱='"+email+'\''+
                ",出生年月='"+birthday+'\''+",备注'"+memo+'\''+' }' +"\n";}}
```

（4）设计MainActivity类，其源代码如下：

```
public class MainActivity extends Activity implements OnClickListener{
    private static final String TAG="MainActivity";
    private Button btnSax1,btnSax2,btnDom,btnPull;
    protected void onCreate(Bundle savedInstanceState){
```

```
        super.onCreate(savedInstanceState);
        setContentView(R.layout.activity_main);
        findView();
        setOnClickListener();}
    private void findView(){
        btnSax1=(Button)findViewById(R.id.btnSax1);
        btnSax2=(Button)findViewById(R.id.btnSax2);
        btnDom=(Button)findViewById(R.id.btnDom);
        btnPull=(Button)findViewById(R.id.btnPull);}
    private void setOnClickListener(){
        btnSax1.setOnClickListener(this);
        btnSax2.setOnClickListener(this);
        btnDom.setOnClickListener(this);
        btnPull.setOnClickListener(this);}
    private void doSax1Click(){  //SAX解析
        Intent intent=new Intent(MainActivity.this,SAX1Activity.class);
        startActivity(intent);}
    private void doSax2Click(){  //SAX解析
        Intent intent=new Intent(MainActivity.this,SAX2Activity.class);
        startActivity(intent);}
    private void doDomClick(){   //DOM解析
        Intent intent=new Intent(MainActivity.this,DOMActivity.class);
        startActivity(intent);}
    private void doPullClick(){  //PULl解析
        Intent intent=new Intent(MainActivity.this, PullActivity.class);
        startActivity(intent);}
   public void onClick(View v){
        switch(v.getId()){
        case R.id.btnSax1:doSax1Click();break;  //SAX解析
        case R.id.btnSax2:doSax2Click();break;  //SAX解析
        case R.id.btnDom:    doDomClick();break;  //DOM解析
        case R.id.btnPull:doPullClick();break;  //PULl解析
}}}
```

（5）继承DefaultHandler实现SAXStudentsHandler类，其源代码如下：

```
public class SAXStudentsHandler extends DefaultHandler{
    private final String TAG="SAXStudentsHandler";
    private List<Student>studentList;    //解析数据存储
    Student students;Boolean currTag=false;
    String currTagVal="";StudentsCallback callback;
    public SAXStudentsHandler(StudentsCallback callback){
        studentList=new ArrayList<Student>();this.callback=callback;}
    public void startDocument()throws SAXException{  //处理文档开始事件
```

```
            super.startDocument();
            studentList=new ArrayList<Student>();}
        public void startElement(String uri, String localName, String qName,
 Attributes attributes)throws SAXException{  //处理标签开始事件
            super.startElement(uri, localName, qName, attributes);
            //遍历students标签
            currTag=true;
            if(localName.equals("student")){students=new Student();
                //遍历解析student标签
                for(int i=0;i<attributes.getLength(); i++){
                    if(attributes.getLocalName(i).equals("id")){
                        students.setId(attributes.getValue(i));
                    } else if(attributes.getLocalName(i).equals("group")){
                        students.setGroup(attributes.getValue(i));}}}}
        public void characters(char[] ch, int start, int length)throws
SAXException{   //处理字符
            super.characters(ch, start, length);
            //student标签保存下获取子标签
            if(currTag){currTagVal=new String(ch, start, length);
                currTag=false;}}
        public void endElement(String uri, String localName, String qName)throws
SAXException{   //处理标签结束
            super.endElement(uri, localName, qName);
            currTag=false;
            //标签结束后，保存子标签的值，并将信息保存到studentList中
            if(localName.equals("student")){studentList.add(students);}
            if(localName.equalsIgnoreCase("name")){
                students.setName(currTagVal);
            } else if(localName.equalsIgnoreCase("sex")){
                students.setSex(currTagVal);
            } else if(localName.equalsIgnoreCase("age")){
                students.setAge(currTagVal);
            } else if(localName.equalsIgnoreCase("email")){
                students.setEmail(currTagVal);
            } else if(localName.equalsIgnoreCase("birthday")){
                students.setBirthday(currTagVal);
            } else if(localName.equalsIgnoreCase("memo")){
                students.setMemo(currTagVal);}}
        public void endDocument()throws SAXException{  //处理文档结束
            super.endDocument();
            if(callback!=null && studentList!=null)
                callback.parseCallback(studentList);  //调用回调}}
```

（6）设计DOMActivity类，采用DOM方法解析XML的源代码如下：

```
private void xmlDomParser1(){
    InputStream inputStream=null;
    try{  //创建factory
   DocumentBuilderFactory factory=DocumentBuilderFactory.newInstance();
     //创建DocumentBuilder对象
     DocumentBuilder builder=factory.newDocumentBuilder();
     AssetManager assetManager=getAssets();  //获取文件
     inputStream=assetManager.open("students.xml");
      //创建Document对象，相当于代表xml表的对象
      Document document=builder.parse(inputStream);
      Element root=document.getDocumentElement();  //获取根节点
      //获取根节点所有student的节点
      NodeList nodeList =root.getElementsByTagName("student");
      /***遍历所有的节点**/
      list=new ArrayList<Student>();
        //先遍历sutdent节点数据
        for(int i=0; i<nodeList.getLength(); i++){
             //student节点
             Element parentelement=(Element)nodeList.item(i);
             //获取student节点内的所有节点
             NodeList childNodeList=parentelement.getChildNodes();
             //获取<Students>标签内的参数
             String group=parentelement.getAttribute("group");
             String id=parentelement.getAttribute("id");
             students=new Student();students.setGroup(group);
             students.setId(id);
             //遍历student节点内的节点
             for(int j=0;j< childNodeList.getLength();j++){
             //多标签加 Node.ELEMENT_NODE判断
        if(childNodeList.item(j).getNodeType()==Node.ELEMENT_NODE){
        //依次赋值
        if("name".equals(childNodeList.item(j).getNodeName())){
        String name=childNodeList.item(j).getFirstChild().getNodeValue();
            students.setName(name);}else if
("sex".equals(childNodeList.item(j).getNodeName())){
        String sex=childNodeList.item(j).getFirstChild().getNodeValue();
            students.setSex(sex);} else if
("age".equals(childNodeList.item(j).getNodeName())){
        String age=childNodeList.item(j).getFirstChild().getNodeValue();
           students.setAge(age);} else if
("email".equals(childNodeList.item(j).getNodeName())){
```

```
        String email=childNodeList.item(j).getFirstChild().getNodeValue();
            students.setEmail(email);} else if
("birthday".equals(childNodeList.item(j).getNodeName())){
String birthday=childNodeList.item(j).getFirstChild().getNodeValue();
            students.setEmail(birthday);}else if
("memo".equals(childNodeList.item(j).getNodeName())
String memo=childNodeList.item(j).getFirstChild().getNodeValue();
            students.setMemo(memo);}}}
            list.add(students);  //添加到list中}
        }catch(IOException e){e.printStackTrace();
        }catch(ParserConfigurationException e){e.printStackTrace();
        }catch(SAXException e){e.printStackTrace();
        }finally{if(inputStream!=null){
            try{inputStream.close();
            } catch(IOException e){e.printStackTrace();}}}}
```

（7）设计PullActivity类，采用PULL方法解析XML数据，其源代码如下：

```
private void xmlPullParser1(){
   InputStream inputStream=null;
   try{  //创建factory
   XmlPullParserFactory factory=XmlPullParserFactory.newInstance();
  //创建 XmlPullParser对象
   XmlPullParser parser=factory.newPullParser();
   AssetManager assetManager=getAssets();  //获得AssetManager对象
   inputStream=assetManager.open("students.xml");  //打开文件
   parser.setInput(inputStream, "UTF-8");  //设置输入流和编码方式
   int eventType=parser.getEventType();
   while(eventType!=XmlPullParser.END_DOCUMENT){  //文档标签
       switch(eventType){
       case XmlPullParser.START_DOCUMENT:  //触发文档标签事件
            studentList=new ArrayList<Student>();break;
       case XmlPullParser.START_TAG:  //触发开始元素事件
            String tagName=parser.getName();  //获取标签名称
            //获取students标签下的参数 id,group
            if(tagName.equals("student")){
                students=new Student();
                students.setId(parser.getAttributeValue(1));
                students.setGroup(parser.getAttributeValue(0));}
                   if(students !=null){
                       if("name".equals(tagName)){
                     students.setName(parser.nextText());}
                 if("sex".equals(tagName)){
                     students.setSex(parser.nextText());}
```

```
                if("age".equals(tagName)){
                    students.setAge(parser.nextText());}
                if("email".equals(tagName)){
                    students.setEmail(parser.nextText());}
                if("birthday".equals(tagName)){
                    students.setBirthday(parser.nextText());}
                if("memo".equals(tagName)){
                    students.setMemo(parser.nextText());}}break;
            case XmlPullParser.END_TAG:  //元素标签结束
                String name=parser.getName();
                if(name.equals("student")){
                    if(students!=null){
                        studentList.add(students);}}break;
            case XmlPullParser.END_DOCUMENT:break;  //文档标签}
               eventType=parser.next();  //获取下一条事件}
        }catch(XmlPullParserException e){e.printStackTrace();
        }catch(IOException e){e.printStackTrace();
        } finally{
            if(inputStream != null){
               try{inputStream.close();
               }catch(IOException e){e.printStackTrace();}}}}
```

（8）案例运行结果如图4-14所示。

图4-14 案例运行结果

4.4 知识扩展

4.4.1 根据经纬度信息在地图上定位

在Android平台上调用地图Map服务，Map插件提供了一个MapView，这个MapView的用法就像普通的ImageView一样，直接在界面布局文件中定义，然后在程序中通过方法来控制该组件即可。

视频
基于位置的服务应用实战

MapView支持的常用方法如下：

（1）MapContrller getController()：获取该MapView关联的MapController。

（2）GeoPoint getMapCenter()：获取该MapView所显示的中心。

（3）Int getMaxZoomLevel()：获取该MapView所支持的最大的放大级别。

（4）List<Overlay>getOverlays()：获取该MapView上显示的全部Overlay（覆盖）。

（5）Projection getProjection()：获取屏幕像素坐标与经纬度坐标之间的投影关系。

（6）int getZoomLevel()：获取该屏幕当前的缩放级别。

（7）setBuiltInZoomControls：设置是否显示内置的缩放控制按钮。

（8）setSatellite：设置是否显示卫星地图。

（9）setTraffic：设置是否显示交通情况。

4.4.2 调用地图地址解析服务

地图Map定位必须根据经度和纬度来完成，如果需要让程序根据地址进行定位，需要先把地址解析成经度和纬度。

（1）地址解析：把普通用户能看到的字符串地址转换成经度和纬度。

（2）反向地址解析：把经度和纬度转换成普通的字符串地址。

在Android中地址解析通过Geocoder工具类实现。该工具类提供了两个方法：

- List<Address>getFromLocation：执行地址解析，把经度和纬度转换为字符串地址值。
- List<Address>getFromLocationName：执行反向地址解析，把字符串地址转换为经度和纬度。

本章小结

本章主要讲述了XML的概念、SAX解析xml文件的方法、DOM解析xml文件的方法、基于位置的服务的实现和基于地图的应用实现。

（1）XML是一种扩展性标识语言，作为承载数据的一个重要角色，被设计用来传输和存储数据。

（2）SAX(Simple API for XML)解析器是一种基于事件的解析器，它的核心是事件处理模式，主要是围绕着事件源以及事件处理器来工作的。

（3）DOM（Document Object Model，文档对象模型）是基于树形结构的节点或信息片段的集合，采用DOM解析器API遍历XML树、检索所需数据。

（4）PULL解析器的运行方式和SAX类似，都是基于事件的模式。

（5）JSON（JavaScript Object Notation）是一种轻量级的数据交换格式。正确JSON的数据格式，才能转换成对应的对象，进行处理。

强化练习

一、填空题

1. XML 非常适合（ ）传输，提供统一的方法来描述和交换独立于应用程序或供应商的结构化数据。

2. SAX 的全称是（ ），既指一种接口，也指一个软件包。

3. SAX 是（ ）的一个标准接口。

4. DOM 的全称是（ ），它在本质上是一种文档平台。

5. 基于位置服务（Location-Based Services，LBS）又称定位服务或基于位置的服务，融合了（ ）、（ ）和（ ）等多种技术，提供了与空间位置相关的综合应用服务。

6. DOM 解析器是属于（ ）驱动。

7. SAX 解析是属于（ ）驱动。

8. PULL 解析是属于（ ）驱动。

9. Android 中实现了 EntityResolver,DTDHandler,ContentHandler,ErrorHandler 接口的内置 SAX 解析类为（ ）。

10. XmlPullParser 的（ ）方法得到了解析文档的事件类型。

11. 在 Android 的 DOM 解析中 Element element=document.getDocumentElement() 方法可以得到 XML 文档中标签（ ）。

12. 可扩展标记语言（EXtensible Markup Language）的缩写为（ ）。

13. Simple API for XML 解析器简称为（ ）。

14. 文档对象模型（Document Object Model）简称为（ ）。

二、单选题

1. Android 内置的 XML 解析为（ ）解析。

A. SAX　　B. DOM　　C. DOM4j　　D. PULL

2. 解析 XML 过程中，将整个文件加载到内存中进行解析方法为（ ）。

A. DOM　　B. SAX　　C. JSON　　D. PULL

3. 下列不属于 PULL 解析 XML 中标签的是（ ）。

A. START_DOCUMENT　　B. START_TAG
C. END_TAG　　D. TEXT _TAG

4. Android 中得到内置 PULL 解析器实例的方法为（ ）。

A. XmlPullParserFactory.newInstance()
B. XmlPullParserFactory.getInstance()
C. XmlPullParserFactory.newPullParser()
D. Xml.newPullParser()

5. DOM 解析是属于（ ）。

A. 文档驱动　　B. 事件驱动
C. 和 Pull 一样　　D. 和 SAX 一样

6. SAX解析属于（　　）。

A. 数据驱动　　B. 文档驱动　　C. 事件驱动　　D. 和DOM一样

7. Android内置PULL解析器用到的类为（　　）。

A. SAXParserFactory　　B. DefaultHandler　　C. XMLReader　　D. XmlPullParser

8. 用DOM解析XML，获得<age>58</age>中的节点age的方法为（　　）。

A. Node age=doc.getElements("age");

B. Node age=doc.getElements("age").item(0);

C. Node age=doc.getElementsByTagName("age");

D. Node age=doc.getElementsByTagName("age").item(0);

9. Android中SAX解析器没有使用的类为（　　）。

A. XmlPullParser　　B. ContentHandler

C. SAXParserFactory　　D. DefaultHandler

10. PULL解析XML不需要用到的类或者接口是（　　）。

A. XmlPullParser　　B. XmlPullParserFactory

C. Xml　　D. SAXParserFactory

11. Android中通过DocumentBuilder抽象类得到DOM解析器实例的方法为（　　）。

A. newInstance();　　B. getInstance();

C. newDocumentBuilder();　　D. builder.newDocument();

12. XmlPullParser的getEventType()方法得到了解析文档的事件类型，下列（　　）不属于它的事件类型。

A. START_DOCUMENT和END_DOCUMENT

B. TEXT

C. START_DOC和END_DOC

D. START_TAG和END_TAG

13. PULL解析器与SAX解析器的主要区别在于（　　）。

A. 主动提取数据可随时终止解析

B. 整个文档树在内存中，支持增删改查等功能

C. 事件驱动

D. 整个文档调入内存，浪费时间和空间

三、简答题或编程题

1. 简述SAX分析器中的几个主要API接口及其作用。
2. 简述SAX和DOM各自的优点和不足。
3. 编程实现基于位置的服务。

第 5 章

Android 多媒体应用

学习目标

- 了解MediaPlayer和MeidaRecorder的生命周期。
- 理解Android系统中多媒体实现的方法。
- 掌握Android中提供的音频相关的类和方法的实现。
- 掌握Android中提供的视频相关的类和方法的实现。
- 掌握Android中提供的摄像头相关的类和方法的实现。

5.1 学习导入

Android应用面向的是普通个人用户，这些用户往往更加关注用户体验，因此为Android应用增加动画、视频、音乐等多媒体功能十分重要。Android平台内置了常用类型媒体的编解码，可被方便地集成到应用程序中。Android平台媒体的访问机制直观简单，通常可使用相同的Intent和Activity机制。因此，为手持设备提供音频录制、播放、视频录制、播放的功能十分重要。

5.2 技术准备

5.2.1 使用多媒体播放器 MediaPlayer 播放音频

MediaPlayer是播放媒体最为广泛使用的类，适合播放比较长且时间要求不高的音频。MediaPlayer已设计用来播放大容量的音频文件，可支持播放操作（开始、停止、暂停等）和查找操作的流媒体。MediaPlayer的生命周期如图5-1所示。MediaPlayer的重要方法包括：

（1）MediaPlayer create（Context context，Uri uri，SurfaceHolder holder）：创建一个多媒体播放器并加载指定uri和holder的多媒体文件。

（2）MediaPlayer create(Context context, Uri uri)：创建一个多媒体播放器并加载指定uri的多媒体文件。

（3）MediaPlayer create(Context context, int resid)：创建一个多媒体播放器并加载指定资源id的多媒体文件。

MediaPlayer的3个方法create()都是静态方法，用于创建MediaPlayer对象。参数context表示上下文；参数uri表示多媒体文件的位置；参数holder表示显示视频的SurfaceHolder对象；参数resid表示存放在本地的多媒体资源ID，以R.raw.×××形式描述。

（4）void setDataSource(String path)：通过文件路径为多媒体播放器指定加载文件。

（5）void setDataSource(Context context, Uri uri)：通过文件uri为多媒体播放器指定加载文件。

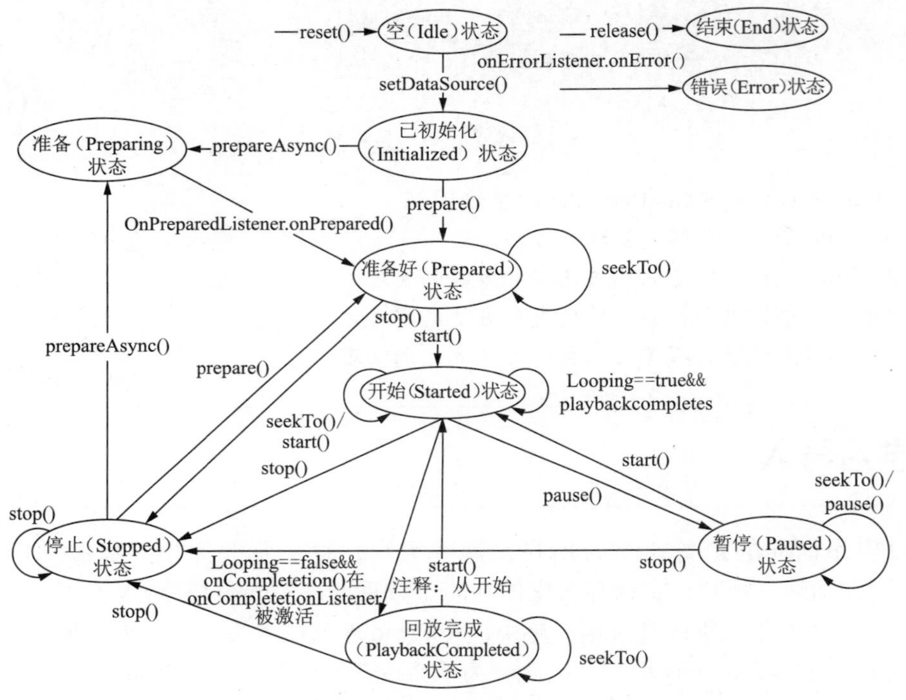

图 5-1　MediaPlayer 的生命周期

（6）void prepare()：同步加载，方法返回时已加载完毕。

（7）void prepareAsync()：准备播放器进行异步加载播放，方法返回时未加载完成，常用于网络文件的加载。加载完成之后，才可以对音频文件进行播放控制。

（8）void start()：开始播放音频。

（9）void stop()：停止播放音频。

（10）void pause()：暂停播放音频。

（11）int getDuration()：返回int，得到歌曲的总时长，以ms（毫秒）为单位。

（12）boolean isLooping()：返回boolean，是否循环播放。

（13）boolean isPlaying()：返回 boolean，是否正在播放。

（14）void release()：无返回值，释放 MediaPlayer 对象。

（15）void reset()：无返回值，重置 MediaPlayer 对象。

（16）void seekTo(int msec)：无返回值，指定歌曲的播放位置，以 ms（毫秒）为单位。

（17）void setLooping(boolean looping)：无返回值，设置是否循环播放。

（18）void setVolume(float leftVolume, float rightVolume)：无返回值，设置音量。

（19）int getCurrentPosition()：获取多媒体当前的播放位置，时间单位为毫秒。

（20）int getVideoWidth()：返回视频的宽度，若没有视频，则返回 0。

（21）void setOnBufferingUpdateListener（MediaPlayer.OnBufferingUpdateListener listener）：注册一个回调事件，当网络流的缓冲状态被改变时调用。

（22）void setOnErrorListener（MediaPlayer.OnErrorListener listener）：注册一个回调事件，在进行异步操作期间若发生错误时调用。

（23）void setAudioStreamType（int streamtype）：设置声音流的类型。参数 streamtype 表示音频流类型，已在 AudioManager 类中定义。

（24）void setAudioStreamType（int streamtype）：设置声音流的类型。参数 streamtype 表示音频流类型，已在 AudioManager 类中定义。

（25）void setOnPreparedListener（MediaPlayer.OnPreparedListener listener）：注册一个回调事件，当多媒体源已经准备好播放时调用。

（26）void setOnSeekCompleteListener（MediaPlayer.OnSeekCompleteListener listener）：注册一个回调事件，当一个 Seek 搜寻操作完成后调用。

视频

使用MediaPlayer播放音频应用实战

（27）void setOnVideoSizeChangedListener（MediaPlayer.OnVideoSizeChangedListener listener）：注册一个回调事件，当视频的大小被更新时调用。

（28）void setOnInfoListener（MediaPlayer.OnInfoListener listener）：注册一个回调事件，当有信息或者警告时调用。

使用 MediaPlayer 播放音乐的步骤如下：

（1）获得 MediaPlayer 实例。既可通过 MediaPlayer mp=new MediaPlayer();获得实例；也可使用 create 的方式，如 MediaPlayer mp=MediaPlayer.create(this, R.raw.demo);获得实例，如果使用 create 方式就不需要调用 setDataSource() 方法。

（2）调用 setDataSource() 方法设置 MediaPlayer 需要播放的文件。

（3）调用 MediaPlayer 常用方法进行播放控制。

- start()：开始或恢复播放。
- stop()：停止播放。
- pause()：暂停播放。

（4）设置不同监听器的方法来对播放器的工作状态进行监听。

为了让 MediaPlayer 装载指定音频文件，MediaPlayer 提供了如下简单的静态方法。static MediaPlayer create(Context context, int resid)：从 resid 资源 ID 对应的资源文件中装载音频文件，并返回新创建的 MediaPlayer 对象。上面这个方法用起来非常方便，但这个方法每次都会返回新创建的 MediaPlayer 对象，如果程序需要使用 MediaPlayer 循环播放多个音频文件，使用 MediaPlayer 的静态 create 方法就不太适合了，此时可通过 MediaPlayer 的 setDataSource() 方法装载指定的音频文件。MediaPlayer

提供了如下方法来指定装载相应的音频文件。

- setDataSource(String path)：指定装载path路径所代表的文件。
- setDataSource(FileDescription fd, long offset, long length)：指定装载fd所代表的文件中从offset开始、长度为length的文件内容。
- setDataSource(FileDescription fd)：指定装载fd所代表的文件。
- setDataSource(Context context, Uri uri)：指定装载Uri所代表的文件。

执行上面的setDataSource()之后，MediaPlayer并未真正去装载那些音频文件，还需要调用MediaPlayer的prepare()方法去准备音频，所谓“准备"，就是让MediaPlayer真正去装载音频文件。因此，使用MediaPlayer对象播放歌曲的代码模板为：

```
try{mPlayer.reset();  //重置MediaPlayer对象，使其处理空闲状态
    //播放前通过DDMS或手机助手将night.mp3复制到SD卡上
    mPlayer.setDatasource("/mnt/sdcard/night.mp3");  //指定加载音乐文件路径
    mPlayer.prepare();  //同步加载，方法返回时已加载完毕
    mPlayer.start();    //开始播放
}catch(IOException e){e.printStackTrace();}
```

除此之外，MediaPlayer还提供了一些绑定事件监听器的方法，用于监听MediaPlayer播放过程中发生的特定事件。绑定事件监听器的方法如下：

- setOnCompletionListener(MediaPlayer.OnCompletionListener listener)：注册一个回调事件，当MediaPlayer已经播放完媒体资源之后调用。
- setOnErrorListener(MediaPlayer.OnErrorListener listener)：为MediaPlayer的播放错误事件绑定事件监听器。
- setOnPreparedListener(MediaPlayer.OnPreparedListener listener)：当MediaPlayer调用prepare()方法时触发该监听器，用于监听MediaPlayer的prepare结束事件。
- setOnSeekListener(MediaPlayer.OnSeekCompletedListener listener)：当MediaPlayer调用seek()方法时触发该监听器。

因此，可在创建一个MediaPlayer对象之后，通过为该MediaPlayer绑定监听器来监听用户所触发的事件。其示例代码如下：

```
MediaPlayer mPlayer=new MediaPlayer();
//1.为MediaPlayer的播放过程中错误事件绑定事件监听器
mPlayer.setOnErrorListener(new OnErrorListener(){
    void onError(MediaPlayer mp, int what, int extra)
   { //针对错误进行相应的处理...
   }});
//2.为MediaPlayer的播放完成事件绑定事件监听器
mPlayer.setOnCompletionListener(new OnCompletionListener(){
  @Override
  public void onCompletion(MediaPlayer mp){
   //setOnCompletionListener监听器实现方法
  }}
```

```
//3.用于监听MediaPlayer的prepare结束事件
mPlayer.setOnPreparedListener(new OnPreparedListener(){
  @Override
  public void OnPrepared(MediaPlayer mp){
  //setOnPreparedListener监听器实现方法
  }}
```

下面简单归纳一下MediaPlayer播放不同来源的音频文件。

1. 播放应用的资源文件

播放应用的资源文件需要两步即可：

- 调用MediaPlayer对象的create(Context context, int resid)方法加载指定的资源文件。
- 调用MediaPlayer对象的start()、pause()、stop()等方法控制播放即可。

其示例代码如下：

```
MediaPlayer mPlayer=MediaPlayer.create(this,R.raw.night);  //创建媒体播放器
mPlayer.start();  //开始播放。注意：先创建res/raw目录并将night.mp3复制到该目录
```

2. 播放应用原始文件

播放应用的原始文件按如下步骤执行：

（1）调用Context的getAssets()方法获取应用的AssetManager。

（2）调用AssetManager对象的openFd(String name)方法打开指定的原始资源，该方法返回一个AssetFileDescriptor对象。

（3）调用AssetFileDescriptor的getFileDescriptor()、getStartOffset()和getLength()方法获取音频文件的FileDesciptor、开始位置、长度等。

（4）创建MediaPlayer对象（或利用已有的MediaPlayer对象），并调用MediaPlayer对象的setDataSource(FileDescriptor fd, long offset, long length)方法装载音频资源。

（5）调用MediaPlayer对象的prepare()方法准备音频。

（6）调用MediaPlayer对象的start()、pause()、stop()等方法控制播放即可。

其示例代码如下：

```
AssetManager am=getAssets();  //获取用的AssetManager对象
AssetFileDescriptor afd=am.openFd("bomb.mp3");  //打开指定音乐文件
MediaPlayer mPlayer=new MediaPlayer();  //创建MediaPlayer对象
//使用MediaPlayer加载指定的声音文件（注意：先将bomb.mp3复制到assets目录）
mPlayer.setDataSource(afd.getFileDescriptor());
mPlayer.prepare();  //准备声音，同步加载，方法返回时已加载完毕
mPlayer.start();  //开始播放
```

3. 播放外部存储器上的音频文件

播放外部存储器上的音频文件按如下步骤执行：

（1）创建MediaPlayer对象（或利用已有的MediaPlayer对象），并调用MediaPlayer对象的setDataSource(String path)方法装载指定的音频文件。

（2）调用MediaPlayer对象的prepare()方法准备音频。

（3）调用MediaPlayer的start()、pause()、stop()等方法控制播放即可。

其示例代码如下：

```
MediaPlayer mPlayer;  //定义MediaPlayer对象
file=new File("/mnt/sdcard/night.mp3");  //获取要播放的文件
mPlayer=MediaPlayer.create(this,Uri.parse(file.getAbsolutePath()));
mPlayer.setDataSource(file.getAbsolutePath());  //重新设置要播放的音频
mPlayer.prepare();  //准备声音，同步加载，方法返回时已加载完毕
mPlayer.start();  //开始播放
```

4. 播放来自网络的音频文件

播放来自网络的音频文件有两种方式：第一种，直接使用MediaPlayer的静态create(Context context, Uri uri)方法；第二种，调用MediaPlayer的setDataSource(Context context, Uri uri)装载指定的Uri对应的音频文件。

以第二种方式播放来自网络的音频文件的步骤如下：

（1）根据网络上的音频文件所在的位置创建Uri对象。

（2）创建MediaPlayer对象（或利用已有的MediaPlayer对象），并调用MediaPlayer对象的setDataSource(Context context , Uri uri)方法装载Uri对应的音频文件。

（3）调用MediaPlayer对象的prepare()方法准备音频。

（4）调用MediaPlayer对象的start()、pause()、stop()等方法控制播放即可。

其示例代码如下：

```
Uri uri=Uri.parse("http://www.zig-cloud.com/test.mp3");
MediaPlayer mPlayer=new MediaPlayer();
mPlayer.setDataSource(this,uri);  //使用MediaPlayer根据Uri来加载指定的声音文件
mPlayer.prepare();  //准备声音
mPlayer.start();  //开始播放
```

MediaPlayer除了调用prepare()方法准备声音之外，还可以调用prepareAsynchronous()方法准备声音，prepareAsync()方法与普通prepare()方法的区别在于，prepareAsync()方法是异步的，它不会阻塞当前的UI线程。

MediaPlayer典型案例：基于MediaPlayer的MP3播放器

该程序界面的布局大纲如图5-2所示。程序的布局代码activity_main.xml如下(先将night.mp3复制到/mnt/sdcard/目录下)：

图 5-2　基于 MediaPlayer 的 MP3 播放器布局大纲

```
<?xml version="1.0"encoding="utf-8"?>
<LinearLayout xmlns:android="http://schemas.android.com/apk/res/android"
    xmlns:tools="http://schemas.android.com/tools"
    android:id="@+id/activity_main"
    android:layout_width="match_parent"
    android:layout_height="match_parent"
    android:orientation="horizontal"
    tools:context="sziit.lihz.mediaplayer_Mp3.MainActivity">
    <Button
        android:id="@+id/btnPlay"
        android:layout_width="wrap_content"
        android:layout_height="wrap_content"
        android:layout_weight="1"
        android:text="播放"/>
    <Button
        android:id="@+id/btnPause"
        android:layout_width="wrap_content"
        android:layout_height="wrap_content"
        android:layout_weight="1"
        android:text="暂停"/>
    <Button
        android:id="@+id/btnReplay"
        android:layout_width="wrap_content"
        android:layout_height="wrap_content"
        android:layout_weight="1"
        android:text="重播"/>
    <Button
        android:id="@+id/btnStop"
        android:layout_width="wrap_content"
        android:layout_height="wrap_content"
        android:layout_weight="1"
        android:text="停止"/>
    <Button
        android:id="@+id/btnExit"
        android:layout_width="wrap_content"
        android:layout_height="wrap_content"
        android:layout_weight="1"
        android:text="退出"/>
</LinearLayout>
```

程序MainActivity源代码如下：

```
public class MainActivity extends Activity implements OnClickListener{
    private MediaPlayer mediaPlayer=null;
```

```
    private Button btnPlay,btnPause,btnReplay,btnStop,btnExit;
    protected void onCreate(Bundle savedInstanceState){
        super.onCreate(savedInstanceState);
        setContentView(R.layout.activity_main);
        mediaPlayer=new MediaPlayer();  //创建MediaPlayer对象
        initMediaPlayer();  //初始化MediaPlayer
        findView();setOnClickListener();}
    private void setOnClickListener(){
        btnPlay.setOnClickListener(this);
        btnPause.setOnClickListener(this);
        btnStop.setOnClickListener(this);
        btnReplay.setOnClickListener(this);
        btnExit.setOnClickListener(this);}
    private void findView(){
        btnPlay=(Button)findViewById(R.id.btnPlay);
        btnPause=(Button)findViewById(R.id.btnPause);
        btnReplay=(Button)findViewById(R.id.btnReplay);
        btnStop=(Button)findViewById(R.id.btnStop);
        btnExit=(Button)findViewById(R.id.btnExit);}
    private void initMediaPlayer(){
        try{  //初始化MediaPlayer
            File file=new File(Environment.getExternalStorageDirectory()
                +"/", "night.mp3");  //设置SD卡上"night.mp3"文件
            mediaPlayer.setDataSource(file.getPath());  //设置音频路径
            mediaPlayer.prepare();  //加载音频，完成准备
    }catch(IOException e){e.printStackTrace();}}
    protected void onDestroy(){  //关闭应用时释放MediaPlayer对象占用的资源
        if(mediaPlayer!=null){
            mediaPlayer.stop();mediaPlayer.release();mediaPlayer=null;}
        super.onDestroy();}
    public void onClick(View v){
        switch(v.getId()){
        case R.id.btnPlay:doPlay();break;
        case R.id.btnPause:doPause();break;
        case R.id.btnReplay:doReplay();  break;
        case R.id.btnStop:doStop();break;
        case R.id.btnExit:finish();break;}}
    private void doReplay(){
        if(mediaPlayer.isPlaying()){
            mediaPlayer.stop();  //停止播放}
        try{mediaPlayer.reset();  initMediaPlayer();
            mediaPlayer.start();
    }catch(IllegalStateException e){e.printStackTrace();}}
```

```
    private void doStop(){
          if(mediaPlayer.isPlaying()){mediaPlayer.stop();
              try{mediaPlayer.prepare();
         }catch(IOException e){e.printStackTrace();}}}
    private void doPause(){
         if(mediaPlayer.isPlaying()){mediaPlayer.pause();  //暂停播放
    }}
    private void doPlay(){
         if(!mediaPlayer.isPlaying()){mediaPlayer.start();  //开始播放
    }}
}
```

该程序的运行结果如图5-3所示。

图5-3　基于MediaPlayer的MP3播放器运行结果

5.2.2　使用音频池SoundPool播放音频

如果应用程序经常需要播放密集、短促的音频，这时MediaPlayer就显得有些不合适了。除了前面介绍的MediaPlayer播放音频之外，Android还提供了SoundPool播放音频，SoundPool使用音频池的概念管理多个短促的音频，例如它可以加载20个音频，以后在程序中按音频的ID进行播放。SoundPool提供了一个构造器，该构造器可以指定它总共支持多少个声音（也就是池的大小）、声音的品质等。构造器如下：

视频

使用音频池SoundPool播放音频

SoundPool(int maxStreams, int streamType, int srcQuality)：第一个参数maxStreams指定支持多少个声音；第二个参数streamType指定声音类型；第三个参数srcQuality指定声音品质。一旦得到了SoundPool对象之后，接下来就可调用SoundPool的多个重载的load()方法加载声音。SoundPool提供了如下4个load()方法：

- int load(Context context, int resId, int priority)：从resId所对应的资源加载声音。
- int load(FileDescriptor fd, long offset, long length, int priority)：加载fd所对应的文件的offset开始、长度为length的声音。
- int load(String path, int priority)：从path对应的文件加载声音。
- int load(AssetFileDescriptor afd, int priority)：从afd所在对应的文件中加载声音。

上面3个方法中都有一个priority参数，该参数目前还没有任何作用，Android建议将该参数设置为1，保持和未来的兼容性。

上面3个方法加载声音之后，都返回该声音的ID，以后程序就可以通过该声音的ID播放指定的声音。SoundPool提供的播放指定声音的方法如下：

int play(int soundID, float leftVolume, float rightVolume, int priority, int loop, float rate)：该方法的第一个参数指定播放哪个声音；leftVolume、rightVolume指定左、右的音量；priority指定播放声音的优先级，数值越大，优先级越高；loop指定是否循环，0为不循环，-1为循环；rate指定播放的比率，

数值范围0.5～2，为正常比率。

为了更好地管理SoundPool加载的每个声音的ID，程序一般会使用HashMap<Integer,Integer>对象管理声音。归纳起来，使用SoundPool播放声音的步骤如下：

（1）调用SoundPool的构造器创建SoundPool的对象。

（2）调用SoundPool对象的load()方法从指定资源、文件中加载声音。最好使用HashMap<Integer, Integer>管理加载声音。

（3）调用SoundPool的play声音播放声音。

SoundPool典型案例：播放音频

该案例示范了如何使用SoundPool播放音频，该程序提供了八个按钮，八个按钮分别用于播放不同的声音。该程序的界面布局UI如图5-4所示，程序SoundPoolActivity类实现代码如下：

```
//从活动基类派生子类，并实现单击监听器接口
public class SoundPoolActivity extends Activity implements OnClickListener{
    private static final String TAG="SoundPoolDemo";  //定义私有静态字符串常量
        private Button btnBomb,btnShot,btnArrow,btnBird;  //声明4个按钮对象
        Button btnPiano,btnTrumpets,btnSiren,btnBell;  //声明4个按钮对象
        private SoundPool soundPool;  //定义一个声音池SoundPool
        private int streamVolume;  //音效的音量
        private HashMap<Integer, Integer>soundMap=new
            HashMap<Integer,Integer>();  //声音映射数组
        private boolean mIsLoadCompelete,isFinishedLoad;
        private TextView tvHint;  //定义提示文本对象
        public void onCreate(Bundle savedInstanceState)  //重写onCreate()方法
        {   Log.d(TAG, "执行onCreate! ");
            super.onCreate(savedInstanceState);  //调用基类onCreate方法
            setContentView(R.layout.activity_main);  //设置活动界面布局
            findView();setOnClickListener();initSoundPool();}
        private void initSoundPool(){  //获得声音设备和设备最大音量
            AudioManager am=(AudioManager)getSystemService(AUDIO_SERVICE);
            streamVolume=am.getStreamMaxVolume(AudioManager.STREAM_MUSIC);
            //设置最多可容纳16个音频流，音频的品质为5
            soundPool=new SoundPool(16,AudioManager.STREAM_SYSTEM,5);
            //load()方法加载指定音频文件返回所加载音频ID，使用HashMap来管理音频流
            soundMap.put(1,soundPool.load(this,R.raw.bomb,1));
            soundMap.put(2,soundPool.load(this,R.raw.shot,1));
            soundMap.put(3,soundPool.load(this,R.raw.arrow,1));
            soundMap.put(4,soundPool.load(this,R.raw.bird,1));
            loadRes(5,R.raw.piano);loadRes(6,R.raw.trumpets);
            loadRes(7,R.raw.siren);loadRes(8,R.raw.bell);
            soundPool.setOnLoadCompleteListener(new SoundPool.OnLoadCompleteListener(){
                public void onLoadComplete(SoundPool soundPool,
                    int sampleId, int status){
```

```
                    mIsLoadCompelete=true;isFinishedLoad=true;}});}
        //把资源中的音效加载到指定的id(播放的时候就对应到该ID播放)
        private void loadRes(int no, int resId){
            soundMap.put(no,soundPool.load(this,resId,1));}
        //循环播放loop次，音乐池中第no号音乐
        private void play(int no, int loop){
    soundPool.play(soundMap.get(no),streamVolume,streamVolume,1,loop,1)}
    private void setOnClickListener(){  //设置按钮单击监听器
        btnBomb.setOnClickListener(this);
        btnShot.setOnClickListener(this);
        btnArrow.setOnClickListener(this);
        btnBird.setOnClickListener(this);   //增加监听器
        btnPiano.setOnClickListener(this); //增加监听器
        btnTrumpets.setOnClickListener(this);
        btnSiren.setOnClickListener(this);
        btnBell.setOnClickListener(this);}
    private void findView(){
        //从布局文件中查找或绑定按钮控件对象
        btnBomb=(Button)findViewById(R.id.btnBomb);
        btnShot=(Button)findViewById(R.id.btnShot);
        btnArrow=(Button)findViewById(R.id.btnArrow);
        btnBird=(Button)findViewById(R.id.btnBird);
        btnPiano=(Button)findViewById(R.id.btnPiano);
        btnTrumpets=(Button)findViewById(R.id.btnTrumpets);
        btnSiren=(Button)findViewById(R.id.btnSiren);
        btnBell=(Button)findViewById(R.id.btnBell);
        tvHint=(TextView)findViewById(R.id.tvHint);}
    //重写OnClickListener监听器接口的方法
    public void onClick(View v)  //重写onClick()方法
    { Log.d(TAG, "执行onClick! ");
      switch(v.getId())  //判断哪个按钮被单击
      {//播放音频资源
      case R.id.btnBomb:tvHint.setText("爆炸声1遍...");
          soundPool.play(soundMap.get(1),1,1,0,0,1);break;
      case R.id.btnShot:tvHint.setText("射击声1遍...");
          soundPool.play(soundMap.get(2),1,1,0,0,1);break;
      case R.id.btnArrow:tvHint.setText("射箭声1遍...");
          soundPool.play(soundMap.get(3),1,1,0,0,1);break;
      case R.id.btnBird:tvHint.setText("播放鸟鸣声1遍...");
         soundPool.play(soundMap.get(4),1,1,0,0,1);
      case R.id.btnPiano:tvHint.setText("播放钢琴声1遍...");
         play(5, 1);break;
      case R.id.btnTrumpets:tvHint.setText("播放号角声2遍...");
```

```
            play(6, 2);break;
        case R.id.btnSiren:tvHint.setText("播放汽笛声3遍...");
            play(7, 3);break;
        case R.id.btnBell:tvHint.setText("播放铃声4遍...");
            play(8,4);break;}}
    protected void onDestroy(){
        for(int i=1;i<=8;i++)soundPool.unload(soundMap.get(i));
        soundPool.release(); super.onDestroy();}}
```

该程序的运行结果如图5-4所示。

图 5-4　声音池 SoundPool 播放音效运行结果

5.2.3　使用 VideoView 和 MediaController 播放视频

为了在Android应用中播放视频，Android提供了VideoView组件，它是一个位于android.widget包下的组件，其作用与ImageView类似，只是ImageView用于显示图片，而VideoView用于播放视频。使用VideoView播放视频的步骤如下：

（1）在界面布局中定义VideoView组件，或在程序中创建VideoView组件。

（2）调用VideoView对象的如下两个方法来加载指定视频。

- setVideoPath(String path)：加载path文件所代表的视频。
- setVideoURI(Uri uri)：加载uri所对应的视频。

（3）调用VideoView对象的start()、stop()、pause()方法控制视频播放。

视频

使用VideoView和MediaController播放视频

实际上与VideoView一起结合使用的还有一个MediaController类，它的作用是提供一个友好的图形控制界面，通过控制界面来控制视频的播放。

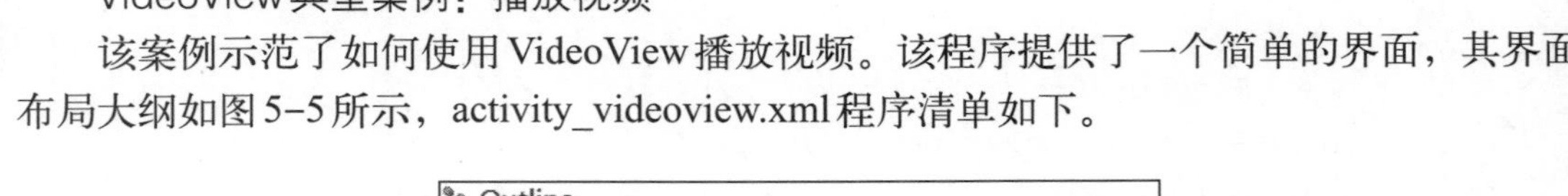

VideoView典型案例：播放视频

该案例示范了如何使用VideoView播放视频。该程序提供了一个简单的界面，其界面布局大纲如图5-5所示，activity_videoview.xml程序清单如下。

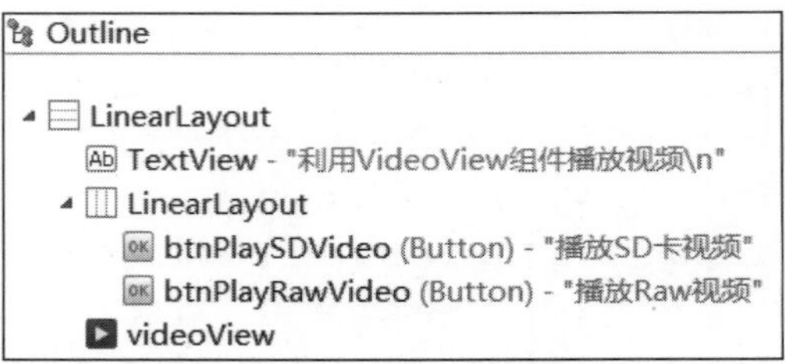

图 5-5　VideoView 播放视频布局大纲

```
<LinearLayout xmlns:android="http://schemas.android.com/apk/res/android"
    android:layout_width="fill_parent"
    android:layout_height="fill_parent"
    android:orientation="vertical">
    <TextView
        android:layout_width="fill_parent"
        android:layout_height="wrap_content"
        android:background="#0000FF"
        android:text="利用VideoView组件播放视频\n"
        android:textColor="#FFFFFF"
        android:textSize="20sp"/>
    <LinearLayout
        android:layout_width="match_parent"
        android:layout_height="wrap_content"
        android:orientation="horizontal">
        <Button
            android:id="@+id/btnPlaySDVideo"
            android:layout_width="wrap_content"
            android:layout_height="wrap_content"
            android:layout_weight="1"
            android:text="播放SD卡视频"/>
        <Button
            android:id="@+id/btnPlayRawVideo"
            android:layout_width="wrap_content"
            android:layout_height="wrap_content"
            android:layout_weight="1"
            android:text="播放Raw视频"/>
    </LinearLayout>
    <VideoView
        android:id="@+id/videoView"
        android:layout_width="match_parent"
        android:layout_height="match_parent"
        android:layout_gravity="center"/>
</LinearLayout>
```

该布局中定义了一个VideoView组件，在程序中使用该组件并结合MediaController控制视频播放，SD卡上视频文件支持多种定位方式。VideoViewActivity类程序代码如下：

```
public class VideoViewActivity extends Activity implements MediaPlayerControl
{   private VideoView videoView;
    private MediaController controller;
    private Button btnPlaySDVideo,btnPlayRawVideo;
    protected void onCreate(Bundle savedInstanceState){
        super.onCreate(savedInstanceState);
```

```
        setContentView(R.layout.activity_videoview);
        videoView=(VideoView)findViewById(R.id.videoView);
        controller=new MediaController(this);
        videoView.setMediaController(controller);  //设置视频控制器
        btnPlaySDVideo=(Button)findViewById(R.id.btnPlaySDVideo);
      btnPlayRawVideo=(Button)findViewById(R.id.btnPlayRawVideo);
      btnPlaySDVideo.setOnClickListener(new OnClickListener(){
    public void onClick(View v){playSDVideo();}});
      btnPlayRawVideo.setOnClickListener(new OnClickListener(){
    public void onClick(View v){playRawVideo();}});}
    public boolean canPause(){
        videoView.canPause();return false;}
    private void playSDVideo(){  //播放SD卡上视频，支持以下几种描述方式
        File file=new File(Environment.getExternalStorageDirectory()
              +"/","movie.mp4");  //设置SD卡上"movie.mp4"文件路径
        videoView.setVideoURI(Uri.parse(file.getAbsolutePath()));
        if(file.exists()){
            //VideoView与MediaController进行关联
            videoView.setVideoPath(file.getAbsolutePath());
            videoView.setMediaController(controller);  //设置视频控制器
            controller.setMediaPlayer(videoView);
            videoView.requestFocus();  //让VideiView获取焦点
            videoView.start();  //开始播放视频
            videoView.setOnCompletionListener(new OnCompletionListener(){
              public void onCompletion(MediaPlayer mp){  //设置播放完成监听器
                videoView.start();  //循环播放}});}}
    private void playRawVideo(){  //播放res/raw下视频文件
        videoView.setVideoURI(Uri.parse("android.resource://"
            +getPackageName()+"/"+R.raw.video));
        videoView.setMediaController(controller);  //设置视频控制器
        controller.setMediaPlayer(videoView);
        videoView.requestFocus();  //让 VideiView 获取焦点
        videoView.start();  //开始播放视频
        videoView.setOnCompletionListener(new OnCompletionListener(){
          public void onCompletion(MediaPlayer mp){  //设置播放完成监听器
            videoView.start();//循环播放
     }});}
    public boolean canSeekBackward(){
       videoView.canSeekForward();return false;}
    public boolean canSeekForward(){videoView.canSeekForward();return false;}
    public int getBufferPercentage(){return 0;}
    public int getCurrentPosition(){return videoView.getCurrentPosition();}
```

```
    public int getDuration(){return videoView.getDuration();}
    public boolean isPlaying(){return videoView.isPlaying();}
    public void pause(){if(videoView.isPlaying()){videoView.pause();}}
    public void seekTo(int pos){videoView.seekTo(pos);}
    public void start(){videoView.start();}}
```

注意：本地视频文件File file的定义有以下几种方法。

```
1.File(Environment.getExternalStorageDirectory()+"/","movie.mp4");
2.File("/mnt/sdcard/movie.mp4");
3.File("file:///mnt/sdcard/movie.mp4");
```

videoView.setVideoURI()调用方法有以下几种应用形式：

```
1.videoView.setVideoURI(Uri.parse("file:///mnt/sdcard/movie.mp4"));
2.videoView.setVideoURI(Uri.parse("/mnt/sdcard/movie.mp4"));
3.videoView.setVideoURI(Uri.parse(file.getAbsolutePath()));
```

或用如下videoView.setVideoPath()方法代替videoView.set VideoURI()方法：

```
videoView.setVideoPath(file.getAbsolutePath());
```

视频资源文件的String uri定义方式如下：

```
Uri.parse("android.resource://"+getPackageName()+
"/"+R.raw.video)
```

在AndroidManifest.xml设置读外部存储的权限READ_EXTERNAL_STORAGE权限，保证/mnt/sdcard/movie.mp4和res/raw/video.3gp视频文件存在的前提下，可正常运行该程序，运行效果如图5-6所示。类似方法也可以用来播放网络视频。

```
<uses-permission android:name="android.permission.
READ_EXTERNAL_STORAGE"/>
```

图 5-6　VideoView 播放视频运行效果

5.2.4　使用 MediaPlayer 与 SurfaceView 播放视频

视频

使用MediaPlayer与SurfaceView播放视频

使用MediaPlayer与SurfaceView播放视频的步骤如下：

（1）创建MediaPlayer对象，并让它加载指定的视频文件。

（2）在界面布局文件中定义SurfaceView组件，或在程序中创建SurfaceView组件，并为SurfaceView的SurfaceHolder添加Callback监听器。

（3）调用MediaPlayer对象的setDisplay(SurfaceHolder sh)将所播放的视频图像输出到指定的SurfaceView组件。

（4）调用MediaPlayer对象的start()、stop()和pause()方法控制视频播放。

播放视频典型案例：使用MediaPlayer和SurfaceView播放视频

该案例使用MediaPlayer和SurfaceView播放视频，同时支持播放SD卡、res/raw和网络上视频，用3个按钮来控制视频播放、暂停和停止。该程序activity_main.xml的界面布局大纲如图5-7所示。

该程序SurfaceViewPlayVideo类实现代码如下：

```
    //从活动基类派生子类，并实现单击监听器接口
    public class SurfaceViewPlayVideo extends Activity
implements OnClickListener
    {private static final String TAG="SurfaceView
PlayVideo";
     private SurfaceView surfaceView;  //声明表面视图包
     private ImageButton play,pause,stop;  //声明3个位
图按钮包
     private MediaPlayer mPlayer;  //声明多媒体播放器包
     private int position;  //记录当前视频的播放位置
     private String path="/mnt/sdcard/movie.mp4";  //设置视频文件SD卡路径
     private Uri uri=Uri
         .parse("android.resource://sziit.lihz.MediaPlayerSurfaceView1/"
            +R.raw.video);
     private static final String videoUrl;  //设置自己的网络视频文件路径
     private int VideoNo=1;  //1:播放SD上视频;2:播放raw资源视频;3:播放网络视频
     public void onCreate(Bundle savedInstanceState)  //重写onCreate()方法
     {    Log.d(TAG, "执行onCreate! ");
          super.onCreate(savedInstanceState);  //调用基类onCreate()方法
          setContentView(R.layout.activity_main);  //设置活动界面布局
          uri=Uri.parse("android.resource://"+getPackageName()+"/"
              +R.raw.video);  //定义资源视频uri
         findView();setOnClickListener();initMedia();}
     private void initMedia(){  //本地的视频需要在手机SD卡根目录添加movie.mp4
         path=Environment.getExternalStorageDirectory().getPath()
         +"/movie.mp4";  //SD卡视频路径
         mPlayer=new MediaPlayer();  //创建MediaPlayer
         surfaceView=(SurfaceView)this.findViewById(R.id.surfaceView);
         surfaceView.getHolder().setKeepScreenOn(true);  //设置播放时打开屏幕
         surfaceView.getHolder().addCallback(new SurfaceListener());}
     private void setOnClickListener(){  //为3个按钮的单击事件绑定事件监听器
         play.setOnClickListener(this);pause.setOnClickListener(this);
         stop.setOnClickListener(this);}
     private void findView(){  //获取界面中的3个按钮
         play=(ImageButton)findViewById(R.id.play);
         pause=(ImageButton)findViewById(R.id.pause);
         stop=(ImageButton)findViewById(R.id.stop);}
     public void onClick(View v){Log.d(TAG,"执行onClick! ");
         try{switch(v.getId())  //根据按钮ID号，分别处理不同按钮单击事件响应{
           case R.id.play:play();break;  //播放视频
           case R.id.pause:
```

```
Outline
◢ RelativeLayout
    surfaceView
  ◢ LinearLayout
      play (ImageButton) - play
      pause (ImageButton) - pause
      stop (ImageButton) - stop
```

图 5-7　播放视频布局大纲

```
            if(mPlayer.isPlaying()){mPlayer.pause();  //暂停播放
            } else{mPlayer.start();}  break;  //开始播放
         case R.id.stop:  //停止按钮被单击
            if(mPlayer.isPlaying()){mPlayer.stop();}break;}
      }catch(Exception e){e.printStackTrace();}}
   private void play()throws IOException{mPlayer.reset();  //视频复位
   mPlayer.setAudioStreamType(AudioManager.STREAM_MUSIC);  //设置音频类型
   //设置需要播放的视频(用户自行修改播放视频)
   switch(VideoNo)
   { case 1:  mPlayer.setDataSource(path);  //播放SD卡上视频
         VideoNo++;break;
     case 2:  //播放res/raw目录资源视频
        mPlayer.setDataSource(getApplicationContext(), uri);
        VideoNo++;break;
     case 3:  //播放网络视频
       mPlayer.setDataSource(getApplicationContext(), Uri.parse(videoUrl ));
VideoNo++;break;}
     if(VideoNo>3){VideoNo=1;}
     //把视频画面输出到SurfaceView
     mPlayer.setDisplay(surfaceView.getHolder());  //设置显示的区域
     mPlayer.prepare();  //准备播放
     WindowManager wManager=getWindowManager();  //获取窗口管理器
     DisplayMetrics metrics=new DisplayMetrics();
     //获取屏幕大小
     wManager.getDefaultDisplay().getMetrics(metrics);
     //设置视频保持纵横比缩放到占满整个屏幕
     surfaceView.setLayoutParams(new LayoutParams(metrics.widthPixels,
         mPlayer.getVideoHeight()*metrics.widthPixels
            /mPlayer.getVideoWidth()));
     mPlayer.start();  //开始播放视频
     //播放完成回调
     mPlayer.setOnCompletionListener(new OnCompletionListener(){
        public void onCompletion(MediaPlayer mp){ //处理完成后事件
            Toast.makeText( getApplicationContext(), "播放完成了", Toast.LENGTH_
SHORT).show();
            mPlayer.start();}});}  //播放后重播
   private class SurfaceListener implements SurfaceHolder.Callback{
       public void surfaceChanged(SurfaceHolder holder, int format, int width,
int height){//构造器
          Log.d(TAG, "执行surfaceChanged! ");}
       public void surfaceCreated(SurfaceHolder holder){
          Log.d(TAG, "执行surfaceCreated! ");
          if(position >0){
```

```
                try{play();  //开始播放
                    mPlayer.seekTo(position);  //并直接从指定位置开始播放
                    position=0;
                } catch(Exception e){e.printStackTrace();}}}
        public void surfaceDestroyed(SurfaceHolder holder){
            Log.d(TAG, "执行surfaceDestroyed! ");}}
    //当其他Activity被打开，暂停播放
    protected void onPause()  //重写onPause()方法
    {   Log.d(TAG, "执行onPause! ");
        if(mPlayer.isPlaying())  //若当前正在播放
        {position=mPlayer.getCurrentPosition();//保存当前的播放位置
        mPlayer.stop();}  //停止播放
        super.onPause();}  //调用基类方法
    protected void onDestroy()  //重写onDestroy()方法
    {   Log.d(TAG, "执行onDestroy! ");
        if(mPlayer.isPlaying())mPlayer.stop();  //停止播放
        mPlayer.release();  //释放资源
        super.onDestroy();}}  //调用基类方法
```

在AndroidManifest.xml中增加INTERNET、MOUNT_UNMOUNT_FILESYSTEMS、READ_EXTERNAL_STORAGE和WRITE_EXTERNAL_STORAGE权限：

```
    <uses-permission android:name="android.permission.MOUNT_
UNMOUNT_FILESYSTEMS"/>
    <uses-permission android:name="android.permission.WRITE_
EXTERNAL_STORAGE"/>
    <uses-permission android:name="android.permission.READ_
EXTERNAL_STORAGE"/>
    <uses-permission android:name="android.permission.
INTERNET"/>
```

该程序的运行结果如图5-8所示。从上面的开发过程不难看出，使用MediaPlayer播放视频要复杂一些，并且需要自己开发控制按钮来控制视频播放，因此一般推荐使用VideoView来播放视频。

图5-8 播放视频案例运行效果

视频

使用Media-Recorder录制音频

5.2.5 使用MediaRecorder录制音频

为了在Android系统中录制音频，Android提供了MediaRecorder类。使用MediaRecorder录制音频的过程很简单，按照如下步骤进行即可：

（1）创建MediaRecorder对象。

（2）调用MediaRecorder对象的setAudioSource()方法设置声音来源，一般传入MediaRecorder. AudioSource.MIC参数指定录制来自麦克风的声音。

（3）调用MediaRecorder对象的setOutputFormat()方法设置所录制的音频文件格式。

（4）调用MediaRecorder对象的setAudioEncoder()、set Audio EncodingBitRate(int bitRate)、setAudioSamplingRate(int samplingRate)方法设置所录制的声音的编码格式、编码位率、采样率等，

这些参数都可以控制所录制的声音的品质、文件的大小。一般来说，声音品质越好，声音文件越大。

（5）调用 MediaRecorder 对象的 setOutputFile(String path) 方法设置录制的音频文件的保存位置。

（6）调用 MediaRecorder 对象的 prepare() 方法准备录制。

（7）调用 MediaRecorder 对象的 start() 方法开始录制。

（8）录制完成，调用 MediaRecoder 对象的 stop() 方法停止录制，并调用 release() 方法释放资源。图 5–9 所示为 MediaRecorder 的状态。

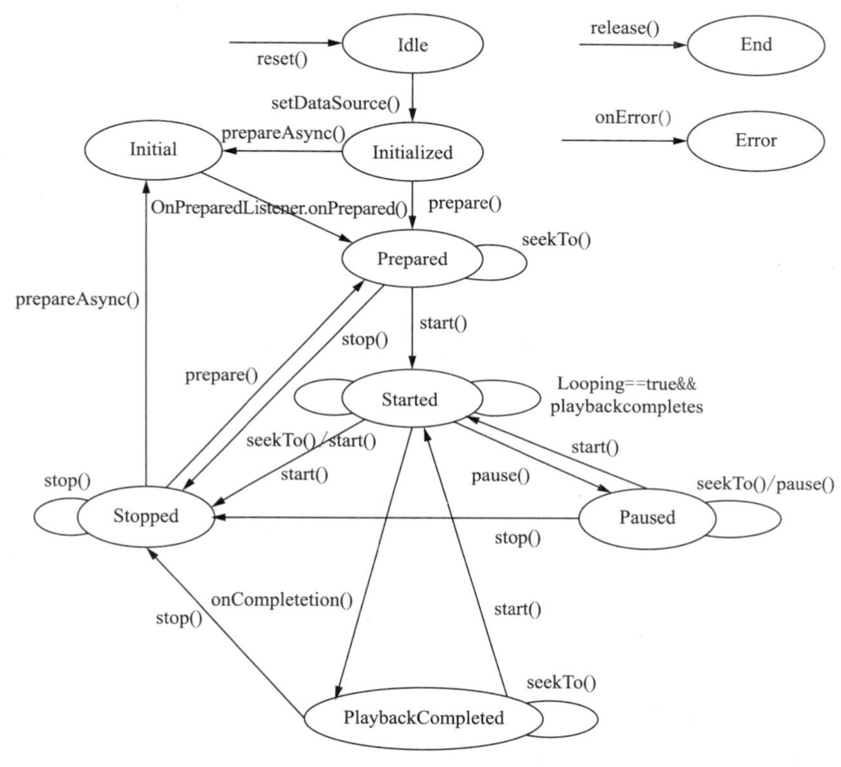

图 5–9　MediaRecorder 的状态

MediaRecorder 的重要方法包括：

（1）void setAudioSource(int audio_source)：设置录音来源。

（2）void setOutputFormat(int output_format)：设置输出格式。

（3）void setAudioEncoder(int audio_encoder)：设置编码方式。

（4）void setOutputFile(String path)：设置输出文件。

（5）void prepare()：让 MediaRecorder 对象处于就绪状态。

（6）void start()：开始录音。

（7）void stop()：停止录音，一旦停止则必须重新配置 MediaRecorder 对象才能再次开始录音。

（8）void release()：释放和 MediaRecorder 对象相关的所有资源。

（9）void reset()：重新启动 MediaRecorder 对象让它处于空闲状态。

MediaRecorder典型案例：迷你录音机

该程序通过MediaRecorder实现声音的录制功能，程序的界面布局大纲如图5-10所示。程序的界面布局文件activity_mediarecorder.xml如下：

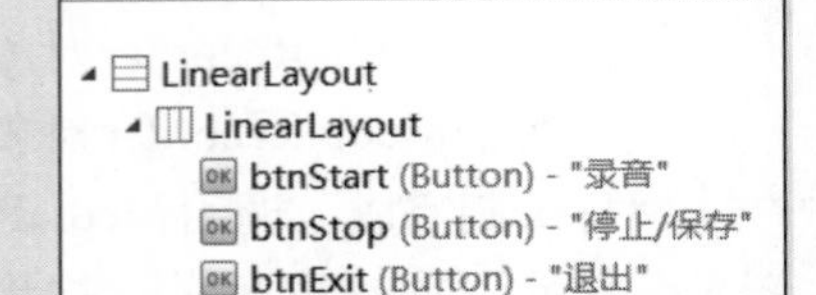

图5-10　录音界面布局大纲

```
<LinearLayout xmlns:android="http://schemas.
android.com/apk/res/android"
    android:layout_width="match_parent"
    android:layout_height="match_parent"
    android:orientation="vertical"
    tools:context=".MediaRecorderActivity">
    <LinearLayout
        android:layout_width="match_parent"
        android:layout_height="wrap_content"
        android:orientation="horizontal">
        <Button
            android:id="@+id/btnStart"
            android:layout_width="wrap_content"
            android:layout_height="wrap_content"
            android:layout_weight="1"
            android:text="@string/start"/>
        <Button
            android:id="@+id/btnStop"
            android:layout_width="wrap_content"
            android:layout_height="wrap_content"
            android:layout_weight="1"
            android:text="@string/stop"/>
        <Button
            android:id="@+id/btnExit"
            android:layout_width="wrap_content"
            android:layout_height="wrap_content"
            android:layout_weight="1"
            android:text="@string/exit"/>
    </LinearLayout>
     <TextView
        android:layout_width="match_parent"
        android:layout_height="wrap_content"
        android:background="#0000ff"
        android:textColor="#ffffff"
        android:textSize="20sp"
        android:text="@string/info"/>
    <ListView
        android:id="@+id/lvFile"
        android:layout_width="match_parent"
```

```
        android:layout_height="wrap_content">
    </ListView>
</LinearLayout>
```

程序MediaRecorderActivity类的Java代码如下：

```
public class MediaRecorderActivity extends Activity implements OnClickListener{
    private static final String TAG="MediaRecorderActivity";
    private MediaPlayer mPlayer;  //录音文件播放
    private MediaRecorder mRecorder;   //录音
    private Button btnStart,btnStop,btnExit;
    private ListView lvFile;
    private String path;  //音频文件保存地址
    private String paths=path;
    private File saveFile;
    private String[] listFile=null;  //所录音的文件
    private DisplayRecorderAdapter drAdapter;
    private AlertDialog alertDialog=null;
    protected void onCreate(Bundle savedInstanceState){
        super.onCreate(savedInstanceState);
        setContentView(R.layout.activity_mediarecorder);
        Log.d(TAG, "onCreate()");findViews();
        mPlayer=new MediaPlayer();initMediaRecorder();
        drAdapter=new DisplayRecorderAdapter();
        mkdir();setOnClickListener();
        if(listFile!=null){lvFile.setAdapter(drAdapter);}}
    private void initMediaRecorder(){
        mRecorder=new MediaRecorder();
        //从麦克风源进行录音
        mRecorder.setAudioSource(MediaRecorder.AudioSource.DEFAULT);
        //设置输出格式
        mRecorder.setOutputFormat(MediaRecorder.OutputFormat.DEFAULT);
        //设置编码格式
        mRecorder.setAudioEncoder(MediaRecorder.AudioEncoder.DEFAULT);}
    private void setOnClickListener(){
        btnStart.setOnClickListener(this);
        btnStop.setOnClickListener(this);
        btnExit.setOnClickListener(this);}
    private void mkdir(){
        if(Environment.getExternalStorageState().equals(
            Environment.MEDIA_MOUNTED)){  //检查是否存在SD卡
          try{
            path=Environment.getExternalStorageDirectory()
                .getCanonicalPath().toString()
```

```
            +"/recordes";  //在SD卡上创建存放录音文件的文件目录
        File files=new File(path);
        if(!files.exists()){files.mkdir();}
        listFile=files.list();
    } catch(IOException e){e.printStackTrace();}}}
  private void findViews(){
      btnStart=(Button)findViewById(R.id.btnStart);
      btnStop=(Button)findViewById(R.id.btnStop);
      btnExit=(Button)findViewById(R.id.btnExit);
      lvFile=(ListView)findViewById(R.id.lvFile);}
  class DisplayRecorderAdapter extends BaseAdapter{
      public int getCount(){return listFile.length;}
      public Object getItem(int arg0){return arg0;}
      public long getItemId(int arg0){return arg0;}
      public View getView(final int postion,View arg1,ViewGroup arg2){
      View views=LayoutInflater.from(MediaRecorderActivity.this).
inflate(R.layout.recorderfilelist,null);
      TextView filename=(TextView)views.findViewById(R.id.tvfilename);
      ImageButton btnPlay=(ImageButton)
  views.findViewById(R.id.listBtnPlay);
        ImageButton btnStop=(ImageButton)
  views.findViewById(R.id.listBtnStop);
       filename.setText(listFile[postion]);
       btnPlay.setOnClickListener(new OnClickListener(){ //播放录音
          public void onClick(View arg0){doPlayClick(postion);}});
       btnStop.setOnClickListener(new OnClickListener(){  //停止播放
          public void onClick(View arg0){
            doStopClick();}});
       return views;}}
  private void doPlayClick(final int postion){
      try{mPlayer.reset();  //重置
          mPlayer.setDataSource(path+"/"+listFile[postion]);
          if(!mPlayer.isPlaying()){
              mPlayer.prepare();  //准备播放
              mPlayer.start();  //开始播放
          } else{mPlayer.pause();  //停止播放}
      } catch(IOException e){e.printStackTrace();}}
  private void doStopClick(){
     if(mPlayer.isPlaying()){mPlayer.stop();  //停止播放
  }}
  public void onClick(View v){
      switch(v.getId()){
      case R.id.btnStart:doStartRecorderClick();break;
```

```
        case R.id.btnStop:doStopRecorderClick();break;
        case R.id.btnExit:finish();break;}}
    private void doStopRecorderClick(){
        if(saveFile.exists()&& saveFile!=null){
            mRecorder.stop();  //停止录音
            mRecorder.release();  //释放录音对象
            //判断是否保存 如果不保存则删除
            new AlertDialog.Builder(this).setTitle("是否保存该录音?")
              .setPositiveButton("确定", null).setNegativeButton("取消",
                new DialogInterface.OnClickListener(){
                  public void onClick(DialogInterface dialog,
                  int which){deleteRecorderFile();}}).show();}
        btnStart.setText("录音");btnStart.setEnabled(true);
        btnExit.setEnabled(true);}
    private void deleteRecorderFile(){saveFile.delete();  //删除文件
        File files=new File(path);  //重新读取 文件
        listFile=files.list();
        drAdapter.notifyDataSetChanged();  //刷新ListView}
    private void doStartRecorderClick(){
        final EditText etfilename=new EditText(this);
        Builder buidler=new Builder(this);
        buidler.setTitle("请输入要保存录音文件名!").setView(etfilename)
            .setPositiveButton("确定", new DialogInterface.OnClickListener(){
              public void onClick(DialogInterface dialog,int which){
                doRecorder(etfilename);}});
        alertDialog=buidler.create();  //创建对话框
        alertDialog.setCanceledOnTouchOutside(false);
        alertDialog.show();  //显示对话框}
    protected void onDestroy(){    //释放资源
      if(mPlayer.isPlaying()){     //如果正在播放
        mPlayer.stop();  //停止播放
        mPlayer.release();  //释放媒体播放器对象资源
    }
      mPlayer.release();     //释放媒体播放器对象资源
      mRecorder.release();  //释放录音器对象资源
      super.onDestroy();}
    private void doRecorder(final EditText etfilename){
      String filename=etfilename.getText().toString();
      try{paths=path+"/"+filename
           +new SimpleDateFormat("yyyyMMddHHmmss").format(System
               .currentTimeMillis())+".amr";
        saveFile=new File(paths);
        initMediaRecorder();  //重新初始化录音对象
```

```
            mRecorder.setOutputFile(saveFile.getAbsolutePath());
            saveFile.createNewFile();  //创建新文件
            mRecorder.prepare();  //准备录音
            mRecorder.start();  //开始录音
            btnStart.setText("正在录音中......");
            btnStart.setEnabled(false);btnExit.setEnabled(false);
            alertDialog.dismiss();  //关闭对话框
            File files=new File(path);  //重新读取文件
            listFile=files.list();
             drAdapter.notifyDataSetChanged();  //刷
新ListView
        } catch(Exception e){e.printStackTrace();
    }}}
```

在AndroidManifest.xml申请录音权限RECORD_AUDIO等：

```
    <uses-permission android:name="android.
permission.RECORD_AUDIO"/>
    <uses-permission android:name="android.
permission.WRITE_EXTERNAL_STORAGE"/>
```

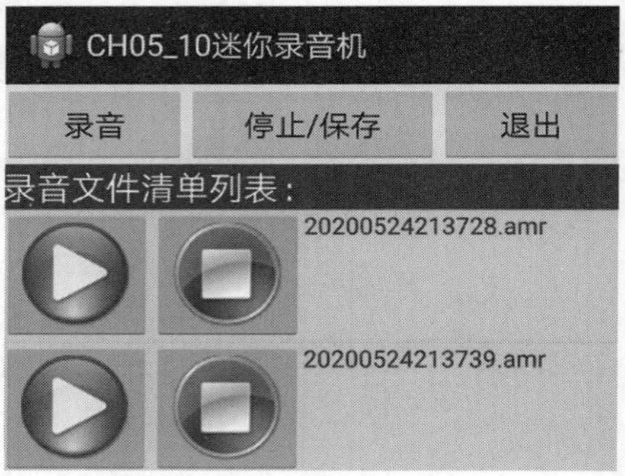

该程序的运行结果如图5-11所示。

图 5-11　录音案例运行主界面

5.2.6　使用手机摄像头Camera拍照

1. Camera类

实现手机摄像头拍照功能有两种方法：一种是利用Intent直接调用系统的照相机程序；另外一种是利用Android提供的Camera类API来定制一个自己的照相机处理程序。Camera类是Android系统实现拍照功能的重要类，能够管理照相机硬件，设置照相机的图像捕捉参数，打开照相机并抓拍照片，也可开始或者停止预览图像。在程序中使用照相机设备时，应在AndroidManifest.xml文件中使用<uses-permission>声明应用程序使用的照相机权限，并使用<uses-feature>声明应用程序用到的照相机特性：

```
    <uses-permission android:name="android.permission.CAMERA"/>
    <uses-feature android:name="android.hardware.camera"/>
    <uses-feature android:name="android.hardware.camera.autofocus"/>
```

Camera类的主要方法包括：

（1）Camera open()：创建一个新照相机对象，可访问照相机设备。该方法为静态方法，可直接调用。

（2）void autoFocus（Camera.AutoFocusCallback cb）：让照相机进行自动对焦，当照相机自动聚焦后，执行回调对象cb中的函数。在回调函数中编写自动对焦的有关处理方法，如聚焦成功后调用MediaActionSound类来播放相应的声音提醒用户。执行该方法要求照相机处于预览状态，即该方法应放在startPreview()和stopPreview()方法间运行。需要注意的是，有些设备可能不支持自动聚焦，可通过getFocusMode()方法进行查询。

（3）void cancelAutoFocus()：取消自动对焦。

（4）Camera.Parameters getParameters()：获得照相机当前的参数。更改Parameters对象的参数后，必须将其作为setParameters（Camera.Parameters）的参数进行传递并执行，才能使参数生效。

（5）void setParameters（Camera.Parameters params)：设置照相机服务的参数。

（6）void reconnect()：当其他进程使用了照相机服务后，重新连接照相机服务，此操作会请求对照相机加锁。若在程序中调用了unlock()，其他的进程可能会使用照相机，若不再使用，则程序必须重新调用reconnect()方法连接照相机。

（7）void release()：断开连接，释放照相机资源。

（8）void setDisplayOrientation（int degrees)：设置照相机预览显示旋转的角度。

（9）void setPreviewDisplay（SurfaceHolder holder)：设置用于照相机预览显示的SurfaceHolder。

（10）void startPreview()：开始在屏幕上捕捉和画出预览帧。在执行该方法之前，应先执行setPreviewDisplay（SurfaceHolder）方法。

（11）void stopPreview()：停止在Surface上捕捉和画出预览帧。

（12）void startSmoothZoom（int value)：开始平滑变焦，参数value表示变焦值。

（13）void stopSmoothZoom()：停止平滑变焦。

（14）void takePicture（Camera.ShutterCallback shutter，Camera.PictureCallback raw，Camera.PictureCallback postview，Camera.PictureCallback jpeg)：以异步方式进行图像捕捉，即拍照。照相机服务将在图像捕捉过程中开始一系列的回调操作。参数shutter所表示的回调发生在图像被捕捉后，可设置播放“咔嚓”声之类操作，以便让用户知道已经拍摄了一张图像。后面有3个PictureCallback接口参数，用于捕捉相片图像数据的回调，分别对应3种形式图像数据。图像数据可在PictureCallback接口的void onPictureTaken（byte[] data，Camera camera）中获得。这3种数据对应的回调函数正好是按照参数放置的先后顺序进行调用。参数raw表示的回调发生在原始图像（未被压缩）可用的时候。参数postview表示的回调发生在已经处理过的图像可用时。参数jpeg表示的回调发生在有压缩图像时。如果应用程序不需要特定的回调操作，可将其参数设置为null。最简单的形式是将所有的参数都设置为null，这样尽管能够捕获照片，但是不能对照片数据进行处理。通常只关心JPEG图像的数据，此时可将前面两个PictureCallback接口的参数直接设置为null。

（15）void takePicture（Camera.ShutterCallback shutter，Camera.PictureCallback raw，Camera.Picture Callback jpeg)：与takePicture（shutter，raw，null，jpeg）等价。

（16）void setErrorCallback（Camera.ErrorCallback cb)：设置照相机在出错时进行回调处理。

（17）void setPreviewCallback（Camera.PreviewCallback cb)：设置照相机的预览回调处理。

（18）void lock()：对照相机进行锁定，避免其他进程调用。照相机对象在默认情况下处于锁定状态，除非之前调用过unlock()方法。在一般情况下，该方法不常用。

（19）void unlock()：对照相机解锁，以便其他进程可以使用。当其他进程使用完后，应调用reconnect()方法。

Camera类方法调用所涉及的主要回调接口包括：

（1）Camera.AutoFocusCallback是一个用于通知照相机自动聚焦完成的回调接口，实现方法为onAutoFocus（boolean success，Camera camera)。该方法在照相机自动聚焦完成时被调用，可将拍照片的代码放在该方法中，如takePicture()方法。其中，参数success为true时表示聚焦成功，若为

false则表示聚焦失败；参数camera表示照相机对象。

（2）Camera.ErrorCallback是一个用于照相机错误通知的回调接口，实现方法为onError（int error，Camera camera）。该方法在照相机出错时调用，其中参数error表示错误代码，取值有CAMERA_ERROR_ UNKNOWN和CAMERA_ERROR_SERVER_DIED；参数camera表示照相机对象。

（3）Camera.PictureCallback是一个用于从照片捕捉中获得图像数据的回调接口，实现方法为onPictureTaken（byte[] data，Camera camera）。该方法在照片被捕捉后图像数据可用时自动调用。其中，参数data是图像数据字节数组，数据的格式依赖于回调的上下文context以及Camera.Parameters参数的设置；参数camera表示照相机对象。

（4）Camera.PreviewCallback用于预览帧被显示时，传送预览帧的副本，实现方法为onPreviewFrame（byte[] data，Camera camera）。该方法在预览帧被显示时自动调用，其中参数data是预览帧的内容，以字节数组形式表示。其格式由ImageFormat定义，并可以通过getPreviewFormat()方法查询获得。如果没有调用过setPreviewFormat（int）方法进行格式设置，那么默认是YCbCr_420_SP（NV21）；参数camera表示照相机对象。

（5）Camera.ShutterCallback是一个用在实际图像捕捉这一时刻的回调接口，实现方法为onShutter()。该方法在照片从传感器中捕捉的一刻即自动调用。可以在该方法中编写按下快门按键发出的声音或者其他的反馈操作。

2. Camera.Parameters类

Camera.Parameters类用于对照相机服务进行设置。照相机的主要参数包括颜色效果、抗条带、闪光灯模式、聚焦模式、情景模式、白平衡、预览大小等。Camera.Parameters类的主要方法如下：

（1）setAntibanding（String antibanding）：设置抗条带或防条带功能，主要用于防止照片场景在渐变时出现严重的一条一条的带状纹理。参数antibanding表示新的抗条带数值。

（2）setColorEffect（String value）：设置颜色效果。参数value表示新的颜色效果。

（3）setExposureCompensation（int value）：设置曝光补偿的索引值。参数value表示曝光补偿的索引值。

（4）setFlashMode（String value）：设置闪光模式。参数value表示闪光模式。

（5）setFocusMode（String value）：设置聚焦模式。参数value表示聚焦模式。

（6）setGpsAltitude（double altitude）：设置GPS的高度值，以便将该值存放在JPEG文件的EXIF头部。EXIF（Exchangeable Image File Format）用于表示图像的信息，例如，拍照当时的解析度、时间、光圈、照相机厂商、照相机型号等，但这些信息不直接显示在图像上。

（7）setJpegQuality（int quality）：设置JPEG图像的质量。参数quality表示质量值，取值范围从1～100。数值越大，则质量越好。

（8）setJpegThumbnailQuality（int quality）：设置JPEG缩略图的质量。

（9）setJpegThumbnailSize（int width，int height）：设置JPEG缩略图的尺寸。

（10）setPictureFormat（int pixel_format）：设置图像的格式。

（11）setPictureSize（int width，int height）：设置相片的尺寸。

（12）setPreviewFormat（int pixel_format）：设置预览相片的图像格式。

（13）setPreviewFrameRate（int fps）：设置预览帧接收的速度。

（14）setPreviewSize（int width，int height）：设置预览相片的尺寸。

（15）setRotation（int rotation）：设置相对于照相机方位顺时针旋转的角度。参数rotation的取值为0、90、180或者270。

（16）setSceneMode（String value）：设置场景模式。场景模式的设置有可能覆盖其他设置的参数，如闪光模式、聚焦模式。

（17）setWhiteBalance（String value）：设置白平衡。参数value表示白平衡的取值。

（18）setZoom（int value）：设置变焦。

3. BitmapFactory类

BitmapFactory类的作用是根据文件、流、字节数据等不同的来源创建Bitmap（BMP）图片。BitmapFactory类提供的都是静态方法，用于创建Bitmap文件。

（1）Bitmap decodeFile（String pathName）和Bitmap decodeFile（String pathName，BitmapFactory.Options opts）：这两种方法都是根据给出的文件路径，将文件解码转换成BMP格式。参数pathName表示完整路径的文件名，参数opts表示Bitmap的参数设置。

（2）Bitmap decodeFileDescriptor（FileDescriptor fd）：根据文件描述符，编码转换成Bitmap文件。参数fd表示包含编码数据的文件描述符。

（3）Bitmap decodeStream（InputStream is）：从输入流中解码转换为BMP文件。参数is为用于编码的数据流。

4. BitmapFactory.Option类

在BitmapFactory加载图片时，需使用BitmapFactory.Options对相关参数进行配置以减少加载的像素。下面介绍其主要属性：

（1）boolean inJustDecodeBounds：若取值为true，解码器返回空，即不生成Bitmap文件，但可计算出原始图片的outwidth和outheight两个属性。这种实现方法不会占用很多内存。有了这两个参数，再通过一定的算法，即可得到一个恰当的inSampleSize。

（2）int inSampleSize：为了节省内存，请求编码器裁剪原始的图像，以便于返回一个更小的图像，取值应大于1。图像大小根据整体取值进行等比缩小，如inSampleSize=4，则表示返回的图像的宽度和高度都相当于原来图像的1/4，也就相当于原来的图像大小的1/16。如果取值小于等于1，则相当于没有对原来的图像做任何改变。

（3）byte[] inTempStorage：用于解码的临时控件，建议取值为16k。

（4）int outHeight：Bitmap文件的高度。

（5）int outWidth：Bitmap文件的宽度。

视频

使用摄像头拍照应用实战

5. 实现拍照程序的步骤

实现拍照程序的步骤如下：

（1）定义一个Activity类实现Callback接口，在OnCreate()方法中获得SurfaceView和Surface Holder对象，并设置SurfaceHolder.Callback回调对象。参考代码如下：

```
//R.id.sv为布局文件中的SurfaceView控件的ID
SurfaceView surfaceView=(SurfaceView)findViewById(R.id.sv);
SurfaceHolder holder=surfaceView.getHolder();
holder.addCallback(this);
```

```
holder.setType(SurfaceHolder.SURFACE_TYPE_PUSH_BUFFERS);
```

（2）在SurfaceHolder.Callback接口的surfaceCreated()方法中，使用Camera的Open()方法打开照相机，获得Camera对象。参考代码如下：

```
mCamera=Camera.open();
```

（3）成功开启照相机后，在SurfaceHolder.Callback接口的surfaceCreated()方法中调用getParameters()方法获得照相机的默认设置。如果有需要可对其返回的Camera.Parameters对象进行参数修改，还可调用setDisplayOrientation()方法设置图像预览显示的角度。参考代码如下：

```
public void setCameraParams(){if(mCamera!=null){return;}
    mCamera=Camera.open();  //创建照相机，打开照相机
    Parameters params=mCamera.getParameters();  //设置照相机参数
    params.setFocusMode(Parameters.FOCUS_MODE_AUTO);  //拍照自动对焦
    params.setPreviewFrameRate(3);  //设置预览帧速率
    params.setPreviewFormat(PixelFormat.YCbCr_422_SP);  //设置预览格式
    params.set("jpeg-quality", 85);  //设置图片质量百分比
    //获取照相机支持图片分辨率
    List<Size>list=params.getSupportedPictureSizes();
    Size size=list.get(0);int w=size.width;int h=size.height;
    params.setPictureSize(w, h);  //设置图片大小
    params.setFlashMode(Camera.Parameters.FLASH_MODE_AUTO);  //设置自动闪光
}
```

（4）在surfaceChanged()方法中，通过setPreviewDisplay为照相机设置SurfaceHolder对象，没有Surface，照相机就不能进行预览。设置成功后调用startPreview()函数开启预览功能，在拍照之前，必须先进行预览。参考代码如下：

```
public void surfaceChanged(SurfaceHolder holder,int format,int width,int
height){try{  //启动照相机
  mCamera.setPreviewDisplay(holder);mCamera.startPreview(); //开始预览
  mPreviewRunning=true;}catch(Exception e){e.printStackTrace();}}
```

（5）若需要自动对焦功能，则在上述surfaceChanged调用完startPreview()函数后，可调用Camera::autoFocus()函数来设置自动对焦回调函数。该步为可选操作，有些设备可能不支持自动对焦功能，可通过Camera::getFocusMode()函数进行查询。参考代码如下：

```
mCamera.autoFocus(new AutoFocusCallback(){  //设置自动对焦
    public void onAutoFocus(boolean success,Camera camera){  //聚集后进行拍照
        mCamera.takePicture(shutterCallback, null,pictureCallback);}});
```

（6）调用takePicture（Camera.ShutterCallback，Camera.PictureCallback，Camera.PictureCallback，Camera.PictureCallback）方法进行拍照，等回调完成后获得图像数据。

（7）每次调用takePicture获取图像后，照相机会停止预览。若需要继续拍照，则在PictureCallback接口的onPictureTaken()方法中再次调用startPreview()方法。

（8）在不需要拍照时，应主动调用stopPreview()方法停止预览功能，并且调用release()方法释放照相机，以便让其他应用程序能够调用照相机。建议在onPause()方法中释放照相机，在onResume()方法中重新打开照相机。如果需要进入照相机的视频拍摄模式，除了需要使用Camera类之外，还需要使用MediaRecorder类，并使用一组方法来按照一定的步骤来进行视频拍摄。

（9）处理按钮拍照事件，设置自动对焦并进行拍照。参考代码如下：

```
public void onClick(View v){
    if(mPreviewRunning){ //判断是否可以进行拍照
      shutter.setEnabled(false);
      mCamera.autoFocus(new AutoFocusCallback(){ //设置自动对焦
       public void onAutoFocus(boolean success,Camera camera){ //聚焦拍照
       mCamera.takePicture(shutterCallback,null,pictureCallback);}});}}
```

（10）定义照相机捕捉图片的回调接口，可从中得到图片数据，将其保存为图片并显示出来。

```
PictureCallback mPictureCallback=new PictureCallback(){
    public void onPictureTaken(byte[] data,Camera camera1){
      if(data!=null){  //判定相片数据是否不为空
         savePicture(data);  //保存图片
         showPicture();  //显示图片
  }}};
public void savePicture(byte[] data){  //保存图片
   try{String imgId=System.currentTimeMillis()+"";  //图片id
      String pathName=Environment.getExternalStorageDirectory().getPath()+
PATH;  //相片保存路径
       File file=new File(pathName);  //创建文件
        if(!file.exists()){file.mkdirs();}
        pathName+="/"+imgId +".jpeg"; //创建文件
        file=new File(pathName);
        if(!file.exists()){file.createNewFile();}  //文件不存在则新建
           FileOutputStream fos=new FileOutputStream(file);
            fos.write(data);fos.close();
        }catch(Exception e){e.printStackTrace();}}
public void showPicture(){  //显示图片
    try{AlbumActivity album=new AlbumActivity();
     bm=album.loadImage(pathName);  //读取相片Bitmap
     iv.setImageBitmap(bm);  //设置到控件上显示
     iv.setVisibility(View.VISIBLE);sv.setVisibility(View.GONE);
     if(mPreviewRunning){  //停止照相机浏览
      mCamera.stopPreview();mPreviewRunning=false;}
      shutter.setEnabled(true);}catch(Exception e){e.printStackTrace();
}}
```

（11）定义相片被捕捉瞬间的回调接口，即快照。

```
ShutterCallback scb=new ShutterCallback(){public void onShutter(){
    Log.d("onShutter","快照回调函数.....");}};
```

（12）实现Surface被销毁时的回调方法。停止照相机预览，并回收照相机资源。

```
public void surfaceDestroyed(SurfaceHolder holder){
   if(mCamera!=null){mCamera.stopPreview();          //停止照相机预览
     mPreviewRunning=false;mCamera.release();      //回收照相机
     mCamera=null;}}
```

MediaStore类是媒体提供者，包含所有内部和外部存储设备可用的媒体元数据。属性ACTION_IMAGE_CAPTURE表示标准的启动照相机程序捕捉照片的Intent动作。属性ACTION_VIDEO_CAPTURE则表示标准的启动照相机程序捕捉视频的Intent动作。

调用camera拍照的最简便方式是调用系统功能，然后通过onActivityResult（）方法获得图像数据，进行自主处理。Camera的uri为android:media.action.IMAGE_CAPTURE，在程序中通过intent跳转到该uri可调用系统的camera硬件。

调用camera录制视频的便捷方式是启动ACTION_VIDEO_CAPTURE捕捉视频，然后从onActivityResult取出视频并显示到VideoView。

Camera典型应用案例：迷你照相机

该程序使用Camera进行拍照，当用户按下拍照键时，该应用会自动对焦，当对焦成功时拍下照片。该程序的界面布局大纲如图5-12所示。程序的界面布局activity_camera.xml代码如下：

图5-12　照相机案例布局大纲

```
<?xml version="1.0"encoding="utf-8"?>
<LinearLayout xmlns:android="http://schemas.
android.com/apk/res/android
"android:layout_width="fill_parent"android:
layout_height="fill_parent"
    android:orientation="horizontal">
    <!-- 照相机视频窗口 -->
    <SurfaceView
        android:id="@+id/surfaceView"
        android:layout_width="fill_parent"
        android:layout_height="fill_parent"
        android:layout_marginLeft="-20.0dip"
        android:layout_weight="1.0"/>
    <!-- 照片 -->
    <ImageView
        android:id="@+id/imageView"
        android:layout_width="fill_parent"
        android:layout_height="fill_parent"
        android:layout_weight="1.0"/>
    <!-- 快照 -->
```

```
<LinearLayout
    android:layout_width="72.0dip"
    android:layout_height="fill_parent"
    android:layout_gravity="center"
    android:background="@drawable/bg_camera_pattern"
    android:orientation="vertical">
    <Button
        android:id="@+id/btnPriview"
        android:layout_width="wrap_content"
        android:layout_height="wrap_content"
        android:layout_weight="1"
        android:text="预览"
        android:textSize="18sp"/>
     <Button
        android:id="@+id/btnRetakePic"
        android:layout_width="wrap_content"
        android:layout_height="wrap_content"
        android:layout_weight="1"
        android:text="重拍"
        android:textSize="18sp"/>
    <ImageView
        android:id="@+id/ivShutter"
        android:layout_width="wrap_content"
        android:layout_height="wrap_content"
        android:layout_gravity="center"
         android:layout_weight="1"
        android:clickable="true"
        android:src="@drawable/capture"/>
    <Button
        android:id="@+id/btnOpenAlbum"
        android:layout_width="wrap_content"
        android:layout_height="wrap_content"
        android:layout_weight="1"
        android:text="相册"
        android:textSize="18sp"/>
    <Button
        android:id="@+id/btnExit"
        android:layout_width="wrap_content"
        android:layout_height="wrap_content"
        android:layout_weight="1"
        android:text="退出"
        android:textSize="18sp"/>
```

```
    </LinearLayout>
</LinearLayout>
```

CameraActivity类的Java代码如下：

```
public class CameraActivity extends Activity implements OnClickListener,
SurfaceHolder.Callback{
    private static final String TAG="CameraActivity";
    private SurfaceView surfaceView;  //照相机视频浏览
    private ImageView imageView; //相片
    private SurfaceHolder surfaceHolder;
    private ImageView ivShutter; //快照按钮
    private Camera mCamera=null; //照相机
    private boolean mPreviewRunning;  //运行照相机浏览
    private static final int MENU_RETAKE_PICTURE=1;
    private static final int MENU_OPEN_ALBUM=2;
    private static final int MENU_EXIT=3;
    private Bitmap bitmap; //相片Bitmap
    private String pathName;
    private MediaPlayer shootMP;
    private Button btnPriview,btnRetakePic,btnOpenAlbum,btnExit;
    protected void onCreate(Bundle savedInstanceState){
        super.onCreate(savedInstanceState);
        setFullScreen();  //设置全屏幕
        setContentView(R.layout.activity_camera);  //设置布局文件
        initView(); //初始化组件
        createAlbumDir();
        setOnClickListener();addCallback();doPreview();}
    private void addCallback(){
        surfaceHolder=surfaceView.getHolder();
        surfaceHolder.addCallback(this);  //设置SurfaceHolder回调事件
        surfaceHolder.setType(SurfaceHolder.SURFACE_TYPE_PUSH_BUFFERS);}
    private void initView(){  //初始化组件
        surfaceView=(SurfaceView)findViewById(R.id.surfaceView);
        imageView=(ImageView)findViewById(R.id.imageView);
        imageView.setVisibility(View.GONE);
        ivShutter=(ImageView)findViewById(R.id.ivShutter);
        btnPriview=(Button)findViewById(R.id.btnPriview);
        btnRetakePic=(Button)findViewById(R.id.btnRetakePic);
        btnExit=(Button)findViewById(R.id.btnExit);
        btnOpenAlbum=(Button)findViewById(R.id.btnOpenAlbum);}
    private void setOnClickListener(){
        ivShutter.setOnClickListener(this);  //设置快照按钮事件
        btnPriview.setOnClickListener(this);
```

```
        btnRetakePic.setOnClickListener(this);
        btnExit.setOnClickListener(this);
        btnOpenAlbum.setOnClickListener(this);}
    private void createAlbumDir(){
        String pathName=Environment.getExternalStorageDirectory()
            .getPath()+"/"+getPackageName();
        File file=new File(pathName);  //创建文件
        if(!file.exists()){
            file.mkdirs();}}
    private void setFullScreen(){  //设置全屏
        requestWindowFeature(Window.FEATURE_NO_TITLE);  //隐藏标题
        getWindow().setFlags(WindowManager.LayoutParams.FLAG_FULLSCREEN,
            WindowManager.LayoutParams.FLAG_FULLSCREEN);}
    public void onClick(View v){  //处理按钮拍照事件，设置自动对焦并进行拍照
        switch(v.getId()){
        case R.id.ivShutter:doShutter();break;  //自动聚焦
        case R.id.btnPriview:doPreview();doReStartTakePicture();break;
        case R.id.btnRetakePic:doReStartTakePicture();break;
        case R.id.btnOpenAlbum:doOpenAlbum();  break;
        case R.id.btnExit:finish();break;}}
    private void doPreview(){addCallback();
        try{ //判断是否运行照相机，运行就停止掉
            if(mPreviewRunning){mCamera.stopPreview();}
            mCamera.setPreviewDisplay(surfaceHolder); //启动照相机
            mCamera.startPreview();  //开始预览
            mPreviewRunning=true;
      } catch(Exception e){e.printStackTrace();}}
    private void doShutter(){
        if(mPreviewRunning){  //判断是否可以进行拍照
          ivShutter.setEnabled(false);
          mCamera.autoFocus(afc);  //设置自动对焦}}
    //自动对焦的接口，对焦完毕则拍照
    private AutoFocusCallback afc=new Camera.AutoFocusCallback(){
        public void onAutoFocus(boolean success, Camera camera){
            if(success){takePicture();}}};
    private void takePicture(){mCamera.takePicture(shutter,raw,jpeg);}
    //未压缩的照片处理接口，作为一个参数传递给takePicture()函数
    private PictureCallback raw=new PictureCallback(){
        public void onPictureTaken(byte[] bytes, Camera camera){
            Log.d(TAG, "未压缩的照片处理接口!");}};
    //照相机图片拍照回调函数
    //参数data: 相片字节数组; camera: 照相机服务对象
    //定义照相机捕捉图片回调接口，可从中得到图片数据，将其保存为图片和显示图片操作
```

```
//JPG格式的照片处理接口，作为一个参数传递给takePicture()函数
private PictureCallback jpeg=new PictureCallback(){
    public void onPictureTaken(byte[] data,Camera camera){
        if(data!=null){  //判断相片数据是否不为空
            savePicture(data);  //保存相片
            showPicture();  //显示相片
            doSaveJpeg(data, camera);}}};
private void doSaveJpeg(byte[] data,Camera camera){
    //根据拍照所得的数据创建位图
    final Bitmap bm=BitmapFactory.decodeByteArray(data,0,data.length);
    //加载/layout/photo_save.xml文件对应的布局资源
    View saveDialog=getLayoutInflater()
        .inflate(R.layout.photo_save, null);
    final EditText photoName=(EditText)saveDialog
        .findViewById(R.id.phone_name);
    //获取saveDialog对话框上的ImageView组件
    ImageView show=(ImageView)saveDialog.findViewById(R.id.isShow);
    show.setImageBitmap(bm);  //显示刚刚拍得的照片
    //使用对话框显示saveDialog组件
    new AlertDialog.Builder(CameraActivity.this).setView(saveDialog)
        .setPositiveButton("保存", new DialogInterface.OnClickListener(){
          public void onClick(DialogInterface dialog, int which){
              //创建一个位于SD卡上的文件
              File file=new File(Environment
                  .getExternalStorageDirectory(),photoName
                  .getText().toString()+".jpg");
              FileOutputStream outStream=null;
              try{//打开指定文件对应的输出流
                 outStream=new FileOutputStream(file);
                 //把位图输出到指定文件中
                 bm.compress(CompressFormat.JPEG, 100, outStream);
                 outStream.close();
            } catch(IOException e){e.printStackTrace();}}

    }).setNegativeButton("取消", null).show();}
//定义相片被捕捉瞬间的回调接口，即快照
ShutterCallback shutter=new ShutterCallback(){
    public void onShutter(){
        Log.d("onShutter", "快照回调函数.....");
        //播放"咔嚓"声提醒用户拍照成功
        shootSound();}};
private void shootSound(){  //以编程方式播放照相机快门声
  AudioManager meng=(AudioManager)getSystemService(Context.AUDIO_SERVICE);
```

```
 int volume=meng.getStreamVolume(AudioManager.STREAM_NOTIFICATION);
      if(volume!= 0){if(shootMP==null)
          shootMP=MediaPlayer.create(getApplicationContext(), Uri.parse("file:///
system/media/audio/ui/camera_click.ogg"));
        if(shootMP!=null)  shootMP.start();}}
    public void surfaceChanged(SurfaceHolder holder, int format, int width,
        int height){  //SurfaceView改变时调用
      try{  //判断是否运行照相机，运行就停止掉
        if(mPreviewRunning){mCamera.stopPreview();}
        mCamera.setPreviewDisplay(holder);  //启动照相机
        mCamera.startPreview();  //开始预览
        mPreviewRunning=true;
    } catch(Exception e){e.printStackTrace();}}
    //SurfaceView创建时调用
    public void surfaceCreated(SurfaceHolder holder){
        setCameraParams();}
    public void setCameraParams(){  //设置Camera参数
      if(mCamera!=null){return;}
      mCamera=Camera.open();  //创建照相机，打开照相机
      Parameters params=mCamera.getParameters();  //设置照相机参数
      params.setFocusMode(Parameters.FOCUS_MODE_AUTO);  //拍照自动对焦
      params.setPreviewFrameRate(3);  //设置预览帧速率
      params.setPreviewFormat(PixelFormat.JPEG);  //设置预览格式
      params.set("jpeg-quality", 100);  //设置图片质量百分比
      //获取照相机支持图片分辨率
      List<Size>list=params.getSupportedPictureSizes();
      Size size=list.get(0);int w=size.width;int h=size.height;
      params.setPictureSize(w, h);  //设置图片大小
      //设置自动闪光灯
      params.setFlashMode(Camera.Parameters.FLASH_MODE_AUTO);}
    //实现Surface被销毁时的回调方法。停止照相机预览，并回收照相机资源
    public void surfaceDestroyed(SurfaceHolder holder){
        if(mCamera!=null){mCamera.stopPreview();  //停止照相机预览
            mPreviewRunning=false;mCamera.release();  //回收照相机
            mCamera=null;}}
    public boolean onCreateOptionsMenu(Menu menu){  //创建菜单
        menu.add(0, MENU_RETAKE_PICTURE, 0,"重拍");
        menu.add(0, MENU_OPEN_ALBUM, 0, "打开相册");
        menu.add(0, MENU_EXIT, 0, "退出");
        return super.onCreateOptionsMenu(menu);}
    //菜单事件
    public boolean onOptionsItemSelected(MenuItem item){
        if(item.getItemId()==MENU_RETAKE_PICTURE){
```

```
            doReStartTakePicture();return true;
        } else if(item.getItemId()==MENU_OPEN_ALBUM){doOpenAlbum();
        } else if(item.getItemId()==MENU_EXIT){finish();}
          return super.onOptionsItemSelected(item);}
      private void doOpenAlbum(){
        Intent intent=new Intent(this, AlbumActivity.class);
        startActivity(intent);}
      private void doReStartTakePicture(){  //重启照相机拍照
      setRequestedOrientation(ActivityInfo.SCREEN_ORIENTATION_PORTRAIT);
      setRequestedOrientation(ActivityInfo.SCREEN_ORIENTATION_LANDSCAPE);}
      public void savePicture(byte[] data){  //保存图片
          try{  //图片id
              String imgId=System.currentTimeMillis()+"";
          String pathName=Environment.getExternalStorageDirectory()
                .getPath()+"/"+getPackageName();
                File file=new File(pathName);  //创建文件
                if(!file.exists()){file.mkdirs();}
                pathName += "/"+imgId+".jpeg";
                file=new File(pathName);  //创建文件
                if(!file.exists()){file.createNewFile();}  //文件不存在则新建
                FileOutputStream fos=new FileOutputStream(file);
                fos.write(data);fos.close();
            } catch(Exception e){e.printStackTrace();}}
      public void showPicture(){  //显示图片
          try{AlbumActivity album=new AlbumActivity();
              bitmap=album.loadImage(pathName);  //读取相片Bitmap
              imageView.setImageBitmap(bitmap);  //设置到控件上显示
              imageView.setVisibility(View.VISIBLE);
              surfaceView.setVisibility(View.GONE);
              if(mPreviewRunning){mCamera.stopPreview();  //停止照相机浏览
                  mPreviewRunning=false;}
              ivShutter.setEnabled(true);
        } catch(Exception e){e.printStackTrace();}}}
```

在AndroidManifest.xml中照相机拍照权限和读写文件权限：

```
    <!-- 照相机拍照权限 -->
    <uses-permission android:name="android.permission.CAMERA"/>
    <uses-feature android:name="android.hardware.camera.autofocus"
android:glEsVersion="4"/>
    <uses-feature android:name="android.hardware.camera"
android:glEsVersion="4"/>
     <!-- 读写文件权限 -->
    <uses-permission android:name="android.permission.WRITE_EXTERNAL_STORAGE"/>
```

```
<uses-permission android:name="android.permission.MOUNT_UNMOUNT_FILESYSTEMS"/>
```

该程序的运行结果如图5-13所示。

图5-13　迷你照相机运行结果

5.2.7　使用 MediaRecorder 录制视频短片

MediaRecorder除了可用于录制音频之外，还可用于录制视频。使用MediaRecorder录制视频与录制音频的步骤基本相同，只是录制视频时不仅需要采集声音，还需要采集图像。为了让MediaRecorder录制时采集图像，应该在调用setAudioSource(int audio_source)方法时再调用setVideoSource(int video_source)方法来设置图像来源。除此之外，还需要调用setOutputFormat()方法设置输出文件格式，之后进行如下步骤：

（1）调用MediaRecorder对象的setVideoEncoder()、setVideoEncodingBitRate(int bitRate)、setVideoFrameRate设置所录制的视频的编码格式、编码位率、每秒多少帧等，这些参数将可以控制所录制的视频的品质、文件的大小。一般来说，视频品质越好，视频文件越大。

视频

控制摄像头录制视频应用实战

（2）调用MediaRecorder对象的setPreviewDisplay(Surface sv)方法设置使用哪个SurfaceView显示视频预览。

（3）之后的步骤基本与录制音频相同。

MediaRecorder典型应用案例：迷你录像机

该案例实现视频的录制，界面中提供了3个按钮用于控制开始、结束录制和退出。该程序的界面布局图大纲如图5-14所示。程序的界面布局代码activity_main.xml如下：

图5-14　录制视频短片案例布局大纲

```
<?xml version="1.0"encoding="utf-8"?>
<LinearLayout xmlns:android="http://schemas.android.com/apk/res/android"
    android:layout_width="fill_parent"
    android:layout_height="fill_parent"
    android:orientation="vertical">
    <SurfaceView
        android:id="@+id/surfaceView1"
        android:layout_width="match_parent"
        android:layout_height="640px"/>
    <LinearLayout
```

```
        android:layout_width="match_parent"
        android:layout_height="wrap_content"
        android:orientation="horizontal">
        <Button
            android:id="@+id/btnRecord"
            android:layout_width="wrap_content"
            android:layout_height="wrap_content"
            android:layout_weight="1"
            android:text="记录"
            android:textSize="26sp"/>
      <Button
            android:id="@+id/btnStop"
            android:layout_width="wrap_content"
            android:layout_height="wrap_content"
            android:layout_weight="1"
            android:text="停止"
            android:textSize="26sp"/>
      <Button
            android:id="@+id/btnExit"
            android:layout_width="wrap_content"
            android:layout_height="wrap_content"
            android:layout_weight="1"
            android:text="退出"
            android:textSize="26sp"/>
    </LinearLayout>
  </LinearLayout>
```

MediaRecorderVideo 类的 Java 代码如下：

```
  public class MediaRecorderVideo extends Activity implements SurfaceHolder.
Callback,OnClickListener
  {   private SurfaceView surfaceView;
      private SurfaceHolder surfaceHolder;
      private Button btnRecord,btnStop,btnExit;
      private MediaRecorder mediaRecorder;
      private File videoFile;  //系统的视频文件
      public void onCreate(Bundle savedInstanceState){
          super.onCreate(savedInstanceState);
          requestWindowFeature(Window.FEATURE_NO_TITLE);
          setContentView(R.layout.activity_main);
          initView();setOnClickListener();createVideoDir();}
      private void setOnClickListener(){
          btnRecord.setOnClickListener(this);
          btnStop.setOnClickListener(this);
```

```
        btnExit.setOnClickListener(this);}
    private void createVideoDir(){
        String pathName=Environment.getExternalStorageDirectory()
            .getPath()+"/"+getPackageName();
        File file=new File(pathName); //创建文件
        if(!file.exists()){file.mkdirs();}}
    private void initView()
    {   findView();
        surfaceHolder=surfaceView.getHolder();
        surfaceHolder.addCallback(this);
        surfaceHolder.setType(SurfaceHolder.SURFACE_TYPE_PUSH_BUFFERS);}
    private void findView(){
        btnRecord=(Button)findViewById(R.id.btnRecord);
        btnStop=(Button)findViewById(R.id.btnStop);
        btnExit=(Button)findViewById(R.id.btnExit);
        surfaceView=(SurfaceView)findViewById(R.id.surfaceView1);}
    private void initRecorder()
    {  //创建保存录制视频的视频文件
        String imgId=System.currentTimeMillis()+"";
        videoFile=new File(Environment.getExternalStorageDirectory()
            .getPath()+"/"+getPackageName()+"/"+imgId+".3gp");
        mediaRecorder=new MediaRecorder();  //获得recorder对象
        //设置预览框
        mediaRecorder.setPreviewDisplay(surfaceHolder.getSurface());
        //设置录制源
        mediaRecorder.setVideoSource(MediaRecorder.VideoSource.CAMERA);
        //设置输出格式
        mediaRecorder.setOutputFormat(MediaRecorder.OutputFormat.THREE_GPP);
        mediaRecorder.setVideoFrameRate(3);  //设置帧率,部分设备无效
        mediaRecorder.setVideoSize(800,480);  //设置视频大小,必须在编码格式之前
        //设置编码格式
        mediaRecorder.setVideoEncoder(MediaRecorder.VideoEncoder.H264);
        mediaRecorder.setOutputFile(videoFile.getAbsolutePath());}
    public void record()
    {  try{mediaRecorder.prepare();mediaRecorder.start();}
       catch(Exception e){e.printStackTrace();}}
    public void onClick(View v){
        switch(v.getId()){
        case R.id.btnRecord:
            initRecorder();  //初始化recorder
            record();  //开始录制
            break;
```

```
            case R.id.btnStop:
                mediaRecorder.stop();  //停止录制
                mediaRecorder.release();  //释放资源
                break;
            case R.id.btnExit:finish();break;}}}
```

在AndroidManifest.xml文件中增加如下CAMERA和WRITE_EXTERNAL_STORAGE权限：

```
<uses-permission android:name="android.permission.CAMERA"/>
<uses-permission android:name="android.permission.WRITE_EXTERNAL_STORAGE"/>
```

该程序的运行结果如图5-15所示。

图5-15　迷你录像机运行结果

5.3 案例——MediaPlayer播放器

在开发一些视频监控软件客户端、网络电视客户端时，要在Android客户端实时显示摄像头或者视频资源的画面，就会涉及多媒体部分的知识。Android中也提供了对流媒体播放的支持，下面的实例将在Android中实现MediaPlayer播放器。该程序的界面布局图大纲如图5-16所示。程序的界面布局代码activity_main.xml如下：

图5-16　MediaPlayer播放器

```
    <?xml version="1.0"encoding="utf-8"?>
    <LinearLayout xmlns:android="http://schemas.android.
com/apk/res/android"
        android:layout_width="match_parent"
        android:layout_height="match_parent"
        android:orientation="vertical">
        <Button
            android:id="@+id/resaudio"
            android:layout_width="match_parent"
            android:layout_height="wrap_content"
            android:background="#0000FF"
            android:text="@string/resaudio"
            android:textSize="26sp"
```

```
        android:textColor="#FFFFFF"/>
    <Button
        android:id="@+id/localaudio"
        android:layout_width="match_parent"
        android:layout_height="wrap_content"
        android:background="#00FF00"
        android:textSize="26sp"
        android:text="@string/localaudio"
        android:textColor="#0000FF"/>
    <Button
        android:id="@+id/localvideo"
        android:layout_width="match_parent"
        android:layout_height="wrap_content"
        android:background="#0000FF"
        android:textSize="26sp"
        android:text="@string/localvideo"
        android:textColor="#FFFFFF"/>
    <Button
        android:id="@+id/streamvideo"
        android:layout_width="match_parent"
        android:layout_height="wrap_content"
        android:background="#00FF00"
        android:textSize="26sp"
        android:text="@string/streamvideo"
        android:textColor="#0000FF"/>
</LinearLayout>
```

播放音频MediaPlayerAudio类的Java代码如下：

```
public class MediaPlayerAudio extends Activity{
    private static final String TAG="MediaPlayerAudio";
    private MediaPlayer mPlayer;
    private static final String MEDIA="media";
    private static final int LOCAL_AUDIO=1;
    private static final int STREAM_AUDIO=2;
    private static final int RES_AUDIO=3;
    private static final int LOCAL_VIDEO=4;
    private static final int STREAM_VIDEO=5;
    private String path;
    private TextView tx;
    public void onCreate(Bundle icicle){
        super.onCreate(icicle);
        tx=new TextView(this);
        setContentView(tx);
```

```
        Bundle extras=getIntent().getExtras();
        playAudio(extras.getInt(MEDIA));}
    private void playAudio(Integer media){
        try{switch(media){
            case RES_AUDIO:doPlayResAudio();
                tx.setText("正在播放资源音频……");break;
            case LOCAL_AUDIO:  doPlayLocalAudio();
                tx.setText("正在播放本地文件音频……");break;}
        }catch(Exception e){Log.e(TAG,"错误: "+e.getMessage(),e);}}
    private void doPlayResAudio(){
        /***:将音频文件上载到res/raw文件夹，并在MediaPlayer.create()方法中提供其
resid*/
        mPlayer=MediaPlayer.create(this, R.raw.night);
        mPlayer.start();}
    private void doPlayLocalAudio()throws IOException{
        /***: 将path变量设置为本地音频文件路径 */
        path="/mnt/sdcard/night.mp3";
        //String sdCard=Environment.getExternalStorageDirectory().getPath();
        //path=sdCard+File.separator+"night.mp3";
        if(path==""){//告诉用户提供音频文件URL
          Toast.makeText(getApplicationContext(),
                "请编辑媒体播放器音频活动,"+"并将路径变量设置为音频文件路径"
                    +"音频文件必须存储在SD卡上。", Toast.LENGTH_LONG)
                .show();}
        mPlayer=new MediaPlayer();mPlayer.setDataSource(path);
        mPlayer.prepare();mPlayer.start();}
    protected void onDestroy(){super.onDestroy();doDestroy();}
    private void doDestroy(){
        if(mPlayer!=null){mPlayer.reset();mPlayer.release();
            mPlayer=null;}}}
```

播放视频MediaPlayerVideo类的Java代码如下：

```
  public class MediaPlayerVideo extends Activity implements
OnBufferingUpdateListener, OnCompletionListener,
OnPreparedListener, OnVideoSizeChangedListener, SurfaceHolder.Callback{
      private static final String TAG="MediaPlayerVideo";
      private int mVideoWidth;
      private int mVideoHeight;
      private MediaPlayer mPlayer;
      private SurfaceView mPreview;
      private SurfaceHolder holder;
      private String path;
      private Bundle extras;
```

```
    private static final String MEDIA="media";
    private static final int LOCAL_AUDIO=1;
    private static final int STREAM_AUDIO=2;
    private static final int RESOURCES_AUDIO=3;
    private static final int LOCAL_VIDEO=4;
    private static final int STREAM_VIDEO=5;
    private boolean mIsVideoSizeKnown=false;
    private boolean mIsVideoReadyToBePlayed=false;
    /****首次创建活动时调用*/
    public void onCreate(Bundle icicle){
        super.onCreate(icicle);
        setContentView(R.layout.mediaplayer_video);
        mPreview=(SurfaceView)findViewById(R.id.surface);
        holder=mPreview.getHolder();holder.addCallback(this);
        holder.setType(SurfaceHolder.SURFACE_TYPE_PUSH_BUFFERS);
        extras=getIntent().getExtras();}
    private void playVideo(Integer Media){doCleanUp();
        try{switch(Media){
            case LOCAL_VIDEO:doPlayLocalVideo();break;
            case STREAM_VIDEO:doPlayStreamVideo();break;}
            execPlayVideo();
        }catch(Exception e){Log.e(TAG, "错误: "+e.getMessage(),e);}}
  private void execPlayVideo()throws IOException{
      //创建新的媒体播放器并设置侦听器
      mPlayer=new MediaPlayer();
      //mMediaPlayer.setDataSource(path);
     Uri uri=Uri.parse("android.resource://sziit.lihz.mediaplayerplayvideoaudio/"+
R.raw.video);
      mPlayer.setDataSource(getApplicationContext(), uri);
      mPlayer.setDisplay(holder);
      mPlayer.prepare();
      mPlayer.setOnBufferingUpdateListener(this);
      mPlayer.setOnCompletionListener(this);
      mPlayer.setOnPreparedListener(this);
      mPlayer.setOnVideoSizeChangedListener(this);
      mPlayer.setAudioStreamType(AudioManager.STREAM_MUSIC);}
  private void doPlayStreamVideo(){
      path="android.resource://sziit.lihz.mediaplayerplayvideoaudio/"+
 R.raw.video;
      if(path==""){
          //告诉用户提供媒体文件URL
          Toast.makeText(getApplicationContext(),
           "请编辑MediaPlayerVideo活动，并将路径变量设置为媒体文件URL",
```

```
                Toast.LENGTH_LONG).show();}}
    private void doPlayLocalVideo(){
        /**将path变量设置为本地媒体文件路径.*/
        path="/mnt/sdcard/sample.3gp";
        String sdCard=Environment.getExternalStorageDirectory().getPath();
        path=sdCard+File.separator+"sample.3gp";
        if(path==""){
            //告诉用户提供媒体文件URL
                Toast.makeText(getApplicationContext(),"请编辑MediaPlayerVideo活动并将
路径变量设置为媒体文件路径。您的媒体文件必须存储在SD卡上。",Toast.LENGTH_LONG).show();}}
      public void onBufferingUpdate(MediaPlayer arg0, int percent){
          Log.d(TAG, "onBufferingUpdate percent:"+percent);}
      public void onCompletion(MediaPlayer arg0){
          Log.d(TAG, "已完成播放！");}
      public void onVideoSizeChanged(MediaPlayer mp,int width,int height){
          Log.v(TAG, "视频尺寸发生改变！");
          if(width==0 || height==0){
              Log.e(TAG, "无效的视频宽度("+width+")或高度("+height+")");
              return;}
          mIsVideoSizeKnown=true;mVideoWidth=width;mVideoHeight=height;
          if(mIsVideoReadyToBePlayed && mIsVideoSizeKnown){
              startVideoPlayback();}}
      public void onPrepared(MediaPlayer mediaplayer){
          Log.d(TAG, "已准备好播放");
          mIsVideoReadyToBePlayed=true;
          if(mIsVideoReadyToBePlayed && mIsVideoSizeKnown){
              startVideoPlayback();}}
      public void surfaceChanged(SurfaceHolder surfaceholder, int i, int j, int k)
{Log.d(TAG, "surfaceChanged called");}
      public void surfaceDestroyed(SurfaceHolder surfaceholder){
          Log.d(TAG, "surfaceDestroyed called");}
      public void surfaceCreated(SurfaceHolder holder){
          Log.d(TAG, "surfaceCreated called");playVideo(extras.getInt(MEDIA));}
      protected void onPause(){
          super.onPause();releaseMediaPlayer();doCleanUp();}
      protected void onDestroy(){
          super.onDestroy();releaseMediaPlayer();doCleanUp();}
      private void releaseMediaPlayer(){
          if(mPlayer!=null){mPlayer.release();mPlayer=null;}}
      private void doCleanUp(){
          mVideoWidth=0;mVideoHeight=0;mIsVideoReadyToBePlayed=false;
          mIsVideoSizeKnown=false;}
      private void startVideoPlayback(){
```

```
        Log.v(TAG, "startVideoPlayback");
        holder.setFixedSize(mVideoWidth, mVideoHeight);
        mPlayer.start();}}
```

在 AndroidManifest.xml 文件中增加如下权限：

```
<uses-permission android:name="android.permission.INTERNET"/>
<uses-permission android:name="android.permission.WRITE_EXTERNAL_STORAGE"/>
<uses-permission
android:name="android.permission.MOUNT_UNMOUNT_FILESYSTEMS"/>
```

该程序的运行结果如图 5-17 所示。

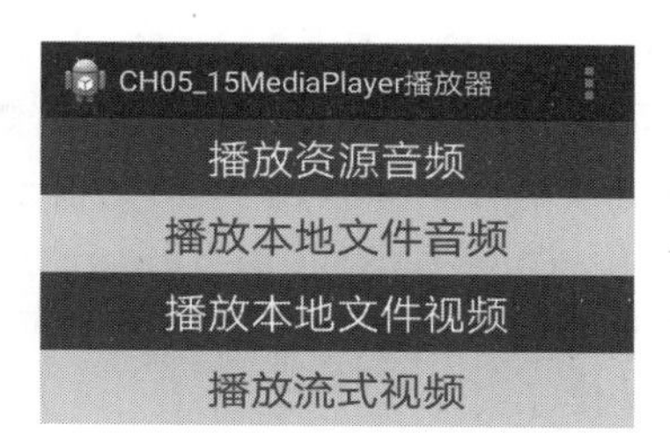

图 5-17 MediaPlayer 播放器运行结果

5.4 知识扩展

5.4.1 传感器知识

（1）加速度传感器（SENSOR_TYPE_ACCELEROMETER）。

（2）磁力传感器（SENSOR_TYPE_MAGNETIC_FIELD）。

（3）方向传感器（SENSOR_TYPE_ORIENTATION）。

（4）陀螺仪传感器（SENSOR_TYPE_GYROSCOPE）。

5.4.2 传感器的典型案例

（1）智能灯。

（2）老人防护仪。

本章小结

音频和视频都是非常重要的多媒体形式，Android 系统为音频、视频等多媒体的播放、录制提供了强大的支持。学习本章需要重点掌握如何使用 MediaPlayer 播放视频和音频、SoundPool 播放音频、MediaRecorder 用来录制音频和视频、利用 Camera 类编写照相机应用程序和利用 Intent 直接调用系统的照相机程序。

强化练习

一、填空题

1. MediaPlayer的3个主要的方法是（　　）、（　　）和（　　）。

2. MediaPlayer存在缺点有（　　）和（　　）。

3. （　　）主要用于播放一些较短的声音片段，与MediaPlayer相比，它的优势在于CPU资源占用量低和反应延迟小。

4. 为了在Android应用中播放视频，Android提供了（　　）组件，它就是一个位于android.widget包下的组件，它的作用与ImageView类似，只是ImageView用于显示图片，而它用于播放视频。

5. 使用VideoView播放视频简单、方便，但是早期开发者更喜欢使用MediaPlayer播放视频。但由于MediaPlayer主要用于播放音频，因此它没有提供图像输出界面，此时就需要借助于（　　）来显示MediaPlayer播放的图像输出。

6. 手持设备一般都提供了传声器硬件，而Android系统可以利用硬件来录制音频。为了在Android系统中录制音频，Android提供了（　　）类。

二、单选题

1. 采用多媒体播放器MediaPlayer播放raw目录下mp3音频文件（　　）。

   A. 直接调用start()方法，无须调用setDataSource方法设置文件源

   B. 需调用setDataSource()方法设置文件源

   C. 直接new MediaPlayer()就行

   D. 需使用MediaPlayer.create方法创建MediaPlayer

2. 用MediaPlayer对象播放音乐时，调用pause()方法实现功能为（　　）。

   A. 暂停音乐　　B. 重播音乐　　C. 快进或快退音乐　　D. 停止播放音乐

3. MediaPlayer播放SD卡音乐前，需要调用（　　）方法完成准备工作。

   A. start　　B. pause　　C. setDataSource　　D. prepare

4. 在播放音乐期间调用pause()和seekTo(0)两个方法将达到的效果为（　　）。

   A. 暂停播放，并释放音乐所占用的缓存

   B. 停止播放，并把播放位置移到音乐开始处

   C. 停止播放，并把播放位置移到音乐末尾

   D. 暂停播放，待下次播放时接着当前暂停的位置继续播放

三、简答题

1. 列举出Android中播放音频相关的类。

2. 列举出Android中播放视频相关的类。

3. 列举出Android中摄像头相关的类。

第6章 Android 系统服务应用

学习目标

- 掌握活动管理器（ActivityManager）的使用。
- 掌握警报管理器（AlarmManager）的使用。
- 掌握音频管理器（AudioManager）的使用。
- 掌握剪贴板管理器（ClipboardManager）的使用。
- 掌握通知管理器（NotificationManager）的使用。

视频

Android系统服务应用

6.1 学习导入

Android系统服务与前面所讲过的服务是不同的，系统服务不仅指服务组件，而且还包括Android系统提供的服务功能。它们的使用方式需要系统提供的特定方式来获取希望得到的服务功能接口，通过这些接口与系统的核心组件进行交互。通过Android系统服务接口可方便地获取系统信息，对系统功能进行集成。Android系统的服务接口都由Context类提供。利用Object android.app.Activity.getSystemService(String name)方法，开发者可通过指定的服务字符串标识获取相应的服务。下面通过表6-1了解在本章的学习中将会用到的服务标识符。

表6-1 Android 系统服务

服务字符串标识	说 明	服务字符串标识	说 明
ACTIVITY_SERVICE	管理 Activity 的服务	NOTIFICATION_SERVICE	通知服务
ALARM_SERVICE	闹钟（警报）服务	POWER_SERVICE	电源管理服务
AUDIO_SERVICE	音量控制服务	SEARCH_SERVICE	搜索服务
CLIPBOARD_SERVICE	剪贴板服务	SENSOR_SERVICE	传感器服务
CONNECTIVITY_SERVICE	连接管理服务	TELTPHONY_SERVICE	电话信息服务

续表

服务字符串标识	说　明	服务字符串标识	说　明
INPUT_METHOD_SERVICE	输入法服务	VIBRATOR_SERVICE	振动器服务
KEYGUARD_SERVICE	键盘锁定服务	WALLPAPER_SERVICE	墙纸服务
LAYOUT_INFLATER_SERVICE	布局填充服务	WIFI_SERVICE	Wi-Fi 服务
LOCATION_SERVICE	定位服务	WINDOW_SERVICE	窗口服务

下面详细讲解如何使用这些服务。

6.2 技术准备

6.2.1 活动管理器（ActivityManager）

视频

Android活动管理器原理及应用实战

ActivityManager是对所有运行中的Activity组件进行管理，在android.app包中。通过ActivityManager可以获取当前设备的配置信息、内存信息、进程错误状态、近期任务、运行中进程、运行中服务和运行中任务信息。通过getSystemService(Context.ACTIVITY_SERVICE)方法获取系统服务ActivityManager对象；在得到系统服务后，调用getDeviceConfigurationInfo()方法获取配置信息接口，在配置信息中包括系统中的多项配置信息，包括输入方式类型、键盘类型、导航方式类型、触摸屏方式类型等内容。

（1）输入方式类型。通过设置配置信息接口的reqInputFeatures属性可以获得当前设备的输入方式，其输入方式的类型在ConfigurationInfo接口中定义。一般有以下两种：

- INPUT_FEATURE_FIVE_WAY_NAV：五向导航键输入。
- INPUT_FEATURE_HARD_KEYBOARD：硬键盘输入。

（2）键盘类型。通过配置信息接口的reqKeyboardType属性可以获得当前设备的键盘类型。键盘类型主要有以下几种：

- KEYBOARD_UNDEFINED：未定义键盘。
- KEYBOARD_NOKEYS：无键键盘。
- KEYBOARD_QWERTY：打字机键盘。
- KEYBOARD_12KEY：十二键盘。

（3）导航方式类型。通过配置信息接口的reqNavigation属性可以获得当前设备的导航方式，有以下几种：

- NAVIGATION_UNDEFINED：未定义导航。
- NAVIGATION_DPAD：面板导航。
- NAVIGATION_TRACKBALL：定位球导航。
- NAVIGATION_WHEEL：滚轮导航。

（4）触摸屏方式类型。触摸屏类型的属性为reqTouchScreen，主要分为：

- TOUCHSCREEN_NOTOUCH：不支持触摸屏。
- TOUCHSCREEN_STYLUS：触摸笔。

- TOUCHSCREEN_FINGER：手指触摸。

除了调用配置信息接口获取系统的配置信息外，还可以通过 ActivityManager 中的 getMemoryInfo() 方法获取系统的内存信息接口。如下代码：

```
MemoryInfo memInfo=new MemoryInfo();service.getMemoryInfo(memInfo);
```

当获取到内存信息接口以后，就可以获取当前的内存信息，其接口的属性如下：

- availMen：可用内存。
- lowMemory：是否低内存。
- threshold：内存阈值。

除此之外，同样可以通过 getProgressesInErrorState() 方法获取系统进程的错误状态信息（ProcessErrorStateInfo）；通过 getRecentTasks() 方法可以获取系统的近期任务信息接口（RecentTaskInfo）。需要注意，在获取 getRecentTasks() 时，需要添加获取任务的使用许可，在 AndroidManifest.xml 文件中添加：

```
<uses-permission android:nam="android.permission.GET_TASKS">
```

ActivityManager 典型应用案例：手持设备活动管理器

程序实例 ActivityManagerTest 为对 ActivityManager 应用的一个简单实例，它在主程序中调用所需要信息，其代码如下：

```
public class ActivityManagerDemo extends Activity{
    private static final String TAG="ActivityManagerDemo";
    private TextView textView1;String strTxt=null;
    ActivityManager amService;  //活动管理器对象
    ActivityManager.MemoryInfo memoryInfo;  //内存信息对象
    ActivityManager.RunningAppProcessInfo runningAppProcessInfo;
    ActivityManager.ProcessErrorStateInfo processErrorStateInfo;
    /**第一次创建活动（activity）时调用 */
    protected void onCreate(Bundle savedInstanceState){
        super.onCreate(savedInstanceState);
        setContentView(R.layout.activity_main);
        textView1=(TextView)findViewById(R.id.textView1);
        //获得ActivityManager服务的对象
        amService=(ActivityManager)
            getSystemService(Context.ACTIVITY_SERVICE);
        //获得配置信息对象
        ConfigurationInfo cfgInfo=amService.getDeviceConfigurationInfo();
        if(cfgInfo.reqInputFeatures==ConfigurationInfo.INPUT_FEATURE_HARD_KEYBOARD)
{strTxt="硬键盘输入";}
      if(cfgInfo.reqInputFeatures==ConfigurationInfo.INPUT_FEATURE_FIVE_WAY_NAV)
{strTxt="五向导航键输入";}
      if(cfgInfo.reqKeyboardType==Configuration.KEYBOARD_UNDEFINED){
          strTxt+=",未定义键盘";}
```

```
        if(cfgInfo.reqKeyboardType==Configuration.KEYBOARD_NOKEYS){
          strTxt+=", 无键键盘";}
        textView1.setText(strTxt);
        getManager();  //01获取对象
        getMemoryInfo();  //02获取内存相关信息
        getRunningAppProcessInfo();  //03可用于判断程序是否前后台
        getTaskDescription();  //04获得任务描述信息
        getProcessErrorStateInfo();}  //05获得进程错误状态信息
    private void getManager(){  /***01获取ActivityManager对象*/
        Log.d(TAG, "01getManager()");
        Log.d(TAG, "getMemoryClass:"+amService.getMemoryClass());
        Log.d(TAG,"getLargeMemoryClass:"+
                amService.getLargeMemoryClass());
        Log.d(TAG,"getLauncherLargeIconSize:"+
                amService.getLauncherLargeIconSize());
        Log.d(TAG,"getLauncherLargeIconDensity:"+
                amService.getLauncherLargeIconDensity());
        Log.d(TAG, "activityManager--还有其他方法没有调用");}
    private void getMemoryInfo(){  /***获取内存相关信息*/
        Log.d(TAG, "02getMemoryInfo()");
        memoryInfo=new ActivityManager.MemoryInfo();
        amService.getMemoryInfo(memoryInfo);
        Log.d(TAG,"getLargeMemoryClass:"+
                amService.getLargeMemoryClass());
        Log.d(TAG, "getMemoryClass:"+amService.getMemoryClass());
        Log.d(TAG, "可用内存:"+(memoryInfo.availMem /1024 /1024)+"M");
        Log.d(TAG, "总内存:"+(memoryInfo.totalMem /1024 /1024)+"M");
        Log.d(TAG, "低内存:"+memoryInfo.lowMemory);
        Log.d(TAG, "阈值:"+memoryInfo.threshold);
        Log.d(TAG, "描述内容:"+memoryInfo.describeContents());}
    private void getRunningAppProcessInfo(){  /***可用于判断程序是否前后台*/
      Log.d(TAG, "03getRunningAppProcessInfo()");
      Log.d(TAG, "RunningAppProcessInfo集合: "
          +amService.getRunningAppProcesses().size());
      for(ActivityManager.RunningAppProcessInfo runningAppProcessInfo : amService.
getRunningAppProcesses()){
      this.runningAppProcessInfo=runningAppProcessInfo;
      ComponentName componentName1=
              runningAppProcessInfo.importanceReasonComponent;
      if(componentName1!=null){
          Log.d(TAG, "getClassName:"+componentName1.getClassName());
          Log.d(TAG, "getPackageName:"+componentName1.getPackageName());
          Log.d(TAG,  "getShortClassName:"+
```

```
                componentName1.getShortClassName());
        Log.d(TAG,"flattenToShortString:"  +
                componentName1.flattenToShortString());
        Log.d(TAG,"flattenToString:"+
                componentName1.flattenToString());
        Log.d(TAG,"toShortString:"+componentName1.toShortString());
        Log.d(TAG,"描述内容:"+componentName1.describeContents());}
        Log.d(TAG, "进程名:"+runningAppProcessInfo.processName);
        Log.d(TAG,  "describeContents():"
                 +runningAppProcessInfo.describeContents());
        Log.d(TAG, "importance:"+runningAppProcessInfo.importance);
        Log.d(TAG, "importanceReasonCode:"
              +runningAppProcessInfo.importanceReasonCode);
        Log.d(TAG, "importanceReasonPid:"
              +runningAppProcessInfo.importanceReasonPid);
        Log.d(TAG, "lastTrimLevel:"+
                runningAppProcessInfo.lastTrimLevel);
        Log.d(TAG, "lru:"+runningAppProcessInfo.lru);
        Log.d(TAG, "uid:"+runningAppProcessInfo.uid);
        for(int i=0; i<runningAppProcessInfo.pkgList.length; i++){
        Log.d(TAG, "runningAppProcessInfo.pkgList:"
              +runningAppProcessInfo.pkgList[i]);Log.d(TAG,"\n");}}}
    private void getTaskDescription(){
       Log.d(TAG, "04getTaskDescription()");
       String label=this.getResources().getString(
          this.getApplicationInfo().labelRes);
       int colorPrimary=
                this.getResources().getColor(R.color.colorPrimary);
       Bitmap icon=BitmapFactory.decodeResource(getResources(),
          R.drawable.ic_launcher);}
    private void getProcessErrorStateInfo(){
       Log.d(TAG, "\n");
       Log.d(TAG, "05getProcessErrorStateInfo()");
       Log.d(TAG,"processErrorStateInfo集合: "
                +amService.getProcessesInErrorState());
       if(amService.getProcessesInErrorState()!= null){
         for(ActivityManager.ProcessErrorStateInfo processErrorStateInfo:
amService.getProcessesInErrorState()){
        this.processErrorStateInfo=processErrorStateInfo;
        Log.d(TAG, "longMsg:"+processErrorStateInfo.longMsg);
        Log.d(TAG, "shortMsg:"+processErrorStateInfo.shortMsg);
        Log.d(TAG, "processName:"+
```

```
            processErrorStateInfo.processName);
        Log.d(TAG, "stackTrace:"+processErrorStateInfo.stackTrace);
        Log.d(TAG, "tag:"+processErrorStateInfo.tag);
        Log.d(TAG, "condition:"+processErrorStateInfo.condition);
        Log.d(TAG,  "describeContents():"
                +processErrorStateInfo.describeContents());
        Log.d(TAG, "pid:"+processErrorStateInfo.pid);
        Log.d(TAG, "uid:"+processErrorStateInfo.uid);}}}}
```

6.2.2 警报管理器（AlarmManager）

视频

报警管理器原理及应用实战

AlarmManager用于访问系统的警报服务，同样位于android.app包中。警报管理器允许用户预定自定义应用程序的运行时间。为了获取AlarmManager，同样需要调用getSystemService()方法，只是需要的参数为Context.ALARM_SERVICE。在获取到AlarmManager对象后，需要设置警报的时区，以便能准确计时，需要调用其setTimeZone()方法。

AlarmManager主要功能方法包括：

（1）void set(int type, long triggerAtMillis, PendingIntent operation)：设置一次性警报。

（2）void setRepeating(int type, long triggerAtMillis, long intervalMillis, PendingIntent operation)：设置周期性警报。

（3）void cancel(PendingIntent operation)：取消报警。

设置警报主要分为两种：一次性警报和周期性警报。一次性警报，顾名思义该提示框只会出现一次。通过警报管理器接口的set()方法确定一个警报时间。该方法包含3个参数：第一个参数type是时间标志，用以确定警报的计时和报警方式。其参数说明如下：

（1）ELAPSED_REALTIME：从系统启动开始计时（包括休眠时间）。

（2）ELAPSED_REALTIME_WAKEUP：从系统启动开始计时（包括休眠时间）并唤醒系统。

（3）RTC：以系统当前的时间戳计时（UTC格式）。

（4）RTC_WAKEUP：以系统当前的时间戳计时（UTC格式）并唤醒系统。

第二个参数triggerAtMillis是警报触发的时间点；第三个参数operation是一个未决意向，用于指明警报的处理方式。设置方式如下：

```
PendingIntent pi=PendingIntent.getBroadcast(context, 0, intent, 0);
am.set(AlarmManager.RTC_WAKEUP, System.currentTimeMillis(), pi);
```

周期性警报与一次性警报的区别就在于该提示框会按照指定的时间间隔出现。周期性警报是通过警报管理器的setRepeating()方法设置，该方法包含4个参数，与前面的设置相比多了一个警报时间的触发间隔时间参数。代码如下：

```
am.setRepeating(AlarmManager.RTC_WAKEUP, System.currentTimeMillis(), 1000 *1 ,
pi);
```

上面介绍了设置警报的知识，当然也可以取消警报设置。相比设置，取消警报比较简单，通过调用AlarmManager.cancel(Intent intent)即可。需要注意在AndroidManifest.xml添加：

```
<uses-permission android:name="android.permission.SET_TIME_ZONE"/>
```

AlarmManager典型应用案例：手持设备警报管理器

该案例利用AlarmManager控件实现手持设备的警报管理器，程序界面布局大纲如图6-1所示。

大纲
LinearLayout
LinearLayout
btnOneTime (Button) - "一次报警"
btnSetDialog (Button) - "设定目标时间报警"
LinearLayout
btnStart (Button) - "重复报警"
btnCancel (Button) - "取消重复报警"

图6-1 AlarmManager案例界面布局大纲

在程序中通过重写一个BroadcastReceiver来对广播事件进行监听。首先注册广播接收器，从而让程序知道有自定义的广播接收器组件，然后就是设置主程序。在主程序中就用到了AlarmManager。主程序的实现代码如下：

```
public class AlarmManagerDemo extends Activity implements OnClickListener{
    private static final String TAG="AlarmManagerDemo";
    private Button btnOneTime,btnSetDialog,btnStart,btnCancel;
    private AlarmListener alarm;
        public void onCreate(Bundle savedInstanceState){  //重写onCreate()方法
            super.onCreate(savedInstanceState);Log.d(TAG, "onCreate()");
            setContentView(R.layout.alarm_manager);findView();
            setOnClickListener(); alarm=new AlarmListener();}
        private void findView(){  //绑定按钮对象
        btnOneTime=(Button)findViewById(R.id.btnOneTime);
          btnSetDialog=(Button)findViewById(R.id.btnSetDialog);
          btnStart=(Button)findViewById(R.id.btnStart);
          btnCancel=(Button)findViewById(R.id.btnCancel);}
    private void setOnClickListener(){  //设置按钮单击监听器
        btnOneTime.setOnClickListener(this);
            btnSetDialog.setOnClickListener(this);
            btnStart.setOnClickListener(this);
            btnCancel.setOnClickListener(this);}
    public void onClick(View v){  //单击事件处理
        switch(v.getId()){
        case R.id.btnOneTime:onetimeTimer();break;  //启动一次报警
        case R.id.btnSetDialog:startSetDialog();break;  //设置目标报警时间
        case R.id.btnStart:startRepeatingTimer();break;  //启动重复报警
        case R.id.btnCancel:cancelRepeatingTimer();break;}}  //取消重复报警
    private void startRepeatingTimer(){  //启动重复报警
        Context context=this.getApplicationContext();
        if(alarm != null){alarm.SetAlarm(context);
        }else{
            Toast.makeText(context,"报警为空",Toast.LENGTH_SHORT).show();}}
    private void startSetDialog(){  //设置目标报警时间
        Intent intent=new Intent(this,TimeActivity.class);
            startActivity(intent);}
    public void cancelRepeatingTimer(){  //取消重复报警
      Context context=this.getApplicationContext();
```

```
        if(alarm!=null){alarm.CancelAlarm(context);
        }else{
          Toast.makeText(context,"报警为空",Toast.LENGTH_SHORT).show();}}
    private void onetimeTimer(){
        Context context=this.getApplicationContext();
        if(alarm!=null){alarm.setOnetimeTimer(context);
        }else{
          Toast.makeText(context,"报警为空",Toast.LENGTH_SHORT).show();}}}
```

AlarmReceiver类派生于BroadcastReceiver，其代码如下：

```
public class AlarmReceiver extends BroadcastReceiver{
    public void onReceive(Context arg0, Intent arg1){  //接收回调函数
        Toast.makeText(arg0, "收到警报!", Toast.LENGTH_LONG).show();
        Intent startIntent=new Intent(arg0, RingtoneService.class);
        arg0.startService(startIntent);  //启动服务}
    public void CancelAlarm(Context context,int RQS_1){  //取消报警
        Intent stopIntent=new Intent(context, RingtoneService.class);
        context.stopService(stopIntent);  //停止服务
        Intent intent=new Intent(context, AlarmReceiver.class);
        PendingIntent sender=PendingIntent.getBroadcast(context,0,intent, 0);
        AlarmManager am=(AlarmManager)
    context.getSystemService(Context.ALARM_SERVICE);
        am.cancel(sender); //取消报警}
    public void SetAlarm(Context context,Calendar targetCal,int RQS_1)
    {
        ArrayList<PendingIntent>intentArray=new ArrayList<PendingIntent>();
        AlarmManager am=(AlarmManager)context.getSystemService(Context.
 ALARM_SERVICE);
        Intent intent=new Intent(context, AlarmReceiver.class);
    PendingIntent pendingIntent=PendingIntent.getBroadcast(context,1,intent,0);
        am.set(AlarmManager.RTC_WAKEUP,targetCal.getTimeInMillis(),
            pendingIntent);
        intentArray.add(pendingIntent);}}
```

设定目标时间报警TimeActivity类代码如下：

```
public class TimeActivity extends Activity implements OnClickListener{
    private static final String TAG="TimeActivity";
    TimePicker timePicker;
    private Button btnStartSetDialog,btnCancleAlarm;
    private TextView tvAlarmPrompt;
    AlarmManager alarmManager;
    private TimePickerDialog timePickerDialog;
    private AlarmReceiver alarmReceiver=new AlarmReceiver();
```

```
static int RQS_1=1;
/**首次创建活动时调用 */
public void onCreate(Bundle savedInstanceState){ //重写onCreate()方法
    super.onCreate(savedInstanceState);Log.d(TAG, "onCreate()");
    setContentView(R.layout.activity_settime);
alarmManager=(AlarmManager)getSystemService(Context.ALARM_SERVICE);
    findViewById();setOnClickListener();}
private void findViewById(){ //绑定控件对象
    tvAlarmPrompt=(TextView)findViewById(R.id.tvAlarmPrompt);
    btnStartSetDialog=(Button)findViewById(R.id.btnStartSetDialog);
    btnCancleAlarm=(Button)findViewById(R.id.btnCancleAlarm);}
private void setOnClickListener(){ //设置按钮单击事件监听器
    btnStartSetDialog.setOnClickListener(this);
    btnCancleAlarm.setOnClickListener(this);}
public void onClick(View v){ //处理单击按钮事件
    switch(v.getId()){
    case R.id.btnStartSetDialog:tvAlarmPrompt.setText("");
        openTimePickerDialog(false);break;
    case R.id.btnCancleAlarm:cancelRepeatingTimer();break;}}
private void openTimePickerDialog(boolean is24r){ //打开时间设置对话框
    Calendar calendar=Calendar.getInstance(); //获得日历实例
    timePickerDialog=new TimePickerDialog(TimeActivity.this,
        onTimeSetListener, calendar.get(Calendar.HOUR_OF_DAY),
        calendar.get(Calendar.MINUTE), is24r);
    timePickerDialog.setTitle("设置报警时间"); //设置对话框标题
    timePickerDialog.show(); //显示对话框}
OnTimeSetListener onTimeSetListener=new OnTimeSetListener(){
    public void onTimeSet(TimePicker view, int hourOfDay, int minute){
        Calendar calNow=Calendar.getInstance();
        Calendar calSet=(Calendar)calNow.clone();
        calSet.set(Calendar.HOUR_OF_DAY, hourOfDay);
        calSet.set(Calendar.MINUTE, minute);
        calSet.set(Calendar.SECOND, 0);
        calSet.set(Calendar.MILLISECOND, 0);
        if(calSet.compareTo(calNow)<=0){
            //今天过了规定的时间，算到明天
            calSet.add(Calendar.DATE, 1);}
        RQS_1+=1;setAlarm(calSet);}};
private void setAlarm(Calendar targetCal){
    tvAlarmPrompt.setText("***\n"+"警报已设置@ "+targetCal.getTime()
        +"\n"+"***");
    alarmReceiver.SetAlarm(this, targetCal, RQS_1);}
private void cancelRepeatingTimer(){
```

```
        Context context=this.getApplicationContext();
        if(alarmReceiver!=null){alarmReceiver.CancelAlarm(context,RQS_1);
        } else{
          Toast.makeText(context,"报警为空",Toast.LENGTH_SHORT).show();}}}
```

最后，在AndroidManifest.xml中声明广播接收器和服务组件即可：

```
<receiver android:name="sziit.lihz.alarmmanager.AlarmReceiver"/>
<service android:name="sziit.lihz.alarmmanager.RingtoneService"/>
```

手持设备警报管理器运行结果如图6-2所示。

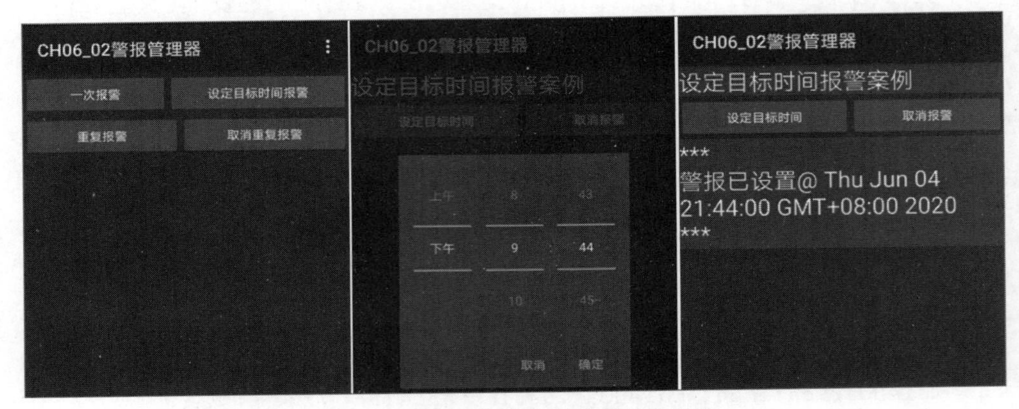

图 6-2　手持设备警报管理器运行结果

AlarmManager应用案例：手持设备壁纸定时切换

该程序在界面中放置3个按钮，第一个按钮用于启动定时更换壁纸，启动ChangeService服务调用WallpaperManager wManager=WallpaperManager.getInstance(this); 改变壁纸wManager.setResource(wallpapers[current++]);第二个按钮用于关闭定时更换壁纸，第三个按钮用于关闭App。该程序的界面布局大纲如图6-3所示，程序的界面布局代码activity_main.xml如下：

图 6-3　定时更换壁纸布局大纲

```
<?xml version="1.0"encoding="utf-8"?>
<LinearLayout xmlns:android="http://schemas.android.com/apk/res/android"
    android:layout_width="fill_parent"
    android:layout_height="fill_parent"
    android:orientation="horizontal">
    <Button
        android:id="@+id/btnStart"
        android:layout_width="wrap_content"
        android:layout_height="wrap_content"
        android:text="@string/start"/>
    <Button
        android:id="@+id/btnStop"
        android:layout_width="wrap_content"
```

```
        android:layout_height="wrap_content"
        android:text="@string/stop"/>
    <Button
        android:id="@+id/btnExit"
        android:layout_width="wrap_content"
        android:layout_height="wrap_content"
        android:text="@string/stop"/>
</LinearLayout>
```

AlarmChangeWallpaper类Java代码如下：

```
public class AlarmChangeWallpaper extends Activity implements OnClickListener{
    private static final String TAG="AlarmChangeWallpaper";
    private AlarmManager aManager;  //定义私有AlarmManager对象
    private Button btnStart, btnStop, btnExit;  //定义私有按钮控件对象
    PendingIntent pi;
    public void onCreate(Bundle savedInstanceState)  //重写onCreate()方法
    {   Log.v(TAG, "onCreate");
        super.onCreate(savedInstanceState);  //调用基类onCreate()方法
        setContentView(R.layout.activity_main);  //设置活动界面布局
        findViewById();setOnClickListener();
        //获得系统服务器对象
        aManager=(AlarmManager)getSystemService(Service.ALARM_SERVICE);
        //指定启动ChangeService组件
        Intent intent=new Intent(this,ChangeService.class);
        //创建PendingIntent对象
    pi=PendingIntent.getService(AlarmChangeWallpaper.this,0,intent, 0);}
    private void findViewById(){  //从布局文件查找按钮控件对象
        btnStart=(Button)findViewById(R.id.btnStart);
        btnStop=(Button)findViewById(R.id.btnStop);
        btnExit=(Button)findViewById(R.id.btnExit);}
    private void setOnClickListener(){  //设置按钮单击事件侦听器
        btnStart.setOnClickListener(this);  //设置启动按钮单击事件侦听器
        btnStop.setOnClickListener(this);  //设置停止按钮单击事件侦听器
        btnExit.setOnClickListener(this);}
    public void onClick(View v){  //初始按钮单击事件函数
        switch(v.getId()){
        case R.id.btnStart:startWakeUp(pi);break;  //启动定时更换壁纸
        case R.id.btnStop:stopWakeUp(pi);break;  //关闭定时更换壁纸
        case R.id.btnExit:finish();break;}}  //关闭应用程序
    private void startWakeUp(final PendingIntent pi){
        //设置每隔5s执行pi代表的组件一次
        aManager.setRepeating(AlarmManager.RTC_WAKEUP, 0, 5000, pi);
        btnStart.setEnabled(false);  //禁用start按钮
```

```
        btnStop.setEnabled(true);  //启动stop按钮
        Toast.makeText(AlarmChangeWallpaper.this, "壁纸定时更换启动成功啦",
            Toast.LENGTH_SHORT).show();
        Log.v(TAG, "壁纸定时更换启动成功啦");}
    private void stopWakeUp(final PendingIntent pi){
        //处理单击事件回调函数
        Log.v(TAG, "stop->onClick");
        btnStart.setEnabled(true);  //启用start按钮
        btnStop.setEnabled(false);  //禁用stop按钮
        aManager.cancel(pi);}}  //取消对pi的调度
```

ChangeService类派生于Service，其程序清单如下：

```
public class ChangeService extends Service  //从服务基类派生子类
{   private static final String TAG="ChangeService";
    //定义定时更换的壁纸资源
    private int[] wallpapers=new int[]{
        R.drawable.shuangta,  R.drawable.lijiang,R.drawable.qiao,
        R.drawable.shui,R.drawable.g,R.drawable.default_wallpaper,
        R.drawable.th,  R.drawable.timg3};
    //定义系统的壁纸管理服务
    private WallpaperManager wManager;
    private int current=0;  //定义当前所显示的壁纸
    //重写onStartCommand()方法
    public int onStartCommand(Intent intent, int flags, int startId)
    {   Log.v(TAG, "onStartCommand");
        //如果到了最后一张，系统重新开始
        if(current >= 8)  current=0;
        try{  //改变壁纸
           wManager.setResource(wallpapers[current++]);}
        catch(Exception e){e.printStackTrace();}
        return START_STICKY;}
    public void onCreate()  //重写onCreate()方法
    {  Log.v(TAG, "onCreate");
       super.onCreate();  //调用基类onCreate()方法
       //初始化WallpaperManager
       wManager=WallpaperManager.getInstance(this);}
    public IBinder onBind(Intent intent)  //重写onBind()方法
    {Log.v(TAG, "onBind");return null;}}
```

在AndroidManifest.xml中增加授予用户修改壁纸的权限：

```
<uses-permission android:name="android.permission.SET_WALLPAPER"/>
```

该程序运行结果如图6-4所示，启动定时更换壁纸后注意观察手机壁纸更换情况。

6.2.3 音频管理器（AudioManager）

图 6-4 定时更换壁纸运行结果

AudioManager 提供了访问音量和响铃模式的控制，定义在 android.media 包中。可以通过 AudioManager 获取和设置音频及音量。同前面其他管理器的使用方法相似，通过 getSystemService() 方法获取音频管理器接口，然后就可以调用 AudioManager 对象的方法实现对音量的设置和管理。

（1）获取音量设置。获取音量设置包括两种类型：获取系统最大音量和获取当前音量值。获取音量的最大值可以通过音量管理器接口的 getStreamMaxVolume() 方法获得指定音频流的音量最大值。该方法仅有的一个参数就是音频流类型：

视频

音频管理器原理及应用实战

- STREAM_VOICE_CALL：呼叫声音。
- STREAM_SYSTEM：系统声音。
- STREAM_RING：用于响铃的音频流。
- STREAM_MUSIC：用于音乐的音频流。
- STREAM_ALARM：用于警报的音频流。

（2）获取当前音量。通过 getStreamVolume() 方法可以获得指定音频流的当前音量设定值。其音频类型上面已经介绍过。

（3）调整音量设置。通过音量管理器接口的 adjustStreamVolume() 方法可以调整指定音频流的当前音量设定值。该方法有 3 个参数：第一个参数是音频流的类型；第二个参数是音量调节的方向；第三个参数是控制标志。第一个参数音频流不再赘述，第二个参数是音量调节的方向。音量调节方向的定义如下：

- ADJUST_LOWER：调低。
- ADJUST_RAISE：调高。
- ADJUST_SAME：不调整。

关于音量控制标志的参数如下：

- FLAG_ALLOW_RINGER_MODES：是否包含响铃模式选项。
- FLAG_PLAY_SOUND：当改变音量的时候是否播放声音。
- FLAG_REMOVE_SOUND_AND_VIBRATE：是否移除队列中的任何声音或震动。
- FLAG_SHOW_UI：是否显示音量调节滑动条。
- FLAG_VIBRATE：是否进入震动响铃模式。

AudioManager 的典型应用案例：音频管理器

该案例利用 AudioManager 控件实现典型的移动设备的音频管理器，程序界面布局大纲如图 6-5 所示。程序的界面布局代码 main_activity.xml 如下：

图 6-5 音频管理器布局大纲

```
<?xml version="1.0"encoding="utf-8"?>
<LinearLayout xmlns:android="http://schemas.android.com/apk/res/android"
    android:orientation="vertical"
    android:layout_width="fill_parent"
    android:layout_height="fill_parent">
  <TextView
```

```
        android:layout_width="fill_parent"
        android:layout_height="wrap_content"
        android:text="@string/app_name"
        android:padding="8sp"
        android:gravity="center_horizontal"/>
     <Spinner android:id="@+id/spnItems"
        android:layout_width="fill_parent"
        android:layout_height="wrap_content"
        android:entries="@array/audio_streams"/>
        <TableLayout
        android:layout_width="fill_parent"
        android:layout_height="wrap_content"
        android:stretchColumns="0,1">
        <TableRow>
          <Spinner android:id="@+id/spnDirections"
            android:layout_width="fill_parent"
            android:layout_height="wrap_content"
            android:entries="@array/directions"/>
          <Button android:id="@+id/btnSet"
            android:layout_width="fill_parent"
            android:layout_height="wrap_content"
            android:text="设置"/>
        </TableRow>
      </TableLayout>
    <SeekBar android:id="@+id/BAR_VOLUME"
        android:layout_width="fill_parent"
        android:layout_height="wrap_content"
        android:max="100"
        android:stepSize="10"
        android:progress="30"/>
    <EditText android:id="@+id/tvContents"
        android:layout_width="fill_parent"
        android:layout_height="fill_parent"
        android:scrollbars="vertical"
        android:editable="false"
        android:textSize="6pt"/>
  </LinearLayout>
```

在主程序中应用AudioManager，代码如下：

```
public class CH06_4 extends Activity implements OnClickListener,
OnItemSelectedListener{private static final String TAG="CH06_4";
  private EditText tvContents=null;
  private Spinner spnItems=null,spnDirections=null;
```

```
    private Button btnSet=null;
    private SeekBar mBarVolume=null;
    private AudioManager mService=null;
    /**首次创建活动时调用 */
    public void onCreate(Bundle savedInstanceState){  //重写onCreate()方法
        super.onCreate(savedInstanceState);
        setContentView(R.layout.main_activity);
        tvContents=(EditText)findViewById(R.id.tvContents);
        spnItems=(Spinner)findViewById(R.id.spnItems);
        spnDirections=(Spinner)findViewById(R.id.spnDirections);
        btnSet=(Button)findViewById(R.id.btnSet);
        mBarVolume=(SeekBar)findViewById(R.id.BAR_VOLUME);
        btnSet.setOnClickListener(this);
        spnItems.setOnItemSelectedListener(this);
        //获取音频服务管理器
        mService=(AudioManager)
          (this.getSystemService(Context.AUDIO_SERVICE));
        spnItems.setSelection(0); //初始化
        spnDirections.setSelection(0);}
    public void onItemSelected(AdapterView<?>parent,View v,int pos,long id)
{//当流类型选择改变时
        int selectedPos=spnItems.getSelectedItemPosition();
        String[] items=
   getResources().getStringArray(R.array.audio_streams);
        final String streamType=items[selectedPos];  int max;
        if(streamType.equalsIgnoreCase("STREAM_VOICE_CALL")){
            max=mService.getStreamMaxVolume(AudioManager.STREAM_VOICE_CALL);
              mBarVolume.setMax(max);mBarVolume.setProgress(max);}
          else if(streamType.equalsIgnoreCase("STREAM_SYSTEM")){
              max=mService.getStreamMaxVolume(AudioManager.STREAM_SYSTEM);
              mBarVolume.setMax(max);mBarVolume.setProgress(max);}
          else if(streamType.equalsIgnoreCase("STREAM_RING")){
              max=mService.getStreamMaxVolume(AudioManager.STREAM_RING);
              mBarVolume.setMax(max);mBarVolume.setProgress(max);}
          else if(streamType.equalsIgnoreCase("STREAM_MUSIC")){
              max=mService.getStreamMaxVolume(AudioManager.STREAM_MUSIC);
              mBarVolume.setMax(max);mBarVolume.setProgress(max);}
          else if(streamType.equalsIgnoreCase("STREAM_ALARM")){
              max=mService.getStreamMaxVolume(AudioManager.STREAM_ALARM);
          mBarVolume.setMax(max);mBarVolume.setProgress(max);}
        getSetting(items[selectedPos]);}
     private void getSetting(String streamType){  //获取设置
        int volume;
```

```
        if(streamType.equalsIgnoreCase("STREAM_VOICE_CALL")){
            volume=mService.getStreamVolume(AudioManager.STREAM_VOICE_CALL);
            mBarVolume.setProgress(volume);clearText();
            printText("STREAM_VOICE_CALL volume: "+volume+
                  "/"+mBarVolume.getMax());}
        else if(streamType.equalsIgnoreCase("STREAM_SYSTEM")){
            volume=mService.getStreamVolume(AudioManager.STREAM_SYSTEM);
            mBarVolume.setProgress(volume);clearText();
            printText("STREAM_SYSTEM volume: "+volume+
                  "/"+mBarVolume.getMax());}
        else if(streamType.equalsIgnoreCase("STREAM_RING")){
            volume=mService.getStreamVolume(AudioManager.STREAM_RING);
            mBarVolume.setProgress(volume);  clearText();
            printText("STREAM_RING volume: "+volume+
                  "/"+mBarVolume.getMax());}
        else if(streamType.equalsIgnoreCase("STREAM_MUSIC")){
            volume=mService.getStreamVolume(AudioManager.STREAM_MUSIC);
            mBarVolume.setProgress(volume);  clearText();
            printText("流音乐音量: "+volume+"/"+mBarVolume.getMax());}
        else if(streamType.equalsIgnoreCase("STREAM_ALARM")){
            volume=mService.getStreamVolume(AudioManager.STREAM_ALARM);
            mBarVolume.setProgress(volume);  clearText();
            printText("流报警音量: "+volume+"/"+mBarVolume.getMax());}}
    public void onNothingSelected(AdapterView<?>parent){
        //TODO Auto-generated method stub}
    public void onClick(View v){  //处理按钮单击事件回调函数
        switch(v.getId()){
          case R.id.btnSet:{doSet();break;}}}
    private void doSet(){  //执行设置
        //TODO Auto-generated method stub
        int selectedPos=spnItems.getSelectedItemPosition();
        int selectedPos2=spnDirections.getSelectedItemPosition();
        String[] items=
  getResources().getStringArray(R.array.audio_streams);
        String[] directions=
  getResources().getStringArray(R.array.directions);
        setSetting(items[selectedPos], directions[selectedPos2]);}
    //调整所选流类型的音量大小
    private void setSetting(String streamType, String direction){

        int direction2=0;  int volume;
        if(direction.equalsIgnoreCase("ADJUST_LOWER")){  //调整方向
             direction2=AudioManager.ADJUST_LOWER;
```

```
            System.out.print("Direction is ADJUST_LOWER");}
        else if(direction.equalsIgnoreCase("ADJUST_RAISE")){
            direction2=AudioManager.ADJUST_RAISE;
            System.out.print("Direction is ADJUST_RAISE");}
        else if(direction.equalsIgnoreCase("ADJUST_SAME")){
            direction2=AudioManager.ADJUST_SAME;
            System.out.print("Direction is ADJUST_SAME");}
        if(streamType.equalsIgnoreCase("STREAM_VOICE_CALL")){
            mService.adjustStreamVolume(AudioManager.STREAM_VOICE_CALL,
               direction2,AudioManager.FLAG_REMOVE_SOUND_AND_VIBRATE);
               volume=mService.getStreamVolume(AudioManager.STREAM_VOICE_CALL);
            mBarVolume.setProgress(volume);clearText();
            printText("STREAM_VOICE_CALL volume: "+volume+
                 "/"+mBarVolume.getMax());}
        else if(streamType.equalsIgnoreCase("STREAM_SYSTEM")){
            mService.adjustStreamVolume(AudioManager.STREAM_SYSTEM,
            direction2,AudioManager.FLAG_REMOVE_SOUND_AND_VIBRATE);
             volume=mService.getStreamVolume(AudioManager.STREAM_SYSTEM);
            mBarVolume.setProgress(volume);clearText();
            printText("STREAM_SYSTEM volume: "+volume+
                 "/"+mBarVolume.getMax());}
        else if(streamType.equalsIgnoreCase("STREAM_RING")){
            mService.adjustStreamVolume(AudioManager.STREAM_RING,
              direction2,AudioManager.FLAG_REMOVE_SOUND_AND_VIBRATE);
            volume=mService.getStreamVolume(AudioManager.STREAM_RING);
            mBarVolume.setProgress(volume);clearText();
            printText("STREAM_RING volume: "+volume+
                 "/"+mBarVolume.getMax());}
        else if(streamType.equalsIgnoreCase("STREAM_MUSIC")){
            mService.adjustStreamVolume(AudioManager.STREAM_MUSIC,
               direction2,AudioManager.FLAG_REMOVE_SOUND_AND_VIBRATE);
            volume=mService.getStreamVolume(AudioManager.STREAM_MUSIC);
            mBarVolume.setProgress(volume);clearText();
            printText("STREAM_MUSIC volume: "+volume+
                 "/"+mBarVolume.getMax());}
        else if(streamType.equalsIgnoreCase("STREAM_ALARM")){
            mService.adjustStreamVolume(AudioManager.STREAM_ALARM,
              direction2,AudioManager.FLAG_REMOVE_SOUND_AND_VIBRATE);
              volume=mService.getStreamVolume(AudioManager.STREAM_ALARM);
            mBarVolume.setProgress(volume);clearText();
            printText("STREAM_ALARM volume: "+volume+
                 "/"+mBarVolume.getMax());}}
    private void clearText(){tvContents.setText("");}
```

```
    private void printText(String text){
        tvContents.append(text);tvContents.append("\n");}};
```

该程序运行结果如图6-6所示。

AudioManager应用案例：手持设备音频控制

该案例提供了4个按钮，分别用于音乐的播放、音乐音量的增加、音乐音量的减少和关闭应用程序。该程序界面布局大纲如图6-7所示。

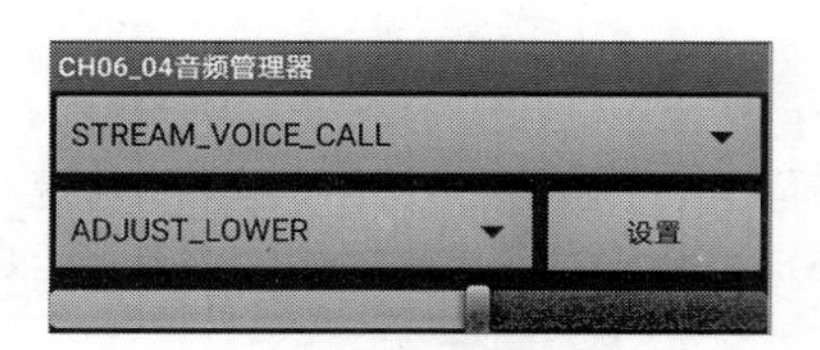

图6-6　音频管理器运行结果

图6-7　音频管理应用案例

程序的界面布局代码如下：

```
<?xml version="1.0"encoding="utf-8"?>
<LinearLayout xmlns:android="http://schemas.android.com/apk/res/android"
    android:orientation="vertical"
    android:layout_width="fill_parent"
    android:layout_height="fill_parent"
    android:gravity="center_horizontal">
<Button
    android:id="@+id/btnPlay"
    android:layout_width="fill_parent"
    android:layout_height="wrap_content"
    android:text="@string/play"/>
<LinearLayout
    android:orientation="horizontal"
    android:layout_width="fill_parent"
    android:layout_height="fill_parent"
    android:gravity="center_horizontal">
<Button
    android:id="@+id/btnUp"
    android:layout_width="wrap_content"
    android:layout_height="wrap_content"
    android:text="@string/up"/>
<Button
    android:id="@+id/btnDown"
    android:layout_width="wrap_content"
    android:layout_height="wrap_content"
    android:text="@string/down"/>
<ToggleButton
```

```
    android:id="@+id/mute"
    android:layout_width="wrap_content"
    android:layout_height="wrap_content"
    android:textOn="@string/normal"
    android:textOff="@string/mute"/>
<Button
    android:id="@+id/btnExit"
    android:layout_width="wrap_content"
    android:layout_height="wrap_content"
    android:text="@string/exit"/>
</LinearLayout>
</LinearLayout>
```

AudioManagerDemo类程序的Java代码如下：

```
public class AudioManagerDemo extends Activity implements OnClickListener
{   private static final String TAG="AudioManagerDemo";
    private Button btnPlay,btnUp,btnDown,btnExit;  //声明4个按钮控件对象
    private ToggleButton mute;  //声明开关按钮对象
    private AudioManager aManager;  //声明音频管理器对象
    private MediaPlayer mPlayer;
    public void onCreate(Bundle savedInstanceState)  //子类重写onCreate()方法
    {   Log.v(TAG, "onCreate");
        super.onCreate(savedInstanceState);  //调用基类onCreate()方法
        setContentView(R.layout.activity_main);  //设置活动界面布局
        //获取系统的音频服务
        aManager=(AudioManager)getSystemService(Service.AUDIO_SERVICE);
        findView();  setOnClickListener();
        //设置按钮状态改变监听器
        mute.setOnCheckedChangeListener(new OnCheckedChangeListener(){
            public void onCheckedChanged(CompoundButton source,
                boolean isChecked){Log.v(TAG, "onCheckedChanged");
        //指定调节音乐的音频，根据isChecked确定是否需要静音
        aManager.setStreamMute(AudioManager.STREAM_MUSIC, isChecked);}});}
  public void onClick(View v){  //处理按钮单击事件回调
       switch(v.getId()){
       case R.id.btnPlay:doPlay();break;  //播放音乐
       case R.id.btnUp:doUp();break;  //调高音量
       case R.id.btnDown:doDown();break;  //调低音量
       case R.id.btnExit:doExit();break;}}  //关闭应用程序
    private void findView(){  //绑定按钮对象
        //获取界面中4个按钮和一个ToggleButton控件
        btnPlay=(Button)findViewById(R.id.btnPlay);
        btnUp=(Button)findViewById(R.id.btnUp);
```

```
        btnDown=(Button)findViewById(R.id.btnDown);
        btnExit=(Button)findViewById(R.id.btnExit);
        mute=(ToggleButton)findViewById(R.id.mute);}
    private void setOnClickListener(){  //设置按钮单击监听器
        btnPlay.setOnClickListener(this);
        btnUp.setOnClickListener(this);
        btnDown.setOnClickListener(this);
        btnExit.setOnClickListener(this);}
    private void doPlay(){Log.v(TAG, "onClick");
        //初始化MediaPlayer对象，准备播放音乐
        mPlayer=MediaPlayer.create(AudioManagerDemo.this,R.raw.night);
        mPlayer.setLooping(true);  //设置循环播放
        mPlayer.start();  //开始播放}
    private void doUp(){Log.v(TAG, "onClick");
        //指定调节音乐的音频，增大音量，而且显示音量图形示意
        aManager.adjustStreamVolume(AudioManager.STREAM_MUSIC,
            AudioManager.ADJUST_RAISE, AudioManager.FLAG_SHOW_UI);}
    private void doDown(){Log.v(TAG, "onClick");
        //指定调节音乐的音频，降低音量，而且显示音量图形示意
        aManager.adjustStreamVolume(AudioManager.STREAM_MUSIC,
            AudioManager.ADJUST_LOWER, AudioManager.FLAG_SHOW_UI);}
    private void doExit(){
        if(mPlayer!=null){mPlayer.stop();  //停止播放
          mPlayer.release();  //释放资源
          mPlayer=null;finish();}}  //关闭应用
    public boolean onKeyDown(int keyCode, KeyEvent event){
        switch(keyCode){
        case KeyEvent.KEYCODE_VOLUME_UP:
           aManager.adjustStreamVolume(AudioManager.STREAM_MUSIC, AudioManager.
ADJUST_RAISE, 0);  return true;
        case KeyEvent.KEYCODE_VOLUME_DOWN:
           aManager.adjustStreamVolume(AudioManager.
STREAM_MUSIC, AudioManager.
ADJUST_LOWER, 0);return true;}
    return super.onKeyDown(keyCode, event);}}
```

视频

剪贴板管理器应用实战

手持设备音频控制运行结果如图6-8所示。

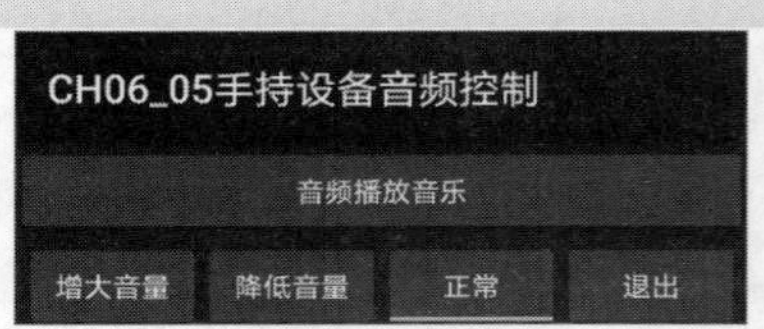

图 6-8　手持设备音频控制运行结果

6.2.4　剪贴板管理器（Clipboard-Manager）

操作 Android 系统剪贴板的类在 android.content 包下，主要包含下面3个类：

（1）ClipboardManager：表示一个剪贴板管理器。

（2）ClipData：剪贴板中保存的所有剪贴数据集（剪贴板可同时复制/保存多条多种数据条目）。

（3）ClipData.Item：剪贴数据集中的一个数据条目。

1. 获取系统剪贴板

通过系统服务接口就可得到ClipboardManager对象，示例代码如下：

```
ClipboardManager clipboard =(ClipboardManager)getSystemService(Context.
CLIPBOARD_SERVICE)
```

2. 复制

获取ClipboardManager对象后，将数据复制到剪贴板的操作如下：

```
//创建一个剪贴数据集，包含一个普通文本数据条目
ClipData clipData=ClipData.newPlainText("","需要复制的文本数据");
clipboard.setPrimaryClip(clipData);  //把数据集设置（复制）到剪贴板
```

3. 粘贴

将剪贴板数据粘贴到EditText etContent对象的操作如下：

```
ClipData clipData=clipboard.getPrimaryClip();  //获取剪贴板的剪贴数据集
if(clipData != null && clipData.getItemCount()>0){
    //从数据集中获取（粘贴）第一条文本数据
    CharSequence text=clipData.getItemAt(0).getText();
    etContent.setText(text);
}
```

4. 剪贴板的数据改变监听

```
//添加剪贴板数据改变监听器
clipboard.addPrimaryClipChangedListener(new ClipboardManager.
OnPrimaryClipChangedListener(){
    public void onPrimaryClipChanged(){
        //剪贴板中的数据被改变，此方法将被回调。使用方法详见下面的案例代码}});
//移除指定的剪贴板数据改变监听器
clipboard.removePrimaryClipChangedListener(listener);
```

下面通过案例来学习ClipboardManager的使用。源代码清单如下所示：

```
public class MainActivity extends Activity implements OnClickListener{
    private ClipboardManager cm;  //系统剪贴板
    private EditText etContent;
    private Button btnCopyClipboard,btnClearClipboard,btnPasete,btnExit;
    private ClipData clipData;  //剪贴数据集
    /***首次创建活动时调用*/
    public void onCreate(Bundle savedInstanceState){  //重写onCreate()方法
        super.onCreate(savedInstanceState);
        setContentView(R.layout.activity_main);
        findView();setOnClickListener();
        //获取系统剪贴板
```

```
        cm=(ClipboardManager)getSystemService(CLIPBOARD_SERVICE);
        //设置事件侦听器
        cm.addPrimaryClipChangedListener(new
  UpdateTextfieldOnClipboardChangedListener());
        updateClipboardContent();}
    private void findView(){
        btnClearClipboard=(Button)findViewById(R.id.btnClear);
        btnCopyClipboard=(Button)findViewById(R.id.btnCopy);
        btnPasete=(Button)findViewById(R.id.btnPasete);
        btnExit=(Button)findViewById(R.id.btnExit);
        etContent=(EditText)findViewById(R.id.etContent);}
    private void setOnClickListener(){
        btnClearClipboard.setOnClickListener(this);
        btnCopyClipboard.setOnClickListener(this);
        btnPasete.setOnClickListener(this);
        btnExit.setOnClickListener(this);}
    private void updateClipboardContent(){ClipData primaryClip;
        if(cm.hasPrimaryClip()){primaryClip=cm.getPrimaryClip();
            etContent.setText(primaryClip.getItemAt(0).coerceToText(getApplicati
onContext()));
        } else{etContent.setText("");}}
    protected void onResume(){
        super.onResume();updateClipboardContent();}
    private class UpdateTextfieldOnClipboardChangedListener
        implements ClipboardManager.OnPrimaryClipChangedListener{
        public void onPrimaryClipChanged(){
            updateClipboardContent();}}
     public void onClick(View v){
        switch(v.getId()){
        case R.id.btnCopy:
            //创建一个剪贴数据集，包含一个普通文本数据条目
            clipData=ClipData.newPlainText("", etContent.getText());
            cm.setPrimaryClip(clipData);  //把数据集设置（复制）到剪贴板
            break;
        case R.id.btnPasete:
            clipData=cm.getPrimaryClip();  //获取剪贴板的剪贴数据集
            if(clipData!=null && clipData.getItemCount()>0){
                //从数据集中获取（粘贴）第一条文本数据
                CharSequence text=clipData.getItemAt(0).getText();
                etContent.setText(text);}  break;
        case R.id.btnClear:
            cm.setPrimaryClip(ClipData.newPlainText("", ""));break;
        case R.id.btnExit:finish();break;}}}
```

剪贴板程序运行结果如图 6-9 所示。

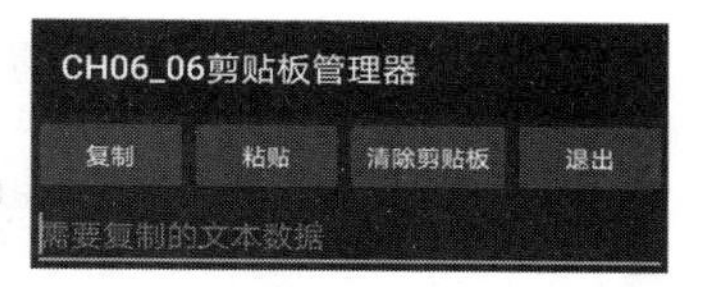

图 6-9　剪贴板程序运行结果

6.2.5　通知管理器（NotificationManager）

NotificationManager 用于通知用户有后台事件发生，其定义于 android.app 包中。当有后台事件发生时，通知管理器会对用户有个提醒。提醒的方式有如下 3 种：

（1）在状态栏中会出现持久的图标，用户可以单击该图标查看通知详情。

（2）屏幕开启或者闪烁。

（3）通过背景灯闪烁、播放声音或者震动的方式。

视频

Android通知管理器应用实战

首先需要获取 NotificationManager 对象，通过 getSystemService() 方法获取通知管理器，然后调用 NotificationManager 对象的 notify() 方法可以发送后台事件通知。在 notify() 方法中包含两个参数：第一个参数是通知 ID；第二个参数是通知实体。这里需要注意，通知的发送也是一种预期行为，用户可以在通知发送的时限内取消发送，所以通知的发送也需要使用未决意向对象。

在发送通知时需要定义一个通知实体，也就是 Notification 对象。注意，实体定义使用了 3 个参数：第一个参数是图标资源 ID；第二个参数是提示文字；第三个参数是通知发送的时间点。下面结合实例进行学习。源代码清单如下。

```
public class CH06_07 extends Activity implements OnClickListener{
    private static final String TAG="CH06_07";
    private Button mBtnRegister,mBtnDo,mBtnUnregister;
    private Button btnExit,btnCreate,btnCurTime;
    private NotificationManager mService=null;
    private Notification mNotification=null;
    public static final int NOTIFICATION_ID=1;
    /** 首次创建活动时调用 */
    public void onCreate(Bundle savedInstanceState){
        super.onCreate(savedInstanceState);
        setContentView(R.layout.main_activity);
        findView();setOnClickListener();init();
        Calendar c=Calendar.getInstance();
        SimpleDateFormat sdf=new SimpleDateFormat("dd:MMMM:yyyy HH:mm:ss a");
        String date=sdf.format(c.getTime());
        int hour=c.get(Calendar.HOUR_OF_DAY);
        btnCurTime.setText(String.valueOf(c.get(Calendar.HOUR_OF_DAY)));
        if(hour==11 || hour==12){
            btnCurTime.setText("午餐时间！");}}
    private void findView(){
        mBtnRegister=(Button)findViewById(R.id.mBtnRegister);
        mBtnDo=(Button)findViewById(R.id.mBtnDo);
        mBtnUnregister=(Button)findViewById(R.id.mBtnUnregister);
        btnCreate=(Button)findViewById(R.id.btnCreate);
```

```
        btnCurTime=(Button)findViewById(R.id.btnCurTime);
        btnExit=(Button)findViewById(R.id.btnExit);}
    private void setOnClickListener(){
        mBtnRegister.setOnClickListener(this);
        mBtnDo.setOnClickListener(this);
        mBtnUnregister.setOnClickListener(this);
        btnCreate.setOnClickListener(this);
        btnCurTime.setOnClickListener(this);
        btnExit.setOnClickListener(this);}
    private void init(){  //初始化通知服务
        mService=(NotificationManager)
            getSystemService(Context.NOTIFICATION_SERVICE);
        mNotification=new Notification(R.drawable.tip, "提醒",
            System.currentTimeMillis());
        setStates(false);}
    public void onClick(View v){
        switch(v.getId()){
        case R.id.mBtnRegister:doRegister();break;
        case R.id.mBtnDo:doNotify();  break;
        case R.id.mBtnUnregister:doUnregister();break;
        case R.id.btnCreate:createNotification();break;
        case R.id.btnExit:finish();break;}}
    private void doRegister(){  //注册通知侦听
        Intent notifyIntent=new Intent(this, RemindAct.class);
        notifyIntent.setFlags(Intent.FLAG_ACTIVITY_NEW_TASK);
        PendingIntent contentIntent=PendingIntent.getActivity(this, 0,
            notifyIntent, 0);
        mNotification.setLatestEventInfo(this.getApplicationContext(),
    "温馨提醒", "该睡觉了！", contentIntent);
        setStates(true);}
    private void setStates(boolean isRegistered){  //设置按钮状态
        //TODO Auto-generated method stub
        mBtnRegister.setEnabled(!isRegistered);
        mBtnUnregister.setEnabled(isRegistered);
        mBtnDo.setEnabled(isRegistered);}
    private void doNotify(){  //发出通知
        mService.notify(NOTIFICATION_ID, mNotification);}
    private void doUnregister(){  //注销通知侦听
        mService.cancel(NOTIFICATION_ID);setStates(false);}
    private void createNotification(){
        Intent intent=new Intent(this, RemindAct.class);
        PendingIntent pi=PendingIntent.getActivity(this, 0,
            intent, 0);
```

```
Notification noti=new Notification.Builder(this)
     .setContentTitle("快点! 午餐时间! ")
     .setContentText("主题").setSmallIcon(R.drawable.ic_launcher)
     .setContentIntent(pi)
     .addAction(R.drawable.image2, "呼叫", pi).build();
NotificationManager notificationManager=(NotificationManager)
      getSystemService(NOTIFICATION_SERVICE);
noti.defaults|=Notification.DEFAULT_SOUND;
noti.flags|=Notification.FLAG_AUTO_CANCEL;
notificationManager.notify(0, noti);}};
```

在 doRegister() 方法中还指定了通知实体所“呈报"的组件，这里指定 RemindAct 组件展现通知内容。该组件是 Activity 组件，与调用它的组件不存在关联，所以在 Intent 中添加了一个标志：FLAG_ACTIVITY_NEW_TASK。setLatestEventInfo() 方法用于设置通知实体的概要信息，同时将通知实体与意向对象绑定。通知实体的概要信息会在单击状态栏图标后展开的界面中显示。最后可以调用 NotificationManager 对象的 cancel() 方法取消通知。程序最终运行结果如图 6-10 所示。

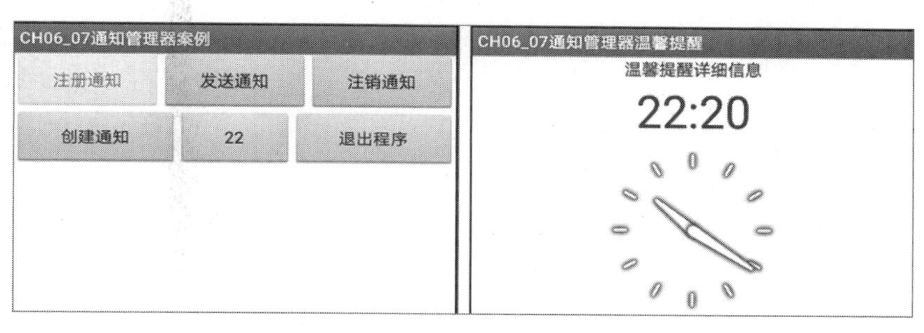

图 6-10　通知管理器运行结果

6.3 案例——网络诊断案例

与 Windows 平台类似，Android 提供了一些常用的系统服务（如访问网络连接状况、GPS 状态等），这些服务在自己开发应用程序时也会用到，下面的实例实现了一个简单网络诊断工具。该程序界面布局大纲如图 6-11 所示。

图 6-11　网络诊断案例布局大纲

程序的界面布局activity_main.xml代码如下：

```
<LinearLayout xmlns:android="http://schemas.android.com/apk/res/android"
    xmlns:tools="http://schemas.android.com/tools"
    android:layout_width="match_parent"
    android:layout_height="match_parent"
    android:orientation="vertical"
    android:paddingBottom="@dimen/activity_vertical_margin"
    android:paddingLeft="@dimen/activity_horizontal_margin"
    android:paddingRight="@dimen/activity_horizontal_margin"
    android:paddingTop="@dimen/activity_vertical_margin"
    tools:context=".MainActivity">
    <LinearLayout
        android:layout_width="match_parent"
        android:layout_height="wrap_content"
        android:orientation="horizontal">
        <Button
            android:id="@+id/network"
            android:layout_width="wrap_content"
            android:layout_height="wrap_content"
            android:layout_weight="1"
            android:onClick="doClick"
            android:text="判断网络是否连接"/>
        <Button
            android:id="@+id/wifi"
            android:layout_width="wrap_content"
            android:layout_height="wrap_content"
            android:layout_weight="1"
            android:onClick="doClick"
            android:text="打开/关闭WIFI"/>
    </LinearLayout>
    <LinearLayout
        android:layout_width="match_parent"
        android:layout_height="wrap_content"
        android:orientation="horizontal">
        <Button
            android:id="@+id/getvoice"
            android:layout_width="match_parent"
            android:layout_height="wrap_content"
            android:layout_weight="1"
            android:onClick="doClick"
            android:text="获取系统的音量"/>
        <Button
```

```
            android:id="@+id/getPackagename"
            android:layout_width="match_parent"
            android:layout_height="wrap_content"
            android:layout_weight="1"
            android:onClick="doClick"
            android:text="获取当前进程包名"/>
    </LinearLayout>
    <Button
        android:id="@+id/btnDiagnostics"
        android:layout_width="match_parent"
        android:layout_height="wrap_content"
        android:text="@string/networkdiagnostics"/>
    <TextView
        android:id="@+id/tvContent"
        android:layout_width="wrap_content"
        android:layout_height="wrap_content"/>
</LinearLayout>
```

主程序MainActivity的Java代码如下：

```
public class MainActivity extends Activity{  /***诊断网络**/
    private DiagositcsAsyncTask mDAsyncTask=new DiagositcsAsyncTask();
    protected void onCreate(Bundle savedInstanceState){
        super.onCreate(savedInstanceState);
        setContentView(R.layout.activity_main);
        findViewById(R.id.btnDiagnostics).setOnClickListener(
              new OnClickListener(){
                   public void onClick(View v){
                        mDAsyncTask=new DiagositcsAsyncTask();
                        mDAsyncTask.execute();}});}
    private void setText(final String str){  /***设置内容**/
        new Handler().post(new Runnable(){
             public void run(){
                  TextView tvContent=(TextView)findViewById(R.id.tvContent);
                  tvContent.setText(String.format("%s\r\n%s",
                  tvContent.getText().toString(), str));}});}
    private class DiagositcsAsyncTask extends
        AsyncTask<String, Integer, String>{  /***诊断网络**/
    protected void onPreExecute(){
      super.onPreExecute();
      setText(getResources().getString(R.string.diagnostics_begin));}
    protected void onProgressUpdate(Integer...values){
      super.onProgressUpdate(values);}
    protected String doInBackground(String...params){
```

```
String res=null;
try{ //判定是否有网络连接
    boolean netState=NetWorkHelper
        .isNetworkAvailable(MainActivity.this);
    //判断MOBILE网络是否可用
    boolean mobileDataState=NetWorkHelper
        .isMobileDataEnable(MainActivity.this);
    //检测是否漫游
    boolean netRoamingState=NetWorkHelper
        .isNetworkRoaming(MainActivity.this);
    //检测Wi-Fi是否可用
    boolean wifiState=NetWorkHelper
        .isWifiDataEnable(MainActivity.this);
    StringBuilder strBuilder=new StringBuilder();
    strBuilder.append(getResources().getString(
            R.string.diagnostics_result))
        .append(netState ? getResources().getString(
            R.string.network_enable): getResources()
        .getString(R.string.network_disable)).append(";")
        .append(mobileDataState ? getResources().getString(
            R.string.mobiledata_enable): getResources()
            .getString(R.string.mobiledata_disable)).append(";")
        .append(netRoamingState ? getResources().getString(
            R.string.wifi_enable): getResources()
            .getString(R.string.wifi_disable)).append(";")
        .append(wifiState ? getResources().getString(
            R.string.roaming_enable): getResources()
                .getString(R.string.roaming_disable).append(";");
    res=strBuilder.toString();
} catch(Exception e){e.printStackTrace();}
   return res;}
protected void onPostExecute(String result){
   super.onPostExecute(result);
   setText(result);
   setText(getResources().getString(R.string.diagnostics_end));}}
public void doClick(View v){
  switch(v.getId()){
  case R.id.network:
      if(isNetWorkConnected(MainActivity.this)==true){
          Toast.makeText(MainActivity.this, "网络已经打开",
Toast.LENGTH_SHORT).show();
    setText("网络已经打开");
} else{
```

```
            Toast.makeText(MainActivity.this, "网络没有打开",
    Toast.LENGTH_SHORT).show();
            setText("网络没有打开");}
            break;
        case R.id.wifi:
            WifiManager wifiManager=(WifiManager)MainActivity.this
                .getSystemService(WIFI_SERVICE);
            if(wifiManager.isWifiEnabled()){
                wifiManager.setWifiEnabled(false);
                Toast.makeText(MainActivity.this, "WIFI已经关闭",
                    Toast.LENGTH_SHORT).show();
                setText("WIFI已经关闭");
        } else{
            wifiManager.setWifiEnabled(true);
            Toast.makeText(MainActivity.this, "WIFI已经打开",
                Toast.LENGTH_SHORT).show();
            setText("WIFI已经打开");}
            break;
        case R.id.getvoice:
           AudioManager mAudioManager=(AudioManager)MainActivity.this
               .getSystemService(AUDIO_SERVICE);
           int max=mAudioManager
               .getStreamMaxVolume(AudioManager.STREAM_SYSTEM);
           int current=mAudioManager
               .getStreamVolume(AudioManager.STREAM_RING);
           Toast.makeText(MainActivity.this,
               "系统的最大音量为:"+max+"当前音量是:"+current,
               Toast.LENGTH_SHORT).show();
           setText("系统的最大音量为:"+max+"当前音量是:"+current);
           break;
        case R.id.getPackagename:
           ActivityManager mActivityManager=(ActivityManager)
           MainActivity.this  .getSystemService(ACTIVITY_SERVICE);
           String packagename=
           mActivityManager.getRunningTasks(1).get(0).topActivity
               .getPackageName();
           Toast.makeText(MainActivity.this, "当前运行的Activity的包名"
           +packagename,Toast.LENGTH_SHORT).show();
           setText("当前运行的Activity的包名"+packagename);}}
      public boolean isNetWorkConnected(Context context){
         if(context!=null){
             ConnectivityManager mConnectivityManager=(ConnectivityManager)
```

```
context.getSystemService(CONNECTIVITY_SERVICE);
        NetworkInfo mNetworkInfo=mConnectivityManager
            .getActiveNetworkInfo();
        if(mNetworkInfo!=null){
          return mNetworkInfo.isAvailable();}}
      return false;}}
```

APN类程序清单如下：

```
public class APN{/***APN定义**/
    String apnId,name,numeric,mcc,mnc,apn,user,server,password;
    String proxy,port,mmsproxy,mmsport,mmsc,authtype,type,current;
    public String getApnId(){return apnId;}
    public void setApnId(String apnId){this.apnId=apnId;}
    public String getName(){return name;}
    public void setName(String name){this.name=name;}
    public String getNumeric(){return numeric;}
    public void setNumeric(String numeric){this.numeric=numeric;}
    public String getMcc(){return mcc;}
    public void setMcc(String mcc){this.mcc=mcc;}
    public String getMnc(){return mnc;}
    public void setMnc(String mnc){this.mnc=mnc;}
    public String getApn(){return apn;}
    public void setApn(String apn){this.apn=apn;}
    public String getUser(){return user;}
    public void setUser(String user){this.user=user;}
    public String getServer(){return server;}
    public void setServer(String server){this.server=server;}
    public String getPassword(){return password;}
    public void setPassword(String password){this.password=password;}
    public String getProxy(){return proxy;}
    public void setProxy(String proxy){this.proxy=proxy;}
    public String getPort(){return port;}
    public void setPort(String port){this.port=port;}
    public String getMmsproxy(){return mmsproxy;}
    public void setMmsproxy(String mmsproxy){this.mmsproxy=mmsproxy;}
    public String getMmsport(){return mmsport;}
    public void setMmsport(String mmsport){this.mmsport=mmsport;}
    public String getMmsc(){return mmsc;}
    public void setMmsc(String mmsc){this.mmsc=mmsc;}
    public String getAuthtype(){return authtype;}
    public void setAuthtype(String authtype){this.authtype=authtype;}
    public String getType(){return type;}
    public void setType(String type){this.type=type;}
```

```
    public String getCurrent(){return current;}
    public void setCurrent(String current){this.current=current;}}
```

APNManager类程序清单如下：

```
public class APNManager{  /***APN控制**/
    private ContentResolver resolver;
    private static final Uri PREFERRED_APN_URI=Uri.parse("content://telephony/
carriers/preferapn");
    private static final Uri APN_TABLE_URI=Uri.parse("content://telephony/
carriers");
    private TelephonyManager tm;
    private Context mContext;
    private static APNManager apnManager=null;
    public static APNManager getInstance(Context context){
        if(apnManager!=null){apnManager=new APNManager(context);}
        return apnManager;}
    private APNManager(Context context){
        resolver=context.getContentResolver();mContext=context;
        tm=(TelephonyManager)context.getSystemService(Context.TELEPHONY_SERVICE);}
    /***判断一个APN是否存在，存在返回id*
     *@param apnNode
     *@return*/
    public int isApnExisted(APN apnNode){
      int apnId=-1;
      Cursor mCursor=resolver.query(APN_TABLE_URI,null,null,null,null);
      while(mCursor!=null && mCursor.moveToNext()){
        apnId=mCursor.getShort(mCursor.getColumnIndex("_id"));
        String name=mCursor.getString(mCursor.getColumnIndex("name"));
        String apn=mCursor.getString(mCursor.getColumnIndex("apn"));
        String type=mCursor.getString(mCursor.getColumnIndex("type"));
        String proxy=mCursor.getString(mCursor.getColumnIndex("proxy"));
        String port=mCursor.getString(mCursor.getColumnIndex("port"));
        String current=mCursor.getString(mCursor.getColumnIndex("current"));
        String mcc=mCursor.getString(mCursor.getColumnIndex("mcc"));
        String mnc=mCursor.getString(mCursor.getColumnIndex("mnc"));
        String numeric=mCursor.getString(mCursor.getColumnIndex("numeric"));
        Log.e("isApnExisted", "info:"+apnId+"_"+name+"_"+apn +
"_"+type+"_"+current+"_"+proxy);  //遍历了所有的APN
        if(/*apnNode.getName().equals(name)*/(apnNode.getApn().equals(apn)
&& apnNode.getMcc().equals(mcc)&& apnNode.getMnc().equals(mnc)&& apnNode.
getNumeric().equals(numeric))&&(type==null||"default".equals(type)||
"".equals(type)))
```

```
            {return apnId;
          } else{apnId=-1;}}
          return apnId;}

        /***设置默认的APN
         *@param apnId
         *@return*/
        public boolean setDefaultApn(int apnId){
          boolean res=false;
          ContentValues values=new ContentValues();
          values.put("apn_id", apnId);
          try{
            resolver.update(PREFERRED_APN_URI, values, null, null);
            Cursor c=resolver.query(PREFERRED_APN_URI, new String[]{"name",
"apn"}, "_id="+apnId, null, null);
            if(c!=null){res=true;  c.close();}
        } catch(SQLException e){e.printStackTrace();}
          return res;}
        public void deleteApn(){  /***删除所有APN*/
          resolver.delete(APN_TABLE_URI, null, null);}
        public APN getDefaultAPN(){
          String id="",apn="",proxy="",name="",port="",type="",mcc="";
          String mnc="",numeric="";  APN apnNode=new APN();
          Cursor mCursor=resolver.query(PREFERRED_APN_URI,null,null, null,null);
          if(mCursor==null){throw new Exception("不存在喜欢的APN ");
            return null;}
          while(mCursor != null && mCursor.moveToNext()){
             id=mCursor.getString(mCursor.getColumnIndex("_id"));
             name=mCursor.getString(mCursor.getColumnIndex("name"));
             apn=mCursor.getString(mCursor.getColumnIndex("apn")).toLowerCase();
             proxy=mCursor.getString(mCursor.getColumnIndex("proxy"));
             port=mCursor.getString(mCursor.getColumnIndex("port"));
             mcc=mCursor.getString(mCursor.getColumnIndex("mcc"));
             mnc=mCursor.getString(mCursor.getColumnIndex("mnc"));
             numeric=mCursor.getString(mCursor.getColumnIndex("numeric"));
             Log.d("getDefaultAPN", "默认Apn信息:"+id+"_"+name+"_"
+apn+"_"+proxy+"_"+proxy);}
           apnNode.setName(name);      apnNode.setApn(apn);
           apnNode.setProxy(proxy);   apnNode.setPort(port);
           apnNode.setMcc(mcc);      apnNode.setMnc(mnc);
           apnNode.setNumeric(numeric);
           return apnNode;}
```

```
    public int getDefaultNetworkType(){
        int networkType=-1;
        ConnectivityManager connectivity=(ConnectivityManager)mContext
            .getSystemService(Context.CONNECTIVITY_SERVICE);
        if(connectivity==null){
    } else{//Wi-Fi 网络优先
        NetworkInfo wifiNetworkInfo=connectivity
            .getNetworkInfo(ConnectivityManager.TYPE_WIFI);
      if(wifiNetworkInfo!=null&& wifiNetworkInfo.getState()==NetworkInfo.
State.CONNECTED){return ConnectivityManager.TYPE_WIFI;}
      NetworkInfo[] info=connectivity.getAllNetworkInfo();
      for(int i=0; i<info.length; i++){
        if(info[i].getState()==NetworkInfo.State.CONNECTED){
          networkType=info[i].getType();  //使用第一个可用的网络
          break;}}}
      return networkType;}
  public int InsetAPN(){  //添加一个 APN
    APN checkApn=new APN();
    checkApn.setName("智能家居物联专用接口");
    checkApn.setApn("ctnet");
    checkApn.setUser("xxx@xxx.vpdn.sd");
    checkApn.setPassword("xxxxxx");
    checkApn.setMcc(getMCC());
    checkApn.setMnc(getMNC());
    checkApn.setNumeric(getSimOperator());
    return addNewApn(checkApn);}
  /***增加新的 APN
  *@param apnNode
  *@return*/
  private int addNewApn(APN apnNode){
    int apnId=-1;
    ContentValues values=new ContentValues();
    values.put("name", apnNode.getName());
    values.put("apn", apnNode.getApn());
    values.put("proxy", apnNode.getProxy());
    values.put("port", apnNode.getPort());
    values.put("user", apnNode.getUser());
    values.put("password", apnNode.getPassword());
    values.put("mcc", apnNode.getMcc());
    values.put("mnc", apnNode.getMnc());
    values.put("numeric", apnNode.getNumeric());
    Cursor c=null;
```

```
        try{
            Uri newRow=resolver.insert(APN_TABLE_URI, values);
            if(newRow!=null){
                c=resolver.query(newRow, null, null, null, null);
                int idindex=c.getColumnIndex("_id");
                c.moveToFirst();
                apnId=c.getShort(idindex);
                Log.d("Robert","New ID: "+apnId+":Inserting new APN succeeded!");
            }
        } catch(SQLException e){e.printStackTrace();}
        if(c!=null)  c.close();
        return apnId;}
    private String getMCC(){
        String numeric=tm.getSimOperator();
        String mcc=numeric.substring(0, 3);
        Log.i("MCC  is", mcc);  return mcc;}
    private String getMNC(){
        String numeric=tm.getSimOperator();
        String mnc=numeric.substring(3, numeric.length());
        Log.i("MNC  is", mnc);  return mnc;}
    private String getSimOperator(){
        String SimOperator=tm.getSimOperator();
        return SimOperator;}
    public String matchAPN(String currentName){
        if("".equals(currentName)|| null==currentName){
             return "";}
        currentName=currentName.toLowerCase();
        if(currentName.startsWith("cmnet")||currentName.startsWith("CMNET"))
            return "cmnet";
        else if(currentName.startsWith("cmwap")||currentName.startsWith("CMWAP"))
            return "cmwap";
        else if(currentName.startsWith("3gwap")||currentName.startsWith("3GWAP"))
            return "3gwap";
        else if(currentName.startsWith("3gnet")||currentName.startsWith("3GNET"))
            return "3gnet";
        else if(currentName.startsWith("uninet")||currentName.startsWith("UNINET"))
            return "uninet";
        else if(currentName.startsWith("uniwap")||currentName.startsWith("UNIWAP"))
            return "uniwap";
        else if(currentName.startsWith("default")|| currentName.
startsWith("DEFAULT"))
            return "default";
        else
```

```
        return "";}
    /***获取APN列表
    *@param context
    *@return */
    public List<APN>getAPNList(){
        String tag="Main.getAPNList()";
        String projection[]={"_id,apn,type,current"};
        Cursor cr=mContext.getContentResolver().query(APN_TABLE_URI, projection,
null,null,null);
        List<APN>list=new ArrayList<APN>();
        while(cr!=null && cr.moveToNext()){
          Log.d(tag,cr.getString(cr.getColumnIndex("_id"))+" "
                  +cr.getString(cr.getColumnIndex("apn"))+" "
                  +cr.getString(cr.getColumnIndex("type"))+" "
                  +cr.getString(cr.getColumnIndex("current")));
          APN a=new APN();
          a.apnId=cr.getString(cr.getColumnIndex("_id"));
          a.apn=cr.getString(cr.getColumnIndex("apn"));
          a.type=cr.getString(cr.getColumnIndex("type"));
          list.add(a);}
        if(cr!=null)cr.close();
        return list;}}
```

NetWorkHelper类的程序清单如下：

```
  public class NetWorkHelper{
      private static String LOG_TAG="NetWorkHelper";
      public static Uri uri=Uri.parse("content://telephony/carriers");
      public static boolean isNetworkAvailable(Context context){
         /***判断是否有网络连接*/
          ConnectivityManager connectivity=(ConnectivityManager)context
.getSystemService(Context.CONNECTIVITY_SERVICE);
      if(connectivity==null){Log.w(LOG_TAG, "无法获得连接管理器");
      } else{
          NetworkInfo[] info=connectivity.getAllNetworkInfo();
          if(info!=null){
            for(int i=0; i<info.length; i++){
              if(info[i].isAvailable()){
                Log.d(LOG_TAG, "网络可用");
                return true;}}}}
        Log.d(LOG_TAG, "网络不可用");
        return false;}
      public static boolean checkNetState(Context context){
        boolean netstate=false;
```

```
        ConnectivityManager connectivity=(ConnectivityManager)context.
etSystemService(Context.CONNECTIVITY_SERVICE);
        if(connectivity!=null){
            NetworkInfo[] info=connectivity.getAllNetworkInfo();
            if(info!=null){
                for(int i=0; i<info.length; i++){
                if(info[i].getState()==NetworkInfo.State.CONNECTED)
                {netstate=true;break;}}}}
        return netstate; }
    public static boolean isNetworkRoaming(Context context){
    /***判断网络是否为漫游*/
        ConnectivityManager connectivity=(ConnectivityManager)context
.getSystemService(Context.CONNECTIVITY_SERVICE);
        if(connectivity==null){
            Log.w(LOG_TAG, "无法获得连接管理器!");
    } else{
        NetworkInfo info=connectivity.getActiveNetworkInfo();
        if(info!=null&& info.getType()==ConnectivityManager.TYPE_MOBILE)
{TelephonyManager tm=(TelephonyManager)context.getSystemService(Context.
TELEPHONY_SERVICE);
            if(tm!=null && tm.isNetworkRoaming()){
              Log.d(LOG_TAG, "网络正在漫游!");return true;
            } else{
                Log.d(LOG_TAG, "网络停止漫游!");}
      } else{Log.d(LOG_TAG, "移动网络不可用! ");}}
      return false;}
    /***判断MOBILE网络是否可用
     *@param context
     *@return
     *@throws Exception*/
    public static boolean isMobileDataEnable(Context context)throws Exception{
        ConnectivityManager connectivityManager=(ConnectivityManager)context
.getSystemService(Context.CONNECTIVITY_SERVICE);
        boolean isMobileDataEnable=false;
        isMobileDataEnable=connectivityManager.getNetworkInfo(
ConnectivityManager.TYPE_MOBILE).isConnectedOrConnecting();
        return isMobileDataEnable;}
    /***判断Wi-Fi 是否可用
    *@param context
    *@return
    *@throws Exception */
    public static boolean isWifiDataEnable(Context context)throws Exception{
```

```
        ConnectivityManager connectivityManager=(ConnectivityManager)context
.getSystemService(Context.CONNECTIVITY_SERVICE);
        boolean isWifiDataEnable=false;
        isWifiDataEnable=connectivityManager.getNetworkInfo(
ConnectivityManager.TYPE_WIFI).isConnectedOrConnecting();
        return isWifiDataEnable;}
        /***设置Mobile网络开关
         *@param context
         *@param enabled
         *@throws Exception*/
    public static void setMobileDataEnabled(Context context, boolean enabled)
throws Exception{
        APNManager apnManager=APNManager.getInstance(context);
        List<APN>list=apnManager.getAPNList();
        if(enabled){
            for(APN apn:list){
                ContentValues cv=new ContentValues();
                cv.put("apn", apnManager.matchAPN(apn.apn));
                cv.put("type", apnManager.matchAPN(apn.type));
                context.getContentResolver().update(uri, cv, "_id=?",new String[]
{apn.apnId });}
        } else{
            for(APN apn:list){
                ContentValues cv=new ContentValues();
                cv.put("apn", apnManager.matchAPN(apn.apn)+"mdev");
                cv.put("type", apnManager.matchAPN(apn.type)+"mdev");
                context.getContentResolver().update(uri, cv, "_id=?",new String[]
  {apn.apnId });}}}}
```

在 AndroidManifest.xml 文件中添加网络和 WIFI 等访问权限：

```
   <uses-permission android:name="android.
permission.INTERNET"/>
   <uses-permission android:name="android.
permission.ACCESS_NETWORK_STATE"/>
   <uses-permission android:name="android.
permission.ACCESS_WIFI_STATE"/>
   <uses-permission android:name="android.
permission.CHANGE_WIFI_STATE"/>
   <uses-permission android:name="android.
permission.GET_TASKS"/>
```

该程序的运行结果如图 6-12 所示。

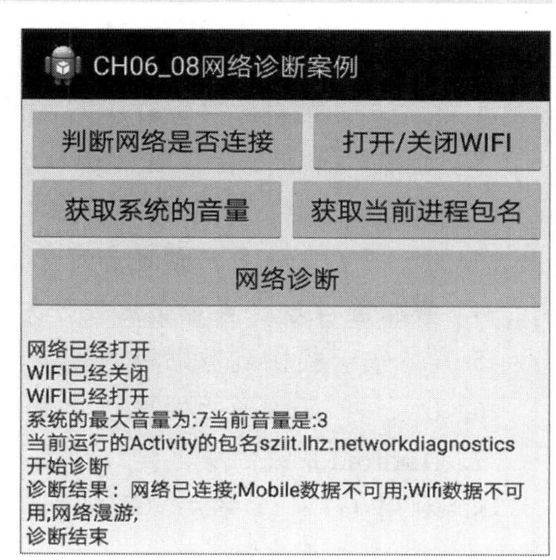

图 6-12　网络诊断案例运行结果

6.4 知识扩展

6.4.1 电话管理器（TelephonyManager）

TelephonyManager是一个管理手机通话状况、电话网络信息的服务类，该类提供了大量的get×××()方法获取电话网络的相关信息。

在程序中获取TelephonyManager非常简单，只需调用如下代码即可：

```
TelephonyManager  tManager=(TelephonyManager)getSystemService(Context.
TELEPHONY_SERVICE);
```

6.4.2 短信管理器（SmsManager）

SmsManager提供了系列send×××Message()方法用于发送短信。实现这个功能非常简单，直接调用sendTextMessage()方法即可。

本章小结

本章讲述Android主要的几个系统服务：活动管理器（ActivityManager）、警报管理器（AlarmManager）、音频管理器（AudioManager）、剪贴板管理器（ClipboardManager）、通知管理器（NotificationManager）的概念、作用及实现。

强化练习

一、填空题

1. （　　）是对所有运行中的Activity组件进行管理，在android.app包中。

2. 通过配置信息接口的reqNavigation属性可以获得当前设备的导航方式，导航方式类型包括（　　）、（　　）、（　　）和（　　）。

3. （　　）允许用户预定自定义应用程序的运行时间。

4. 警报管理器设置警报主要分为两种：（　　）和（　　）。

5. （　　）提供到剪贴板的接口，用于设置或获取全局剪贴板中的文本。

6. （　　）又称通知管理器，用于通知用户有后台事件发生。

7. Android系统的服务接口都由（　　）类提供。

8. 利用（　　）方法，开发者可通过指定的服务字符串标识获取相应服务。

二、编程题

1. 用AlarmManager时钟服务编写3s后一次性报警、每隔3s周期性报警和取消周期性报警应用程序。

2. 在界面布局文件中设置3个按钮，分别为“发送通知”、“删除通知”和“退出应用”，利用NotificationManager在状态栏显示系统通知服务。

3. 编程实现通过调用AudioManager对象的方法实现对音量的设置和管理。

4. 编程实现设置和获取剪贴板中的文字内容。

5. 编程实现通知管理。

综合项目实训篇

本部分在 Android 核心理论知识的基础上，通过 3 个企业真实的、涉及当前热门的物联网技术和移动互联技术等方面的典型综合项目，基于软件工程化思想，按照企业项目实施的关键节点，详细论述每个项目的设计与实现，以提高学生综合应用多种核心知识和技能，解决复杂工程问题的素养和能力。

本篇包含以下 3 章：

第 7 章　基于移动端 GPS 和传感器的运动打卡 APP 项目

第 8 章　Struggle 车牌识别系统 APP 项目

第 9 章　基于 Android 智能仓储系统项目

第7章

基于移动端GPS和传感器的运动打卡APP项目

学习目标

- 掌握软件界面的设计和实现、页面与页面之间数据交互应用技巧。
- 掌握嵌入式SQLite数据库的设计和实现。
- 掌握多媒体录音、拍照、音频播放和视频播放功能。
- 掌握GPS系统服务访问、数据获取存储与地图应用技术。

视频

基于移动端GPS和传感器的运动打卡APP项目

7.1 项目概述

基于移动端GPS和传感器的运动打卡APP项目（足步天下APP）是一款用户在日常生活中用来记录自己运动情况的Android手机软件，用户可以用此软件实时记录自己的运动轨迹、运动幅度、运动量，在运动过程中可以留下自己的运动照、运动趣事小视频、运动语音，并上传分享自己运动的乐趣，另外用户还可以进行打卡，根据用户的运动目标、运动量来分析，软件会给出合理建议的功能。该项目的主要特点如下：

（1）在设计上，用Eclipse编制友好的用户界面；在开发效率和程序功能上具有明显的优越性和通用性；在传统运动APP上，添加了自身APP的新元素。除了精准记录轨迹、里程、配速、步数等数据，自动统计分段里程的配速信息，能够直观展示用户运动全过程外，还可以针对系统用户产生的运动数据，能够分析用户的兴趣爱好，并制订运动计划，根据运动的数据进行与学校的体育课成绩直接挂钩，以此来激励高校学生的运动。

（2）在数据存储技术实现上，同时支持嵌入式数据库和MySQL数据库访问技术，在本地手机和后台记录运动打卡数据，支持insert、delete、update、query等常用数据库操作。

（3）采用GPS定位技术获得用户的经度、纬度、高度、速度、方向等信息，采用Baidu Map提供地图服务。

7.2 项目设计

7.2.1 项目总体功能需求

1. 功能性需求

基于移动端GPS和传感器的运动打卡APP实现项目的主要功能性需求包括：制订运动计划和路线、运动检测、管理后台、运动社区、数据管理。

（1）制订运动计划和路线：学校可以制定固定运动路线，个人可以调整，最终生成每个学生周和月度运动计划，包含跑步、步行、爬山、游泳等多个类别。学生每天按照计划在约定的时间内进行锻炼，学生可提前根据天气和个人时间调整计划，但必须保证足够的运动量。

（2）运动监测：基于GPS和传感器技术，精准记录轨迹、里程、配速、步数等数据，自动统计分段里程的配速信息，能够直观展示学生运动全过程。在运动开始、运动结束记录打卡，在运动过程中随机要求2~3次打卡，会提前5分种语音提示，需要在5分钟内手工打卡，必须打卡时记录定位信息、录音、拍摄照片或视频，以证明是学生自己在某位置打卡。

（3）管理后台：管理端功能包括学校管理、学生管理，数据的录入支持批量导入；对于课程管理，包含课程信息管理和成绩管理，可以通过管理后台向管理的学生分配作业，可以限定运动区域、时间、及格线等属性；针对学生的成绩后台可以手动修改策略，和其他成绩进行综合评定；对应的运动数据有实时排行榜，在后台用户可以查看相应名次（需要考虑数据权限的区域级别）；同时也支持学校针对该学校发布相应的通告信息和知识推送。

（4）运动社区：基于学生的运动成绩，能够有相应的积分，相应积分能够确定不同级别，能够在社区中进行消费；并且在社区中，学生之间可以相互交流，社区信息也包含学校发布的知识专题、赛事活动，学生可以在社区中选择自主参与，针对赛事并有实时进度和排名。

（5）数据管理：对于用户运动数据支持本地缓存，不需要运动过程中实时联网上传数据；数据能够与其他系统对接，实现共享；基于运动数据进行分析，提供智能化的运动计划，并监督学生按时执行，记录执行效果。

视频

足步天下软件详细设计

2. 非功能性需求

人机交互界面友好、管理后台支持响应、能在各平台上流畅运行。

该项目功能需求示意图如图7-1所示。

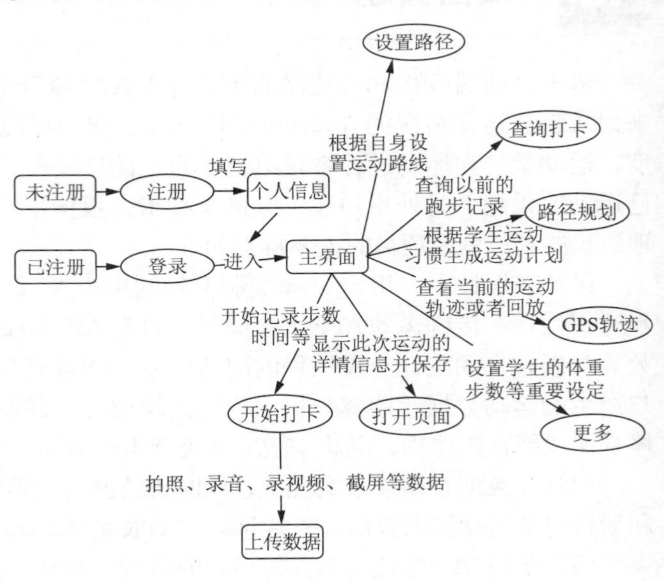

图7-1 项目功能需求示意图

7.2.2 项目总体设计

1. 软件登录界面设计

首先，软件设计的第一步就是进行注册/登录，保存用户信息，完成后系统会把客户端的用户信息上传到服务器中，利用数据库把用户信息分类，接着就是用户的运动信息的保存，实现对用户数据信息的分析保存处理。

2. 计步器功能设计

（1）利用智能手机传感器感知身体的运动功能，分析人体行走时加速度的变化状况，选取了与行走动作最为相关的3个特征：幅度特征，以表示运动时的力度变化范围；角度特征，以反映人体行走运动时方向的变化情况；距离特征，以表示行走运动时人体的位移变化。进一步对3个特征分别设计了特征计算方法。

（2）通过调用手机加速度传感器进行各个方向加速度的采集；之后对加速度进行取模，得出加速度变化曲线，判断加速度波峰、波谷变化的差值；最后，如果差值大于设置的阈值则记步，否则取消该步，最后显示步数结果。

（3）显示步数、计算路程、用时，消耗的卡路里和速率等功能。通过传感器计算出用户的运动数据，能够自动统计分段里程的配速信息，能够直观展示学生运动的全过程。

（4）计步器的设置。用户根据自己的情况设置计步器，包括计步的灵敏性，步长、身高、体重、运动类型、期望步速等。

3. GPS实时定位功能设计

（1）通过该模块可以实现定位、导航等基本功能，可以获取用户当前的经纬度、海拔高度、运动方向和速度等数据。采用Location Provider提供GPS Provider或NETWORK Provider实现定位。GPS Provider使用GPS定位方式进行手机定位；NETWORK Provider则采用网络和基站实现定位。

（2）GPS定位功能可以选择利用网络来定位还是卫星定位，实时得到用户的位置信息，实现卫星图、路况图、热力图等多元化GPS功能。

4. GPS轨迹功能设计

基于GPS定位功能获得的GPS数据信息，可以对GPS数据信息进行分析，实现运动轨迹动态展现功能。

5. 查询GPS数据功能设计

通过GPS定位的功能延伸到查询GPS数据的功能，对于实时定位的数据进行本地的上传，实现本地查看数据信息，清晰、有条理地以列表的方式列出。对于其他功能也有延伸，如查询数据得到的经纬度、海拔、速度、定位的时间、数据类型等。

6. 开始打卡功能设计

（1）在运动开始和结束都要进行运动记录打卡，在运动中也可实行随机打卡，点击“开始打卡”后，语音提示操作成功，数据信息得到分段保存，实现实时数据的打卡数据处理。

（2）在语音提示完后，用户要进行拍摄环境照片、拍摄正面照片、录制录音、录制视频的提示操作，为记录用户的实时数据证明，达到智能化监督的效果特点。

7. 拍摄照片功能设计

基于“开始打卡”功能，实现用户运动过程中的拍照功能，记录运动点滴，并将拍摄的照片保存到本地SQLite嵌入式数据库和远程后台云数据库中，实现在线预览打卡图片功能。

8. 录音功能设计

基于开始打卡功能，用户可以在线录制音频数据、在线播放，并把录制的音频保存到本地SQLite嵌入式数据库和远程后台云数据库中，便于音频回放。

9. 录制小视频功能设计

基于开始打卡功能，录制小视频，在限定的时间内录制视频，记录运动的真实情况，并把录制的小视频保存到本地SQLite嵌入式数据库和远程后台云数据库中，便于视音回放。

10. 打卡页面功能设计

对于“开始打卡”后的各项语音提示进行的操作，该程序对于GPS数据打卡功能所需的程序进行了统计，基于用户保存本地或者选择上传。将GPS数据、时间、信息源、计步器数据、打卡功能的程序结合起来，加上地图界面的截图功能，可对用户每次的打卡进行查询回放。打卡页面功能如图7-2所示。

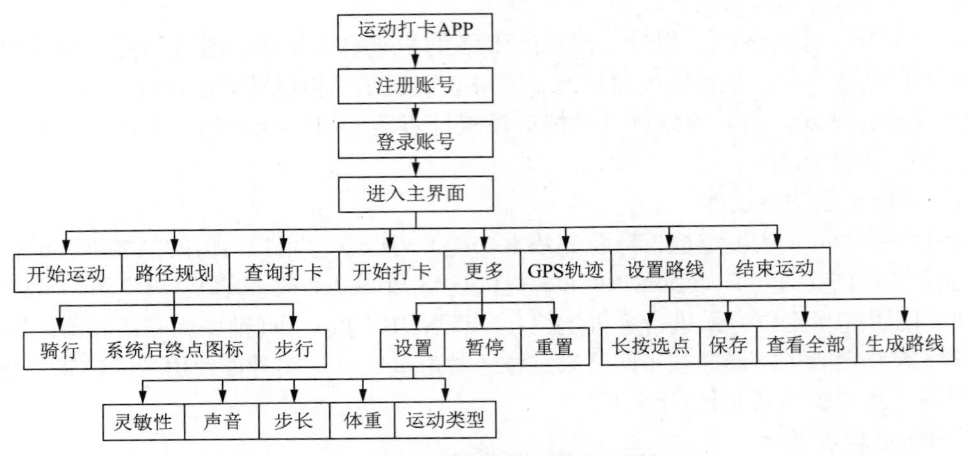

图7-2 打卡页面功能设计框架图

11. 计步查询功能设计

基于传感器计步器功能，对传感器记录的所有数据以列表的形式进行罗列，标注清晰，可供用户进行查询回看。

12. 路径规划功能设计

结合地图显示功能，实现步行路径规划功能。项目整体设计框架如图7-3所示。计步器功能设计框图如图7-4所示。

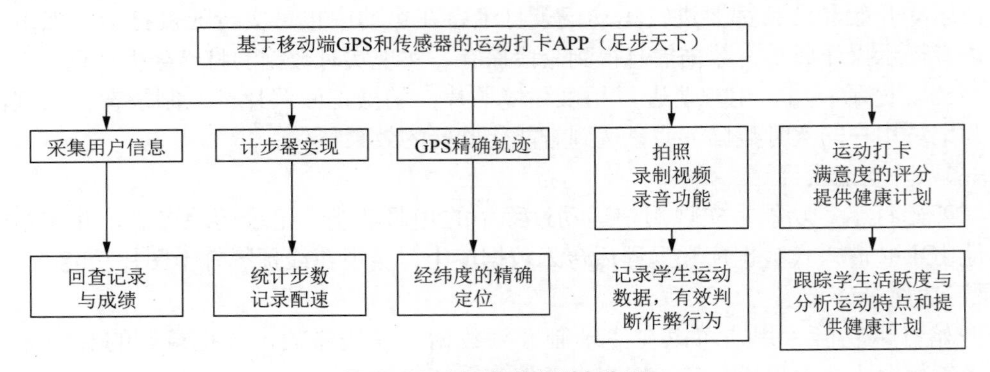

图7-3 项目整体设计框架图

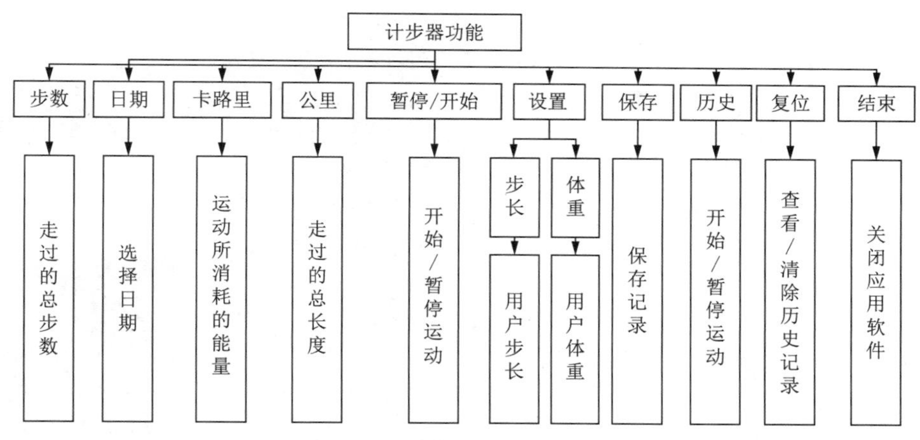

图7-4　计步器功能设计框图

7.3 必备的技术和知识点

本章必备的技术和知识点包括：

（1）界面编程与视图组件。

（2）布局管理器。

（3）TextView及其子类。

（4）ImageView及其子类。

（5）AdapterView及其子类。

（6）ViewAnimator及其子类。

（7）对话框和菜单。

（8）基于监听的事件处理。

（9）Handle消息传递机制。

（10）嵌入式SQLite数据库基本操作。

（11）使用MediaPlayer播放音频和视频。

（12）使用MediaRecorder录制音频。

（13）控制摄像头Camera拍照、录制视频。

（14）GPS定位和地图服务。

7.4 项目实施

7.4.1　闪屏页面

实现软件的欢迎闪屏活动页面和动画品牌宣传页面功能。先设计activity_splash.xml布局代码，然后编写SplashActivity类功能代码，其运行效果如图7-5所示，其源代码如下：

```
    public class SplashActivity extends Activity{
        //定义助手对象，处理消息循环(用于在欢迎界面到闪屏幕之间传递消
        //息的助手)
        Handler WelcomeHandler=new Handler()  //接受信息界面跳转
        { public void handleMessage(Message msg)  //重写方法
            { switch(msg.what){
                case 0:gotoSplashActivity(); break;}}};
                //跳转到主界面
        protected void onCreate(Bundle savedInstanceState){
            super.onCreate(savedInstanceState);
            requestWindowFeature(Window.FEATURE_NO_TITLE);
            //隐藏标题
            getWindow().setFlags(WindowManager.LayoutParams.
FLAG_FULLSCREEN,
                WindowManager.LayoutParams.FLAG_FULLSCREEN);
            gotoWelcomeSurfaceView();}  //程序跳转到欢迎界面
        public void gotoWelcomeSurfaceView(){  /***功能描述:
欢迎界面*/
            WelcomeSurfaceView mView=new WelcomeSurfaceView(this);
            setContentView(mView);}
        private void gotoSplashActivity(){  /***功能描述: 切换到闪屏活动页面 */
            setContentView(R.layout.activity_splash); //设置闪屏活动页面布局
            //程序已经启动，直接跳转到运行界面
            if(StepService.FLAG || GPSPedometerMainActivity.GetStepValue()>0)
            {  //创建一个新的Intent，指定当前应用程序上下文和要启动的StepActivity类
                Intent intent=new Intent(SplashActivity.this,
                    LoginActivity.class);
                startActivity(intent); //传递这个intent给startActivity
                        this.finish();} else{
                        new CountDownTimer(1000L, 1000L){
                          public void onFinish(){
                              Intent intent=new Intent();  //启动界面淡入淡出效果
                              intent.setClass(SplashActivity.this,
                                  LoginActivity.class);
                              try{Thread.sleep(1000);  //延迟1000
                              }catch(InterruptedException e){e.printStackTrace();}
                                startActivity(intent);
                    overridePendingTransition(R.anim.fade_in,R.anim.fade_out);
                        finish();}
                public void onTick(long paramLong){}}.start();}}}
```

图 7-5　闪屏页面

视频

足步天下运动打卡APP软件使用说明书

7.4.2　注册 / 登录页面

实现账号和密码输入文本框、实现登录按键功能、实现密码记录功能和自动登录功能。先设计activity_login.xml界面布局，运行结果如图7-6所示。然后编程RegisterActivity类代码如下：

图 7-6　注册 / 登录页面图

```
public class RegisterActivity extends Activity
implements OnClickListener{
   private static final String TAG="RegisterActivity";
      private EditText et_phone;
      //用户名（手机号码）输入编辑框控件
      private EditText et_code;  //手机验证码输入编辑框控件
      private EditText et_pswd,et_pswdAgine;
      private Button btn_code,btn_show,btn_register;
      private String strPhone,strPswd,strPswdAgine;
      private boolean showPswd=false; /***是否显示密码*/
      protected void onCreate(Bundle savedInstanceState)
      {  //重写onCreate()
         super.onCreate(savedInstanceState);  //调用基类onCreate()方法
         requestWindowFeature(Window.FEATURE_NO_TITLE);  //隐藏标题栏
         setContentView(R.layout.activity_register);initView();}
      private void initView(){  /***功能描述: 初始化视图*/
         RelativeLayout titleBar=(RelativeLayout)
            findViewById(R.id.layout_titlebar);
         TextView center_tx=(TextView)titleBar
            .findViewById(R.id.tv_titlebar_title);
         center_tx.setText(getString(R.string.register));  //"注册"
         Button rightBtn=(Button)titleBar
            .findViewById(R.id.btn_titlebar_right);
         rightBtn.setText(getString(R.string.login));  //"登录"
         Button leftBtn=(Button)
           titleBar.findViewById(R.id.btn_titlebar_back);
         leftBtn.setText(getString(R.string.return_main));  //"返回"
         et_phone=(EditText)findViewById(R.id.et_mobile);
         et_code=(EditText)findViewById(R.id.et_code);
         et_pswd=(EditText)findViewById(R.id.et_pswd);
         et_pswdAgine=(EditText)findViewById(R.id.et_pswdAgine);
         btn_code=(Button)findViewById(R.id.btn_verify);
         btn_show=(Button)findViewById(R.id.btn_show);
         btn_register=(Button)findViewById(R.id.btn_register);
         btn_show.setText(getString(R.string.display));  //"显示"
         btn_code.setOnClickListener(this);
         btn_show.setOnClickListener(this);
         btn_register.setOnClickListener(this);
```

```
        rightBtn.setOnClickListener(this);
        leftBtn.setOnClickListener(this);}
    public void onClick(View v){
        switch(v.getId()){
        case R.id.btn_register:  //注册用户
            strPhone=et_phone.getText().toString().trim();
            strPswd=et_pswd.getText().toString().trim();
            strPswdAgine=et_pswdAgine.getText().toString().trim();
            if(strPhone==null||strPhone.trim().equals("")){
                //"请输入手机号"
                MyToast.showToast(this.getApplicationContext(),
                    getString(R.string.telephone_prompt));
      } else if(isMobileNumber(strPhone)){
          //"请输入正确的手机号"
          MyToast.showToast(this.getApplicationContext(),
              getString(R.string.input_right_telephone));
      } else if(strPswd==null||strPswd.trim().equals("")){
          //"请输入密码"
          MyToast.showToast(RegisterActivity.this,
              getString(R.string.password_prompt));
      } else if(strPswdAgine==null||strPswdAgine.trim().equals("")){
          //"请再次输入密码"
          MyToast.showToast(this.getApplicationContext(),
              getString(R.string.password_reprompt));
      } else if(!strPswdAgine.equals(strPswd)){
          et_pswdAgine.setText("");
          MyToast.showToast(this.getApplicationContext(),
              getString(R.string.password_dismatch_reprompt));
      } else{
          new Thread(new Runnable(){
              public void run(){getRegisterInfo(strPswd,strPhone);}});
          MyToast.showToast(RegisterActivity.this,
              getString(R.string.register_sucess));  //"注册成功"
          Intent intent=new Intent(RegisterActivity.this,
              LoginActivity.class);
          startActivity(intent);RegisterActivity.this.finish();}
          break;
      case R.id.btn_show:
          showPswd=!showPswd;setPswdVisibility(showPswd);break;
      case R.id.btn_titlebar_right:
      case R.id.btn_titlebar_back:
          Intent intent=new Intent(RegisterActivity.this,
              LoginActivity.class);
```

```
        startActivity(intent);RegisterActivity.this.finish();break;}}
    /***功能描述：从后台服务器（通过网络）获取注册信息*/
    private void getRegisterInfo(String pswd, String phoneNo){
        HttpClient mHttpClient=new DefaultHttpClient();
        HttpPost mHttpPost=new HttpPost(
          NetRequestAddress.REQUEST_ADDR_REGISTER);
        //组装数据放到HttpEntity中发送到服务器
        ArrayList<BasicNameValuePair>dataList=new ArrayList<BasicNameValuePair>();
        dataList.add(new BasicNameValuePair("password", pswd));
        dataList.add(new BasicNameValuePair("phone", phoneNo));
        HttpEntity mHttpEntity=null;
        try{//对参数进行编码操作
           mHttpEntity=new UrlEncodedFormEntity(dataList, "UTF-8");
        } catch(UnsupportedEncodingException e){e.printStackTrace();}
        //生成一个post请求对象
        mHttpPost.setEntity(mHttpEntity);
        //向服务器发送POST请求并获取服务器返回的结果
        HttpResponse mResponse=null;
        try{mResponse=mHttpClient.execute(mHttpPost);
        } catch(ClientProtocolException e){e.printStackTrace();
        } catch(IOException e){e.printStackTrace();}
        //获取响应的结果信息
        String result="";
        try{result=EntityUtils.toString(mResponse.getEntity(),"utf-8");
          MyLog.d(TAG, "注册result = "+result);
          if(result != null && !result.trim().equals("")){
            parseResultData(result);}
        } catch(ParseException e){e.printStackTrace();
        } catch(IOException e){e.printStackTrace();}}
        /***功能描述：解析返回结果
        *@param result*/
    private void parseResultData(String result){
       try{JSONObject sellerObject=new JSONObject(result);
         String status=sellerObject.getString("status");
         String info="";
         if(sellerObject.has("info")){
            info=sellerObject.getString("info");}
         if(status != null && status.equals("success")){
            MyToast.showToast(RegisterActivity.this,
                getString(R.string.register_sucess));  //"注册成功"
            Intent mIntent=new Intent(RegisterActivity.this,
              LoginActivity.class);
          startActivity(mIntent);RegisterActivity.this.finish();
```

```
    } else{MyToast.showToast(RegisterActivity.this, info);}
  } catch(JSONException e){e.printStackTrace();}}
  /***功能描述：判断字符串是否为手机号码*
   *@param strParam
   *@return*/
  public boolean isMobileNumber(String strParam){
      Pattern p=Pattern.compile("[1][34578]\\d{9}");
      Matcher m=p.matcher(strParam);return m.matches();}
  /***功能描述：设置密码的可见性*
   *@param isShow*/
  private void setPswdVisibility(boolean isShow){
      if(isShow){btn_show.setText("隐藏");
        et_pswd.setInputType(InputType.TYPE_TEXT_VARIATION_PASSWORD);
  } else{btn_show.setText("显示");
      et_pswd.setInputType(InputType.TYPE_CLASS_TEXT
          |InputType.TYPE_TEXT_VARIATION_PASSWORD);}}
  /***功能描述：处理回退按钮，返回登录界面 */
  public boolean onKeyDown(int keyCode, KeyEvent event){
      if(keyCode==KeyEvent.KEYCODE_BACK
          && event.getAction()==KeyEvent.ACTION_DOWN){
          Intent intent=new Intent(RegisterActivity.this,
              LoginActivity.class);
          startActivity(intent);  RegisterActivity.this.finish();}
      return super.onKeyDown(keyCode, event);}}
```

7.4.3 主页面

程序主界面activity_main.xml采用上下滚屏方式，分区各类信息，包括显示计步器信息区、GPS定位信息区、计步器打卡功能区、功能命令按钮区、百度地图控制按钮区、百度地图显示区、用户积分信息区等。

（1）显示计步器信息区包括：总步数、行程(单位：公里或英里)、总用时（单位毫秒，按照时：分：秒格式显示)、已消耗卡路里数、步行配速（单位：步数/分钟)、步行速度（单位：公里/小时)。

（2）GPS定位信息区包括：位置经纬度、海拔高度、运动方向、运动速度、位置信息来源、GSP时间等。

（3）计步器打卡功能区包括：环境拍照、名称显示和相片查看；运动者自拍照片、名称显示和相片查看；声音录制、名称显示和声音播放；视频录制、名称显示和视频播放；获取地图截图、名称显示和地图查看等。

（4）功能命令按钮区包括：开始运动(计步)、结束运动(计步)、查看记录；开始打卡(上传数据)、GPS定位、查看GPS、查询打卡、设置路径、打卡页面、GPS轨迹、自我定位、路径规划、计步查询。

（5）百度地图控制按钮区包括：截图、普通地图、卫星地图、路况地图、城市热力图；长按地图获取经纬度信息、输入名称，按“保存”按钮保存该收藏点数据。生成收藏点轨迹（利用收藏点技术来设定学生运动路径）、收藏点轨迹跟踪、跟踪复选按钮；生成运动轨迹（利用运动采集到的GPS数据来设定学生运动路径）、运动轨迹跟踪、跟踪复选按钮；更新当前位置复选按钮、地图显示模式（普通、跟随、罗盘）。

主程序GPSPedometerMainActivity类的核心数据成员如下：

```
private SharedPreferences mSettings;
private PedometerSettings mPedometerSettings;
private TextView mStepValueView;
private TextView mPaceValueView;
private TextView mDistanceValueView;
private TextView mSpeedValueView;
private TextView mCaloriesValueView;
TextView mDesiredPaceView;
private int mStepValue;
private int mPaceValue;
private float mDistanceValue;
private float mSpeedValue;
private int mCaloriesValue;
private float mDesiredPaceOrSpeed;
//定义文本框控件
private TextView tv_week_day;  //星期
private TextView tv_date;  //日期
private TextView tvTimer;  //运行时间
private Button btn_querygps;  //查询GPS
private Button btnGpsLocation;  //GPS手动定位
private boolean isStartPunchClock=true;
private static long use_time=0;  //运动时间
private static long startTime=System.currentTimeMillis();  //开始时间
//处理GPS定位
private LocationManager mLocationManager;  //定义位置管理器对象
private Location mLocation;  //定义位置对象
private Criteria mCriteria;  //定义标准对象
private String mProvider;  //定义GPS数据提供者
private TextView tvLongitude;  //显示GPS经度文本视图控件
private TextView tvLatitude;  //显示GPS纬度文本视图控件
private TextView tvHigh;  //显示GPS海拔文本视图控件
private TextView tvDirection;  //显示GPS方向文本视图控件
private TextView tvSpeed;  //显示GPS速度文本视图控件
private TextView tvGpsTime;  //显示GPS时间文本视图控件
private TextView tvInfoType;  //显示GPS数据来源文本视图控件
//用于存储GPS数据的SQLite数据库
private DBGps mDBGps=new DBGps(this);
```

```
SQLiteDatabase mDB=null;
gpsdata my_gpsdata=new gpsdata();
private Cursor mCursor=null;
/*处理录制视频和播放视频 */
public static final int VIDEO_CAPTURED=5;
Button btnCaptureVideo;
Button btnPlayVideo;
TextView tvVideoName;
Uri videoFileUri;
String videoPath=null;
Button btnPrtMapScreen;  //截取地图屏幕
TextView tvMapScreenName;
Button btnShowMapScreen;
//调用系统照相机变量
public static final int UI_SYSTEM_CAMERA_BACK=1;  //调用系统照相机返回标识
public static final int UI_PHOTO_ZOOM_BACK=2;  //图片剪裁后返回标识
public static final int UI_SYSTEM_CAMERA_BACK_SELFIE=3;  //系统照相机返回标识
public static final int UI_PHOTO_ZOOM_BACK__SELFIE=4;  //图片剪裁后返回标识
public static final String IMAGE_UNSPECIFIED="image/*";
//照片存放路径及照片名
private String pathPhoto;
private String EnvironmentName;
private String SelfieName;
private String PrtScrMapName;
//处理拍照和显示照片(环境照片)
private Button btnTakeEnvironPhoto;
private Button btnShowEnvironPhoto;
private TextView tvImageEnvironName;
//处理拍照和显示照片(正面照片)
private Button btnTakeSelfiePhoto;
private Button btnShowSelfiePhoto;
private TextView tvImageSelfieName;
//处理录音和播放声音
private Button btnRecordSound;
private Button btnPlaySound;
private TextView tvSoundName;
private Button btnBegPunchClock;  //开始打卡和上传按钮
private Button btnPunchClock;      //打卡按钮
private Button btnQueryPCData;    //查询打开数据
  private Button btnQuerySensorStep;  //查询传感器计步数据
  private Button btnGpsPath;  //查询轨迹
  private Button btnLocate;  //自我定位
  private Button btnRoutePlan;  //路径规划
```

```
private Button btnSetPath;  //设置运动路径
private Button btnStartWalk;
private boolean isStartWalking=false;
private Button btnEndWalking;
private Button btnTrackList;
//录音控件
private MediaRecorder mRecorder=null;  //录音类
private MediaPlayer mPlayer=null;  //播放类
private boolean mStartRecording=true;
private boolean mStartPlaying=true;
private static String mFileName=null;
private File file; //录音后文件
//照片文件
private File photoEnvironFile;  //环境照片文件
private File photoSelfieFile;  //正面照片文件
private File mapScreenFile;  //地图截图文件
private MediaPlayer mMediaPlayer; //播放音乐的对象(背景音乐)
//暂停图片和注册图片
private ImageView ivpausesong;
private boolean imagebflag=true;
punchclockdata pcdata;
sensorstepdata ssdata;
//定义GPS跟踪显示地图对象
private MapView bMainMapView;
private BaiduMap mMainBaiduMap;
private UiSettings mUiSettings;
//收藏地图坐标点界面控件相关
private EditText locationText;
private EditText nameText;
private View mPop;
private View mModify;
EditText mdifyName;
//定位相关
LocationClient mLocClient;
public MyLocationListenner myListener=new MyLocationListenner();
private LocationMode mCurrentMode;
BitmapDescriptor mCurrentMarker;
```

主程序GPSPedometerMainActivity类中实现GPS定位功能核心代码如下：

```
//提醒用户打开手机GPS定位功能
if(!DeviceTools.isGpsOpen(this.getApplicationContext())){//…}
/***功能描述：初始化定位*/
private void initLocation(){
```

```
    mLocationManager=(LocationManager)
        getSystemService(Context.LOCATION_SERVICE);
    if(mLocationManager.isProviderEnabled(LocationManager.GPS_PROVIDER))
    {     mCriteria=new Criteria();
          mCriteria.setAccuracy(Criteria.ACCURACY_FINE);  //高精度
          mCriteria.setAltitudeRequired(true);  //显示海拔
          mCriteria.setBearingRequired(true);  //显示方向
          mCriteria.setSpeedRequired(true);  //显示速度
          mCriteria.setCostAllowed(false);  //不允许有花费
          mCriteria.setPowerRequirement(Criteria.POWER_LOW);  //低功耗
          Provider=mLocationManager.getBestProvider(mCriteria, true);
          //位置变化监听,默认1秒一次,距离10m以上
          mLocationManager.requestLocationUpdates(mProvider, gps_min_time,
              gps_min_distance, locationListener);
    } else showInfo(null, -1);}
/***功能描述: 定义位置侦听器对象*/
private final LocationListener locationListener=new LocationListener(){
    public void onLocationChanged(Location arg0){
        my_gpsdata=getLastPosition();
        showInfo(my_gpsdata, 2);}
    public void onProviderDisabled(String arg0){showInfo(null,-1);}
    public void onProviderEnabled(String arg0){}
    public void onStatusChanged(String arg0,int arg1,Bundle arg2){}};
/***功能描述: 获得最近GPS数据 */
private gpsdata getLastPosition(){
    gpsdata result=new gpsdata();
    mLocation=mLocationManager.getLastKnownLocation(mProvider);
    if(mLocation != null){
        result.longitude=(int)(mLocation.getLongitude()*1E6);
        result.latitude=(int)(mLocation.getLatitude()*1E6);
        result.high=mLocation.getAltitude();
        result.direct=mLocation.getBearing();
        result.speed=mLocation.getSpeed();Date d=new Date();
        result.gpstime=CommonUtils.GetCurGpsTimeString();d=null;}
    return result;}
/***功能描述: 显示GPS信息*/
private void showInfo(gpsdata cdata, int infotype){
    if(infotype==-1){tvLongitude.setText("GPS功能已关闭");
       tvLatitude.setText("");tvHigh.setText("");
       tvDirection.setText("");tvSpeed.setText("");
       tvGpsTime.setText(""); tvInfoType.setText("");
    } else{if(cdata==null)    return;
       tvLongitude.setText(String.format("经度:%d", cdata.longitude));
```

```
        tvLatitude.setText(String.format("纬度:%d", cdata.latitude));
        tvHigh.setText(String.format("海拔:%f", cdata.high));
        tvDirection.setText(String.format("方向:%f", cdata.direct));
        tvSpeed.setText(String.format("速度:%f", cdata.speed));
        tvGpsTime.setText(String.format("GPS时间:%s", cdata.gpstime));
        cdata.infotype=infotype;
        switch(infotype){
        case 1:tvInfoType.setText("信息源:手动获取更新");break;
        case 2:tvInfoType.setText("信息源:位置改变更新");break;}
        //将GPS数据写到SQLite数据库
        if(cdata.longitude!=0&&cdata.latitude!=0&&cdata.high!=0
                && cdata.gpstime!=null){
            mDBGps.insert_GpsData(cdata);}
        use_time=tempTime+System.currentTimeMillis()-startTime;}}
```

主程序GPSPedometerMainActivity类的实现计步器功能核心实现代码如下：

```
mStepValueView=(TextView)findViewById(R.id.step_value);
mPaceValueView=(TextView)findViewById(R.id.pace_value);
mDistanceValueView=(TextView)findViewById(R.id.distance_value);
mSpeedValueView=(TextView)findViewById(R.id.speed_value);
mCaloriesValueView=(TextView)findViewById(R.id.calories_value);
mDesiredPaceView=(TextView)findViewById(R.id.desired_pace_value);
Handler mUseTimerhandler=new Handler(){  //Handler对象用于更新当前步数
    //这个方法是从父类/接口继承过来的，需要重写一次
    public void handleMessage(Message msg){
    super.handleMessage(msg);
    mStepValueView.setText(mStepValue+"");  //显示当前步数
    //显示路程
    mDistanceValueView.setText(formatDouble((double)mDistanceValue));
    //显示卡路里
    mCaloriesValueView.setText(formatDouble((double)mCaloriesValue));
    //显示速度
    mSpeedValueView.setText(formatDouble((double)mSpeedValue));
    //显示当前运行时间
    tvTimer.setText(CommonUtils.getFormatTime(use_time));
    mPaceValueView.setText(""+(int)mPaceValue);
    pcdata.pc_distance=mDistanceValue;
    pcdata.pc_calories=mCaloriesValue;
    pcdata.pc_velocity=mSpeedValue;
    pcdata.pc_usetime=use_time;
    pcdata.pc_step_counter=mStepValue;
    ssdata.ss_distance=mDistanceValue;
    ssdata.ss_calories=mCaloriesValue;
```

```
    ssdata.ss_velocity=mSpeedValue;
    ssdata.ss_usetime=use_time;
    ssdata.ss_step_counter=mStepValue;
    ssdata.ss_time=CommonUtils.GetCurGpsTimeString();
    mDBGps.insert_ssd(ssdata);  //将计步器数据写到SQLite数据库}};
    private StepService.ICallback mCallback=new StepService.ICallback(){
    public void stepsChanged(int value){
        mHandler.sendMessage(mHandler.obtainMessage(STEPS_MSG,value,0));}
    public void paceChanged(int value){
        mHandler.sendMessage(mHandler.obtainMessage(PACE_MSG,value,0));}
    public void distanceChanged(float value){
        mHandler.sendMessage(mHandler.obtainMessage(DISTANCE_MSG,
          (int)(value *1000),0));}
    public void speedChanged(float value){
         mHandler.sendMessage(mHandler.obtainMessage(SPEED_MSG,
          (int)(value *1000), 0));}
    public void caloriesChanged(float value){
         mHandler.sendMessage(mHandler.obtainMessage(CALORIES_MSG,
          (int)(value), 0));}};
    private static final int STEPS_MSG=1;
    private static final int PACE_MSG=2;
    private static final int DISTANCE_MSG=3;
    private static final int SPEED_MSG=4;
    private static final int CALORIES_MSG=5;
    private Handler mHandler=new Handler(){
        public void handleMessage(Message msg){
            switch(msg.what){
            case STEPS_MSG:
                mStepValue=(int)msg.arg1;
                sStepValue=mStepValue;
                mStepValueView.setText(""+mStepValue);
                pcdata.pc_step_counter=mStepValue;
                ssdata.ss_step_counter=mStepValue;
                ssdata.ss_time=CommonUtils.GetCurGpsTimeString();
                mDBGps.insert_ssd(ssdata);  //将计步器数据写到SQLite数据库
                break;
        case PACE_MSG:
          mPaceValue=msg.arg1;
          if(mPaceValue<=0){
             mPaceValueView.setText("0");
             pcdata.pc_pacevalue=0;
             ssdata.ss_pacevalue=0;
             ssdata.ss_time=CommonUtils.GetCurGpsTimeString();
```

```
        mDBGps.insert_ssd(ssdata);  //将计步器数据写到SQLite数据库
    } else{
        mPaceValueView.setText(""+(int)mPaceValue);
        pcdata.pc_pacevalue=mPaceValue;
        ssdata.ss_pacevalue=mPaceValue;
        ssdata.ss_time=CommonUtils.GetCurGpsTimeString();
        mDBGps.insert_ssd(ssdata);}  //将计步器数据写到SQLite数据库
      break;
    case DISTANCE_MSG:
        mDistanceValue=((int)msg.arg1)/1000f;
        if(mDistanceValue <= 0){
            mDistanceValueView.setText("0");
            pcdata.pc_distance=0;
            ssdata.ss_distance=0;
            ssdata.ss_time=CommonUtils.GetCurGpsTimeString();
            mDBGps.insert_ssd(ssdata);  //将计步器数据写到SQLite数据库
        } else{
               mDistanceValueView
                    .setText((""+(mDistanceValue+0.000001f))
                         .substring(0, 5));
               pcdata.pc_distance=mDistanceValue;
               ssdata.ss_distance=mDistanceValue;
               ssdata.ss_time=CommonUtils.GetCurGpsTimeString();
               mDBGps.insert_ssd(ssdata);}  //将计步器数据写到SQLite数据库
             break;
        case SPEED_MSG:
            mSpeedValue=((int)msg.arg1)/1000f;
            if(mSpeedValue<=0){
                mSpeedValueView.setText("0");
                pcdata.pc_velocity=0;
                ssdata.ss_velocity=0;
                ssdata.ss_time=CommonUtils.GetCurGpsTimeString();
                mDBGps.insert_ssd(ssdata);  //将计步器数据写到SQLite数据库
        } else{
                mSpeedValueView.setText((""+(mSpeedValue+0.000001f))
        .substring(0, 4));
    pcdata.pc_velocity=mSpeedValue;
    ssdata.ss_velocity=mSpeedValue;
    ssdata.ss_time=CommonUtils.GetCurGpsTimeString();
    mDBGps.insert_ssd(ssdata);}  //将计步器数据写到SQLite数据库
  break;
case CALORIES_MSG:mCaloriesValue=msg.arg1;
  if(mCaloriesValue <= 0){mCaloriesValueView.setText("0");
```

```
                pcdata.pc_calories=0;
                ssdata.ss_calories=0;
                ssdata.ss_time=CommonUtils.GetCurGpsTimeString();
                mDBGps.insert_ssd(ssdata); //将计步器数据写到SQLite数据库
            } else{
                mCaloriesValueView.setText(""+(int)mCaloriesValue);
                pcdata.pc_calories=(int)mCaloriesValue;
                ssdata.ss_calories=(int)mCaloriesValue;
                ssdata.ss_time=CommonUtils.GetCurGpsTimeString();
                mDBGps.insert_ssd(ssdata);} //将计步器数据写到SQLite数据库
              break;
             default:
                 super.handleMessage(msg);}}};
      private void savePaceSetting(){
      mPedometerSettings.savePaceOrSpeedSetting(mMaintain,
         mDesiredPaceOrSpeed);}
      private StepService mService;
      private ServiceConnection mConnection=new ServiceConnection(){
          public void onServiceConnected(ComponentName className, IBinder service)
{mService=((StepService.StepBinder)service).getService();
          mService.registerCallback(mCallback);
          mService.reloadSettings();   }
        public void onServiceDisconnected(ComponentName className){
          mService=null;}};
    private void startStepService(){
        if(!mIsRunning){Log.i(TAG,"[SERVICE]Start");mIsRunning=true;
          StartThread(); //开始计步器计时
          startService(new Intent(GPSPedometerMainActivity.this,
              StepService.class));}}
    private void bindStepService(){Log.i(TAG,"[SERVICE] Bind");
        bindService(
          new Intent(GPSPedometerMainActivity.this, StepService.class),
          mConnection, Context.BIND_AUTO_CREATE
              +Context.BIND_DEBUG_UNBIND);}
    private void unbindStepService(){Log.i(TAG,"[SERVICE]Unbind");
        unbindService(mConnection);}
    private void stopStepService(){Log.i(TAG,"[SERVICE]Stop");
        if(mService!=null){Log.i(TAG,"[SERVICE]stopService");
          stopService(new Intent(GPSPedometerMainActivity.this,
              StepService.class));}
        mUseTimerhandler.removeCallbacks(thread);mIsRunning=false;}
```

主程序GPSPedometerMainActivity类的百度地图功能核心实现代码如下：

```
    //在使用SDK各组件之前初始化context信息，传入ApplicationContext
    SDKInitializer.initialize(getApplicationContext());
    //注册 SDK 广播监听者
    IntentFilter iFilter=new IntentFilter();iFilter.addAction(SDKInitializer.SDK_
BROADTCAST_ACTION_STRING_PERMISSION_CHECK_OK);
    iFilter.addAction(SDKInitializer.SDK_BROADTCAST_ACTION_STRING_PERMISSION_CHECK_
ERROR);
    iFilter.addAction(SDKInitializer.SDK_BROADCAST_ACTION_STRING_NETWORK_ERROR);
    mReceiver=new SDKReceiver();registerReceiver(mReceiver, iFilter);
    bMainMapView=(MapView)findViewById(R.id.bMainMapView);  //获取地图控件引用
    bMainMapView.setMapCustomEnable(true);
    mMainBaiduMap=bMainMapView.getMap();
    mMainBaiduMap.setOnMapLongClickListener(this);  //设置地图长按事件侦听器
    mMainBaiduMap.setOnMarkerClickListener(this);  //设置标记单击事件侦听器
    mMainBaiduMap.setOnMapClickListener(this);  //设置地图单击事件侦听器
    /***定位SDK监听函数*/
    public class MyLocationListenner implements BDLocationListener{
        @Override
        public void onReceiveLocation(BDLocation location){
            //map view 销毁后不再处理新接收的位置
            if(location==null||bMainMapView==null){return;}
            mCurrentLat=location.getLatitude();
            mCurrentLon=location.getLongitude();
            //记录当前实时GPS位置
            mRealTimePolylines.add(new LatLng(mCurrentLat, mCurrentLon));
            if(isStartWalking==true)insertGeoPoint(location);
            if(mRealTimePolylines.size()>2){
                PolylineOptions mRealTimePolylineOptions=new
PolylineOptions().points(mRealTimePolylines).width(10).color(Color.RED);
                mPolyline=(Polyline)mMainBaiduMap
                    .addOverlay(mRealTimePolylineOptions);}
                mCurrentAccracy=location.getRadius();
                locData=new MyLocationData.Builder()
                    .accuracy(location.getRadius())
                  //此处设置开发者获取到的方向信息，顺时针0~360。
                  .direction(mCurrentDirection)
                  .latitude(location.getLatitude())
                  .longitude(location.getLongitude()).build();
          mMainBaiduMap.setMyLocationData(locData);
          if(isChangeMapCenter)  //是否实时更新地图中心
          {  isFirstLoc=false;
             LatLng ll=new LatLng(location.getLatitude(),
                 location.getLongitude());
```

```
            MapStatus.Builder builder=new MapStatus.Builder();
            builder.target(ll).zoom(19.0f);
            mMainBaiduMap.animateMapStatus(MapStatusUpdateFactory
                .newMapStatus(builder.build()));}}
    public void onReceivePoi(BDLocation poiLocation){}}
```

主程序运行结果如图 7-7 所示。

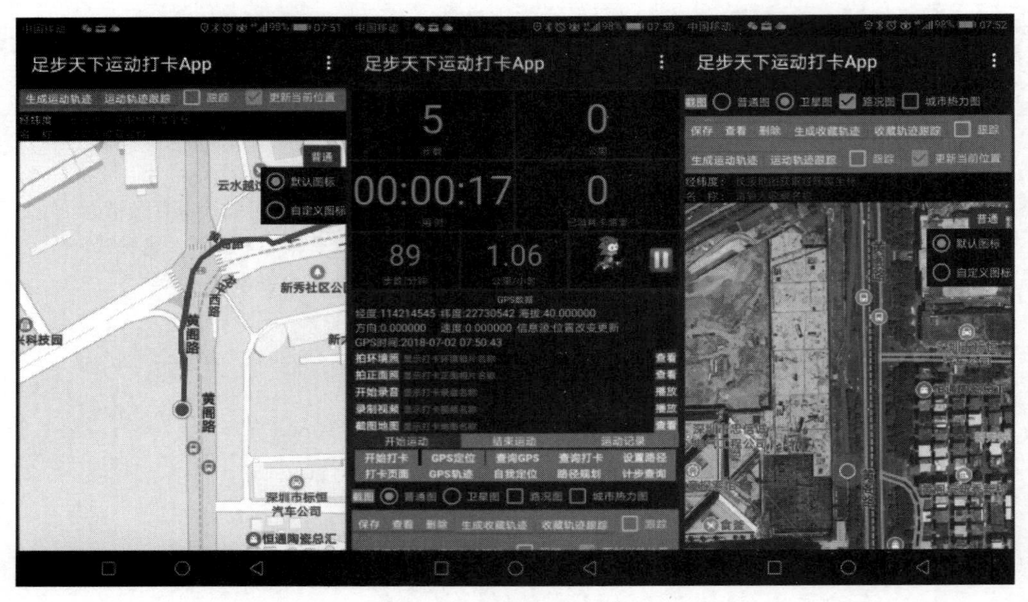

图 7-7　主程序运行结果

7.4.4　运动打卡功能页面

实现足步天下运动打卡功能，包括运动拍照、录音、录视频和照片、录音和视频回放功能等。先设计运动打卡布局activity_punch_clock.xml，然后编码实现PunchClockActivity类，运动打卡运行结果如图 7-8 所示。其核心代码如下：

```
public class PunchClockActivity extends Activity implements OnClickListener{
    protected void onCreate(Bundle savedInstanceState){
        super.onCreate(savedInstanceState);
        requestWindowFeature(Window.FEATURE_NO_TITLE);
        setContentView(R.layout.activity_punch_clock);
        initModel();initView();}
    public void onClick(View v){
      switch(v.getId()){
      case R.id.btnTakeEnvPhoto:
          invokSystemCamera(newEnvName, UI_SYSTEM_CAMERA_BACK);break;
      case R.id.EnvImage:  //放大显示图片
      case R.id.EnvImageName:ShowEnvImage();break;
      case R.id.btnTakeMyPhoto:
```

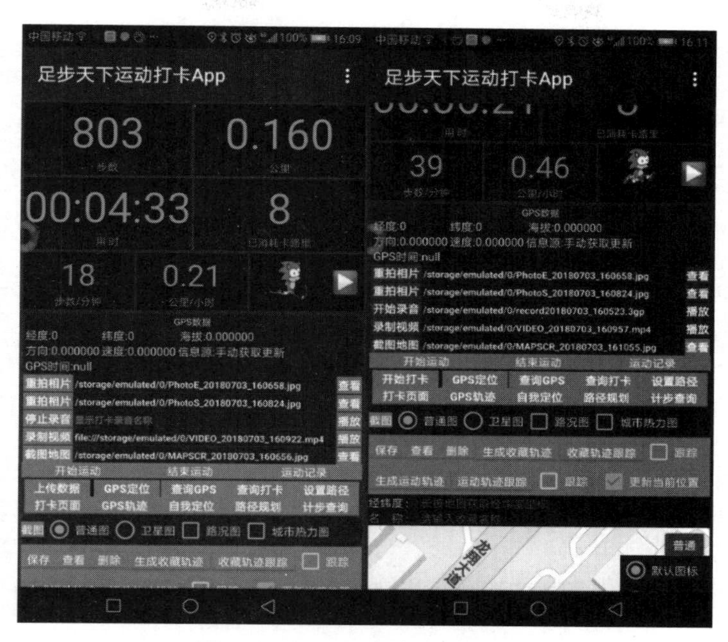

图 7-8　运动打卡功能运行结果

```
        invokSystemCamera(newMyName, UI_SYSTEM_CAMERA_MYBACK);break;
    case R.id.MyImage:  //放大显示图片
    case R.id.MyImageName:ShowMyImage();break;
    case R.id.record_bt:  RecordMyMusic();break;
    case R.id.play_bt:
    case R.id.record_name:PlayMyRecordMusic();break;
    case R.id.video_bt:CaptureMyVideo();break;
    case R.id.playvideo_bt:
    case R.id.video_name:
        if(!video_name.getText().toString().trim().equals("")){
           PlayMyVideo();}break;
    case R.id.gps_bt:
        showInfo(getLastPosition(), 1);break;
    case R.id.save_bt:
        if(isValidData()){
            my_gpsdata=getLastPosition();
            dbgps.insert_pcd(my_gpsdata, pcdata);
            dbgps.insert_GpsData(my_gpsdata);
            SetTextNull();
    } else{DisplayPromptInfo();}break;
    case R.id.query_pcd_bt:
        QueryPunchClockData();break;
    case R.id.exit_bt:PunchClockActivity.this.finish();break;}}
private void RecordMyMusic(){
```

```
    onRecord(mStartRecording);
    if(mStartRecording){record_bt.setText("停止录音");} else{
       record_bt.setText("开始录音");}
    mStartRecording=!mStartRecording;}
private void PlayMyRecordMusic(){
    if(!(record_name.getText().toString().trim().equals(""))){
       onPlay(mStartPlaying);
       if(mStartPlaying){mPlayButton.setText("停止");
    } else{mPlayButton.setText("开始");}
       mStartPlaying=!mStartPlaying;}}
private void QueryPunchClockData(){  //GPS打卡按钮
   Intent intent=new Intent(PunchClockActivity.this,
    QueryPunchClockDataActivity.class);startActivity(intent);}
private void ShowEnvImage(){  //显示打卡环境相片
    if(!(EnvImageName.getText().toString().trim().equals(""))){
       Intent picture=new Intent(PunchClockActivity.this,
          PhotoActivity.class);
       picture.putExtra("photo", path+newEnvName);
       startActivity(picture);}}
private void CaptureMyVideo(){  //录制视频
   Intent captureVideoIntent=new Intent(
       android.provider.MediaStore.ACTION_VIDEO_CAPTURE);
   videoFileUri=getOutputMediaFileUri();  //创建一个文件Uri来存储视频
   if(!(videoFileUri.toString().trim().equals(""))){
     video_name.setText(videoFileUri.toString().trim());
     //设置视频文件名
     captureVideoIntent.putExtra(MediaStore.EXTRA_OUTPUT,videoFileUri);
     //设置视频质量为“高”
     captureVideoIntent.putExtra(MediaStore.EXTRA_VIDEO_QUALITY,1);}
       startActivityForResult(captureVideoIntent, VIDEO_CAPTURED);}
private void PlayMyVideo(){
    if((!(videoFileUri.toString().trim().equals("")))
          &&(!(video_name.getText().toString().trim().equals("")))){
        Intent intent=new Intent(Intent.ACTION_VIEW);
        String type="video/mp4";
        Uri name=Uri.parse(videoFileUri.toString());
        intent.setDataAndType(name,type);startActivity(intent);}}
/**创建用于存储视频的文件Uri */
private Uri getOutputMediaFileUri(){  //获得SD卡路径
   File sdDir=null;
   boolean sdCardExist=Environment.getExternalStorageState().equals(
     Environment.MEDIA_MOUNTED); //判断sd卡是否存在
   if(sdCardExist){
```

```
        sdDir=Environment.getExternalStorageDirectory();} //获取根目录
    if(sdDir!=null){  //获得当前时间
        String timeStamp=new SimpleDateFormat("yyyyMMdd_HHmmss")
            .format(new Date());
        File videoFile=new File(sdDir.getPath()+File.separator
            +"VIDEO_"+timeStamp+".mp4");
        return Uri.fromFile(videoFile);
    } else{Log.d(TAG, "SD卡不存在!!!");return null;}}
private void ShowMyImage(){
    if(!(MyImageName.getText().toString().trim().equals(""))){
        Intent picture=new Intent(PunchClockActivity.this,
            PhotoActivity.class);
        picture.putExtra("photo", path+newMyName);
        startActivity(picture);}}
private void onPlay(boolean start){
    if(start){startPlaying();} else{
        stopPlaying();}}
private void startPlaying(){  //开始播放
    mPlayer=new MediaPlayer();
    try{
        mPlayer.setDataSource(mFileName); //设置播放路径
        mPlayer.prepare();  //准备
        mPlayer.start();  //开始
        mPlayer.setOnCompletionListener(new OnCompletionListener(){
            public void onCompletion(MediaPlayer mp){
                mPlayButton.setText("开始播放");
                mStartPlaying=!mStartPlaying;}});
    } catch(IOException e){MyLog.d("prepare()failed");}}
private void stopPlaying(){  //停止播放
    mPlayer.stop();mPlayer.release();mPlayer=null;}
private void onRecord(boolean start){
    if(start){startRecord();} else{stopRecord();}}
private void startRecord(){  //开始录音
    mRecorder=new MediaRecorder();
    //设置音源为Micphone
    mRecorder.setAudioSource(MediaRecorder.AudioSource.MIC);
    //设置封装格式
    mRecorder.setOutputFormat(MediaRecorder.OutputFormat.THREE_GPP);
    //设置编码格式
    mRecorder.setAudioEncoder(MediaRecorder.AudioEncoder.AMR_NB);
    mRecorder.setOutputFile(mFileName);file=new File(mFileName);
    try{file.createNewFile();mRecorder.prepare();
} catch(IOException e){e.printStackTrace();}
```

```
        mRecorder.start();}
    private void stopRecord(){  //停止录音
        mRecorder.stop();  mRecorder.release();  mRecorder=null;
        record_name.setText(mFileName.toString());
        pcdata.pc_sound=mFileName.toString();}
    private void invokSystemCamera(){/***调用系统照相机方法*/
        Intent cameraIntent=new Intent(MediaStore.ACTION_IMAGE_CAPTURE);
        if(cameraIntent.resolveActivity(getPackageManager())!= null){
          //判断一个activity是否存在于系统中
            photoFile=new File(Environment.getExternalStorageDirectory(),
                newEnvName);
            if(photoFile != null){
                cameraIntent.putExtra(MediaStore.EXTRA_OUTPUT,Uri.fromFile(photoFile));
                startActivityForResult(cameraIntent, UI_SYSTEM_CAMERA_BACK);}}}
  /***调用系统照相机方法*/
  private void invokSystemCamera(String newName,int Ui_System_Camera_Back)
  {   Intent cameraIntent=new Intent(MediaStore.ACTION_IMAGE_CAPTURE);
      if(cameraIntent.resolveActivity(getPackageManager())!= null){
      //判断一个activity是否存在于系统中
          photoFile=new File(Environment.getExternalStorageDirectory(),
              newName);
          if(photoFile!=null){
  cameraIntent.putExtra(MediaStore.EXTRA_OUTPUT,Uri.fromFile(photoFile));
  startActivityForResult(cameraIntent, Ui_System_Camera_Back);}}}
    @Override
    public void onActivityResult(int requestCode, int resultCode, Intent data){
        super.onActivityResult(requestCode, resultCode, data);
        if(resultCode==RESULT_OK){
            switch(requestCode){
            case UI_SYSTEM_CAMERA_BACK:  //设置文件保存路径，这里放在根目录下
              StartEnvPhotoZoom();  break;
            case UI_PHOTO_ZOOM_BACK:PhotoEnvZoomBack();break;
            case UI_SYSTEM_CAMERA_MYBACK:StartMyPhotoZoom();break;
            case UI_PHOTO_ZOOM_MYBACK:PhotoMyZoomBack();break;
            case VIDEO_CAPTURED:
                videoFileUri=data.getData();playvideo_bt.setEnabled(true);
                video_name.setText(videoFileUri.getPath());
                pcdata.pc_video=videoFileUri.getPath();break;}}}
    private void PhotoEnvZoomBack(){
       File file=new File(path+newEnvName);
       if(file.exists()){
          Bitmap photo=BitmapFactory.decodeFile(path+newEnvName);
          EnvImage.setImageBitmap(photo);
```

```
        btnTakeEnvPhoto.setText("重拍相片");}}
    private void PhotoMyZoomBack(){
        File file=new File(path+newMyName);
        if(file.exists()){
            Bitmap photo=BitmapFactory.decodeFile(path+newMyName);
            MyImage.setImageBitmap(photo);btnTakeMyPhoto.setText("重拍相片");}}
    private void StartEnvPhotoZoom(){
        File picture=new File(Environment.getExternalStorageDirectory()
            +File.separator+newEnvName);
        startPhotoZoom(Uri.fromFile(picture), newEnvName, UI_PHOTO_ZOOM_BACK);
        EnvImageName.setText(path+newEnvName);
        pcdata.pc_environphoto=path+newEnvName;}
    private void StartMyPhotoZoom(){
        File picture=new File(Environment.getExternalStorageDirectory()
            +File.separator+newMyName);
        startPhotoZoom(Uri.fromFile(picture), newMyName, UI_PHOTO_ZOOM_MYBACK);
        MyImageName.setText(path+newMyName);
        pcdata.pc_selfiephoto=path+newMyName;}}
```

7.4.5 SQLite 嵌入式数据库 DBGps 功能实现

实现运动打卡 SQLite 嵌入式数据库功能。DBGps 类核心代码如下：

```
public class DBGps{
    private static final String DATABASENAME="GpsStepData.db";
    private static final int VERSION=1; //版本
    //定义保存运动轨迹的GPS数据表
    public static final String GPS_TABLENAME="gps_geopoint";  //gps数据表
    public static final String GEO_ID="id"; //编码ID
    public static final String GEO_LONGITUDE="longitude";  //经度
    public static final String GEO_LATITUDE="latitude"; //纬度
    public static final String GEO_HIGH="high"; //海拔高度
    public static final String GEO_DIRECT="direct"; //方向
    public static final String GEO_SPEED="speed"; //速度
    public static final String GEO_GPSTIME="gpstime"; //速度
    public static final String GEO_INFOTYPE="infotype";  //数据来源
    //创建保存的GPS采集路径
    public final static String SPORT_GPS_TABLENAME="SportGPSGeopoint";
    public final static String SPORTGPS_ID="id";
    public final static String SPORTGPS_LONGITUDE="longitude";
    public final static String SPORTGPS_LATITUDE="latitude";
  private final Context ct;
  private SQLiteDatabase db;
  private SQLiteDB sdb;
```

```
private static class SQLiteDB extends SQLiteOpenHelper{
    public SQLiteDB(Context context){
        super(context, DATABASENAME, null, 1);}
    public void onCreate(SQLiteDatabase sdb){  //建gps表
        String gps_sql="CREATE TABLE "+GPS_TABLENAME+"("+GEO_ID
            +"INTEGER PRIMARY KEY  AUTOINCREMENT,"+GEO_LONGITUDE
            +"INTEGER,"+GEO_LATITUDE+"INTEGER,"+GEO_HIGH
            +"DOUBLE,"+GEO_DIRECT+"DOUBLE,"+GEO_SPEED
            +"DOUBLE,"+GEO_GPSTIME+"DATE,"+GEO_INFOTYPE
            +"INTEGER);";
        sdb.execSQL(gps_sql);
        //建运动打卡表
        String punch_clock_sql="CREATE TABLE "+PUNCH_CLOCK_TABLENAME
            +"("+PUNCH_CLOCK_ID
            +"INTEGER PRIMARY KEY  AUTOINCREMENT,"
            +PUNCH_CLOCK_LONGITUDE+"INTEGER,"
            +PUNCH_CLOCK_LATITUDE+"INTEGER,"+PUNCH_CLOCK_HIGH
            +"DOUBLE,"+PUNCH_CLOCK_DIRECT+"DOUBLE,"
            +PUNCH_CLOCK_SPEED+"DOUBLE,"+PUNCH_CLOCK_GPSTIME
            +"DATE,"+PUNCH_CLOCK_INFOTYPE+"INTEGER,"
            +PUNCH_CLOCK_TIME+"DATE,"+PUNCH_CLOCK_USE_TIME
            +"INTEGER,"+PUNCH_CLOCK_DISTANCE+"DOUBLE,"
            +PUNCH_CLOCK_CALORIES+"DOUBLE,"+PUNCH_CLOCK_VELOCITY
            +"DOUBLE,"+PUNCH_CLOCK_PACEVALUE+"DOUBLE,"
            +PUNCH_CLOCK_STEP_COUNTER+"INTEGER,"
            +PUNCH_CLOCK_ENVIRONPHOTO+"VARCHAR(56),"
            +PUNCH_CLOCK_SELFIEPHOTO+"VARCHAR(56),"
            +PUNCH_CLOCK_SOUND+"VARCHAR(56),"
            +PUNCH_CLOCK_VIDEO  +"VARCHAR(56),"
               +PUNCH_CLOCK_MPACREEN  +"VARCHAR(56));";
        sdb.execSQL(punch_clock_sql);
        //创建GPS位置信息表
        String geop_sql="CREATE TABLE "+
              GEOP_TABLENAME+"("+
            GEOP_ID+"INTEGER PRIMARY KEY AUTOINCREMENT,"+
            GEOP_LONGITUDE+"DOUBLE,"+
             GEOP_LATITUDE  +"DOUBLE,"+
            GEOP_TRACKNAME+"TEXT,"+
            GEOP_STEP_VALUE+"INTEGER,"+
            GEOP_DISTANCE_VALUE+"DOUBLE,"+
            GEOP_USE_TIME+"INTEGER,"+
            GEOP_CALORIES_VALUE+"DOUBLE,"+
            GEOP_PACE_VALUE+"DOUBLE,"+
```

```
                GEOP_SPEED_VALUE+"DOUBLE,"+
                GEOP_TIME  +"TEXT);";
            sdb.execSQL(geop_sql);
            //创建轨迹跟踪表
            String track_sql="CREATE TABLE "+
            TRACK_TABLENAME+"("+
            TRACK_ID+"INTEGER PRIMARY KEY  AUTOINCREMENT,"  +
            TRACK_NAME+"TEXT,"+
            TRACK_CREATETIME+"TEXT,"  +
            TRACK_DIST+"TEXT,"+
            TRACK_SPEED+"TEXT,"+
            TRACK_COUNT+"TEXT,"  +
            TRACK_STEP_VALUE+"INTEGER,"+
            TRACK_DISTANCE_VALUE+"DOUBLE,"+
            TRACK_USE_TIME+"INTEGER,"+
            TRACK_CALORIES_VALUE+"DOUBLE,"+
            TRACK_PACE_VALUE+"DOUBLE,"+
            TRACK_SPEED_VALUE+"DOUBLE,"+
            TRACK_ENDTIME+"TEXT);";
            sdb.execSQL(track_sql);
            //创建GPS采集数据运动路径表
            String sportgps_sql="CREATE TABLE "+
              SPORT_GPS_TABLENAME+"("+
              SPORTGPS_ID+"INTEGER PRIMARY KEY  AUTOINCREMENT,"  +
              SPORTGPS_LONGITUDE+"INTEGER,"  +
              SPORTGPS_LATITUDE+"INTEGER);";
              sdb.execSQL(sportgps_sql);}
      public void onUpgrade(SQLiteDatabase sdb, int oldVersion, int
newVersion){  //删除表GPS_TABLENAME
          String sql1="DROP TABLE IF EXISTS "+GPS_TABLENAME;
          sdb.execSQL(sql1);
          String sql2="DROP TABLE IF EXISTS "+PUNCH_CLOCK_TABLENAME;
          sdb.execSQL(sql2); //删除表PUNCH_CLOCK_TABLENAME
          String sql3="DROP TABLE IF EXISTS "+SENSOR_STEP_TABLENAME;
          sdb.execSQL(sql3); //删除表SENSOR_STEP_TABLENAME
          String sql4="DROP TABLE IF EXISTS "+TRACK_TABLENAME;
          sdb.execSQL(sql4);  //删除表GEOP_TABLENAME
          String sql5="DROP TABLE IF EXISTS "+SET_MOVE_PATH_TABLENAME;
        sdb.execSQL(sql5); //删除表GEOP_TABLENAME
        String sql6="DROP TABLE IF EXISTS "+SPORT_GPS_TABLENAME;
        sdb.execSQL(sql6);onCreate(sdb);}}
    public DBGps(Context context){  //初始化数据库
        ct=context;  sdb=new SQLiteDB(ct);}
```

```
    public SQLiteDatabase openDB(){  //打开数据库
        return db=sdb.getWritableDatabase();}
    public void closeDB(){sdb.close();}
    /**将GPS数据插入数据库GpsStepData.db的数据表gps_geopoint中**/
    public boolean addGpsData(gpsdata cdata){boolean result=true;
        try{
          String StrSql=String.format("INSERT INTO "+GPS_TABLENAME+"("
              +GEO_LONGITUDE+","+GEO_LATITUDE+","+GEO_HIGH+","
              +GEO_DIRECT+","+GEO_SPEED+","+GEO_GPSTIME+","
              +GEO_INFOTYPE+")"
              +"values(%d,%d,%.1f,%.1f,%.1f,' %s' ,%d)", cdata.longitude,
              cdata.latitude, cdata.high, cdata.direct, cdata.speed,
              cdata.gpstime, cdata.infotype);
          db.execSQL(StrSql); result=true;
        } catch(Exception e){result=false;}
        return result;}
    public boolean addSensorStepData(sensorstepdata ssdata){
        boolean result=true;
        try{  String StrSql=String.format("INSERT INTO "
              +SENSOR_STEP_TABLENAME+"("+SENSOR_STEP_TIME+","
              +SENSOR_STEP_USE_TIME+","+SENSOR_STEP_DISTANCE+","
              +SENSOR_STEP_CALORIES+","+SENSOR_STEP_VELOCITY+","
              +SENSOR_STEP_PACEVALUE+","+SENSOR_STEP_COUNTER+","
              +SENSOR_STEP_STAR_NUMBER+")"
              +"values( '%s' ,%d,%.1f,%.1f,%.1f,%.1f,%d,%d)",
              ssdata.ss_time, ssdata.ss_usetime, ssdata.ss_distance,
              ssdata.ss_calories, ssdata.ss_velocity,
              ssdata.ss_pacevalue, ssdata.ss_step_counter,
              ssdata.ss_star_number);
          db.execSQL(StrSql);  result=true;
        } catch(Exception e){result=false;}
        return result;}
    public Cursor select(String table){  /***查询数据表**/
        SQLiteDatabase db=sdb.getReadableDatabase();
        Cursor cursor=db.query(table, null, null, null, null, null, null);
        return cursor;}
    public Cursor select_pcd(){  /***查询打卡数据*/
        SQLiteDatabase db=sdb.getReadableDatabase();
        Cursor cursor=db.query(PUNCH_CLOCK_TABLENAME, null, null, null, null,
null, null);return cursor;
    public long insert_pcd(gpsdata cdata, punchclockdata pcdata){
        SQLiteDatabase db=sdb.getReadableDatabase();
        ContentValues cv=new ContentValues();
```

```
        cv.put(PUNCH_CLOCK_LONGITUDE, cdata.longitude);
        cv.put(PUNCH_CLOCK_LATITUDE, cdata.latitude);
        cv.put(PUNCH_CLOCK_HIGH, cdata.high);
        cv.put(PUNCH_CLOCK_DIRECT, cdata.direct);
        cv.put(PUNCH_CLOCK_SPEED, cdata.speed);
        cv.put(PUNCH_CLOCK_GPSTIME, cdata.gpstime);
        cv.put(PUNCH_CLOCK_INFOTYPE, cdata.infotype);
        cv.put(PUNCH_CLOCK_TIME, pcdata.pc_time);
        cv.put(PUNCH_CLOCK_USE_TIME, pcdata.pc_usetime);
        cv.put(PUNCH_CLOCK_DISTANCE, pcdata.pc_distance);
        cv.put(PUNCH_CLOCK_CALORIES, pcdata.pc_calories);
        cv.put(PUNCH_CLOCK_VELOCITY, pcdata.pc_velocity);
        cv.put(PUNCH_CLOCK_PACEVALUE, pcdata.pc_pacevalue);
        cv.put(PUNCH_CLOCK_STEP_COUNTER, pcdata.pc_step_counter);
        cv.put(PUNCH_CLOCK_ENVIRONPHOTO, pcdata.pc_environphoto);
        cv.put(PUNCH_CLOCK_SELFIEPHOTO, pcdata.pc_selfiephoto);
        cv.put(PUNCH_CLOCK_SOUND, pcdata.pc_sound);
        cv.put(PUNCH_CLOCK_VIDEO, pcdata.pc_video);
        cv.put(PUNCH_CLOCK_MPACREEN, pcdata.pc_mapscreen);
        try{long row=db.insert(PUNCH_CLOCK_TABLENAME, null, cv);
            return row;} catch(Exception e){return -1;}}
    public long insert_GpsData(gpsdata cdata){
        SQLiteDatabase db=sdb.getReadableDatabase();
        ContentValues cv=new ContentValues();
        cv.put(GEO_LONGITUDE, cdata.longitude);
        cv.put(GEO_LATITUDE, cdata.latitude);
        cv.put(GEO_HIGH, cdata.high);
        cv.put(GEO_DIRECT, cdata.direct);
        cv.put(GEO_SPEED, cdata.speed);
        cv.put(GEO_GPSTIME, cdata.gpstime);
        cv.put(GEO_INFOTYPE, cdata.infotype);
        try{long row=db.insert(GPS_TABLENAME, null, cv);
            return row;}catch(Exception e){return -1;}}}
```

7.4.6　查询 GPS 页面

实现GPS 查询功能。用户在首页中点击查询GPS按钮，可以对运动信息的点进行查询，运行结果如图 7-9 所示。QueryGPSActivity 类核心功能代码如下：

```
public class QueryGPSActivity extends Activity{
    private Button btn_refreshgps;  //更新查询GPS信息按钮
    private Button btn_return_main;  //返回主菜单按钮
    private Button exportgps,importgps;private ListView listView;
    private DBGps dbgps=new DBGps(this);SQLiteDatabase db=null;
```

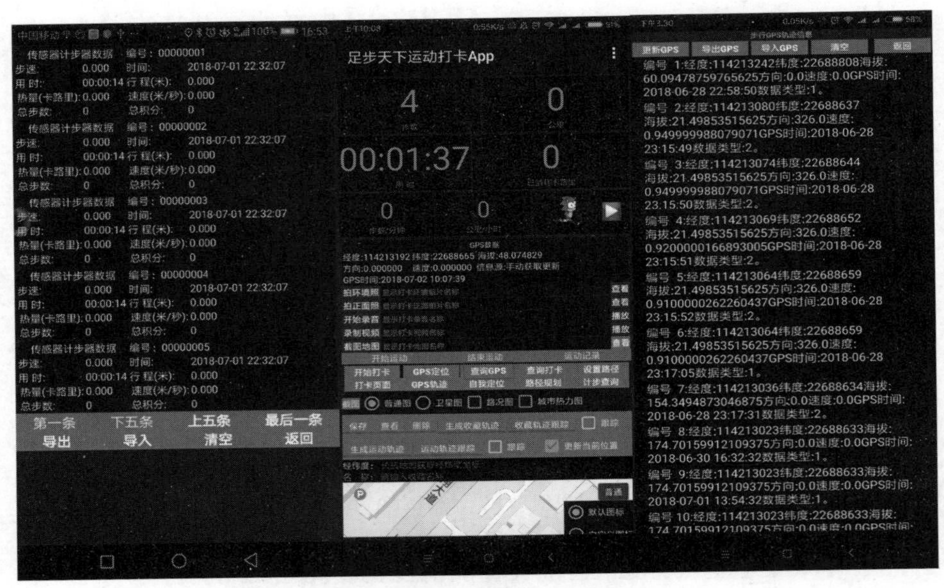

图 7-9　查询 GPS 功能运行结果

```
    gpsdata my_gpsdata=new gpsdata();
    String EXPORT_GPS_FILE_NAME="/myexportgps.txt";
    String myimportgps="/myimportgps.txt";
  /**当活动被第一次创建时调用 */
  public void onCreate(Bundle savedInstanceState){
      super.onCreate(savedInstanceState);
       requestWindowFeature(Window.FEATURE_NO_TITLE);
      setContentView(R.layout.activity_gps_tracker);
      btn_refreshgps=(Button)this.findViewById(R.id.refreshgps);
      btn_return_main=(Button)this.findViewById(R.id.return_main);
      exportgps=(Button)this.findViewById(R.id.exportgps);
      importgps=(Button)this.findViewById(R.id.importgps);
      listView=(ListView)findViewById(R.id.show);
      db=dbgps.openDB();  QueryGPS();  my_gpsdata=new gpsdata();}
  /***功能描述：遍历GPS数据*/
  protected List<String>converCursorToListString(Cursor cursor){
      List<String>result=new ArrayList<String>();
      while(cursor.moveToNext()){  //遍历Cursor结果集
         //将结果集中的数据存入ArrayList中
        StringBuffer sb=new StringBuffer(256 *4 *10);
        sb.append(String.format("编号%3d:", cursor.getInt(cursor.
        getColumnIndex(DBGps.GEO_ID))));  sb.append("经度:");
        sb.append(String.valueOf(cursor.getInt(cursor.
        getColumnIndex(DBGps.GEO_LONGITUDE))));sb.append("纬度:");
        sb.append(String.valueOf(cursor.getInt(cursor.
```

```
            getColumnIndex(DBGps.GEO_LATITUDE))));sb.append("海拔:");
            sb.append(String.valueOf(cursor.getDouble(cursor.
            getColumnIndex(DBGps.GEO_HIGH))));sb.append("方向:");
            sb.append(String.valueOf(cursor.getDouble(cursor.
            getColumnIndex(DBGps.GEO_DIRECT))));sb.append("速度:");
            sb.append(String.valueOf(cursor.getDouble(cursor.
            getColumnIndex(DBGps.GEO_SPEED))));sb.append("GPS时间:");
            sb.append(String.valueOf(cursor.getString(cursor.
            getColumnIndex(DBGps.GEO_GPSTIME))));  sb.append("数据类型:");
            sb.append(String.valueOf(cursor.getInt(cursor.
            getColumnIndex(DBGps.GEO_INFOTYPE))));sb.append("。");
            result.add(sb.toString());}return result;}
    /***功能描述: 将GPS数据导出到文件**/
    protected void writeGpsToFile(Cursor cursor,String filename){
        while(cursor.moveToNext()){ //遍历Cursor结果集
            StringBuffer sb=new StringBuffer(60);
            sb.append(String.format("%3d", cursor.getInt(cursor.
            getColumnIndex(DBGps.GEO_ID))));sb.append(",");
            sb.append(String.valueOf(cursor.getInt(cursor.
            getColumnIndex(DBGps.GEO_LONGITUDE))));sb.append(",");
            ……sb.append(String.valueOf(cursor.getInt(cursor.
            getColumnIndex(DBGps.GEO_INFOTYPE))));sb.append("\n");
            MyFile.write(filename, sb.toString());
            MyFile.write(myimportgps, sb.toString());}}
    private void QueryGPS(){
        try{Cursor cursor=null;
            cursor=db.rawQuery("select *from "+DBGps.GPS_TABLENAME,null);
            List<String>gpsinfo=converCursorToListString(cursor);
            ArrayAdapter<String>adapter=new ArrayAdapter<String>(this, android.
R.layout.simple_list_item_1, gpsinfo);
          listView.setAdapter(adapter); //填充ListView
      } catch(SQLiteException se){}}
    private void exportGPS(){
       try{Cursor cursor=null;
            cursor=db.rawQuery("select *from "+DBGps.GPS_TABLENAME, null);
            EXPORT_GPS_FILE_NAME="/"+"EXPORT_GPS_DATA_"+CommonUtils.
GetCurTimeString()+".txt";
          writeGpsToFile(cursor, EXPORT_GPS_FILE_NAME);
      } catch(SQLiteException se){}}
    private void importGPS(){read(myimportgps);}
    public void onClick(View view){
        switch(view.getId()){
        case R.id.refreshgps:db=dbgps.openDB();QueryGPS();break;
```

```
        case R.id.exportgps:  DoExportGps();  break;
        case R.id.importgps:  DoImportGps();  break;
        case R.id.cleargps:db=dbgps.openDB();clearGPS();QueryGPS();break;
        case R.id.return_main:dbgps.closeDB();db.close();finish();break;}}
    private void DoImportGps(){
        importgps.setEnabled(false);db=dbgps.openDB();
        Thread t=new Thread(new Runnable(){
            public void run(){importGPS();}});
        QueryGPS();  importgps.setEnabled(true);}
    private void DoExportGps(){
        exportgps.setEnabled(false);  db=dbgps.openDB();
        Thread t=new Thread(new Runnable(){
          public void run(){exportGPS();}});
        t.start();exportgps.setEnabled(true);}}
```

7.4.7 步行轨迹跟踪功能页面

实现步行轨迹跟踪功能。用户可以查看自己运动所经过的运动轨迹、生成轨迹图、可导出数，运行结果如图7-10所示。TrackerActivity类核心功能代码如下：

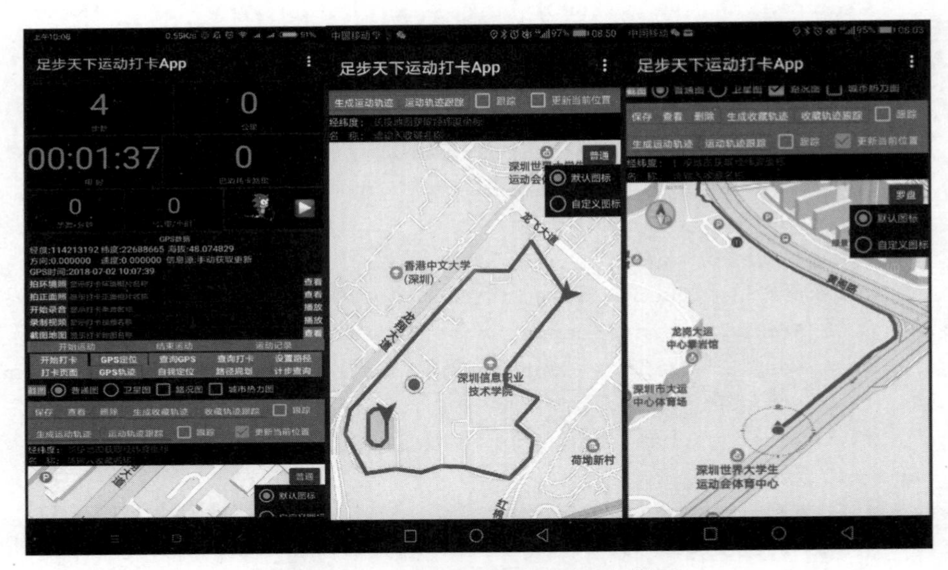

图7-10　步行轨迹跟踪功能运行结果

```
    public class TrackerActivity extends Activity implements
OnMapLongClickListener, OnMarkerClickListener, OnMapClickListener,
SensorEventListener{private com.baidu.mapapi.map.MapView mMapView;
    private BaiduMap mBaiduMap;  private Polyline mPolyline;
    private Marker mMoveMarker;  private Handler mHandler;
    Button btnMode;  CheckBox checkBoxDynThread;
    //通过设置间隔时间和距离可以控制速度和图标移动的距离
```

```
    private static final int TIME_INTERVAL=80;
    private static final double DISTANCE=0.00002;
    private DBGps mDBGps=new DBGps(this);  private SQLiteDatabase mDB=null;
    private Cursor mCursor=null;private Cursor mSportGpsCursor=null;
    protected void onCreate(Bundle savedInstanceState){
        super.onCreate(savedInstanceState);
        setContentView(R.layout.activity_track_detail);
        sportgps=new SportGPS();mCurrentMode=LocationMode.NORMAL;
        btnMode=(Button)findViewById(R.id.btnMode);btnMode.setText("普通");
        btnTrackExport=(Button)findViewById(R.id.btnTrackExport);
        btnTrackImport=(Button)findViewById(R.id.btnTrackImport);
        btnQuerySelecedPath=(Button)findViewById(R.id.btnQuerySelecedPath);
        trackName=getIntent().getStringExtra("trackName");
        android.view.View.OnClickListener btnClickListener=new android.view.View.
OnClickListener(){
            public void onClick(View v){
                switch(mCurrentMode){
                case NORMAL:btnMode.setText("跟随");
                     mCurrentMode=LocationMode.FOLLOWING;
                     mBaiduMap.setMyLocationConfiguration(
            new MyLocationConfiguration(mCurrentMode, true, mCurrentMarker));
                     MapStatus.Builder builder=new MapStatus.Builder();
                     builder.overlook(0);
                     mBaiduMap.animateMapStatus(MapStatusUpdateFactory.newMapStatus
(builder.build())); break;
            case COMPASS:btnMode.setText("普通");
                mCurrentMode=LocationMode.NORMAL;
                mBaiduMap.setMyLocationConfiguration(
        new MyLocationConfiguration(mCurrentMode, true, mCurrentMarker));
            MapStatus.Builder builder1=new MapStatus.Builder();
            builder1.overlook(0);
            mBaiduMap.animateMapStatus(MapStatusUpdateFactory.newMapStatus(builder1.
build()));  break;
            case FOLLOWING:btnMode.setText("罗盘");
                mCurrentMode=LocationMode.COMPASS;
                mBaiduMap.setMyLocationConfiguration(new
    MyLocationConfiguration(mCurrentMode, true, mCurrentMarker));break;}}};
        btnMode.setOnClickListener(btnClickListener);
        checkBoxDynThread=(CheckBox)findViewById(R.id.checkBoxDynThread);
        checkBoxDynThread.setChecked(isThreadRunning);
        mBaiduMap.setMyLocationEnabled(true); //开启定位图层
        mLocClient=new LocationClient(this); //定位初始化
        mLocClient.registerLocationListener(myListener);
```

```
        LocationClientOption option=new LocationClientOption();
        option.setOpenGps(true); //打开gps
        option.setCoorType("bd0911"); //设置坐标类型
        option.setScanSpan(1000);  mLocClient.setLocOption(option);
        mLocClient.start();}
   /***查询全部GPS轨迹数据***/
   private Cursor QuerySelcectGpsTrackData(String trackName){
      try{Cursor cursor=mDB.query(DBGps.GEOP_TABLENAME, null, DBGps.GEOP_
TRACKNAME+"=?", new String[]{trackName}, null, null, null);
         return cursor;} catch(SQLiteException se){return null;}}
   /***绘制GPS轨迹曲线***/
   private void drawSelectedGpsTrack(Cursor cursor){mPolylines.clear();
      while(cursor.moveToNext()){
         mLongitude=cursor.getDouble(cursor.getColumnIndex
                    (DBGps.GEOP_LONGITUDE));
         mLatitude=cursor.getDouble(cursor.getColumnIndex
                   (DBGps.GEOP_LATITUDE));
         mPolylines.add(new LatLng(mLatitude, mLongitude));}
      if(mPolylines.size()<=2)  return;
      PolylineOptions polylineOptions=new PolylineOptions().
         points(mPolylines).width(10).color(Color.RED);
      mPolyline=(Polyline)mBaiduMap.addOverlay(polylineOptions);
      OverlayOptions markerOptions;
      markerOptions=new MarkerOptions().flat(true).anchor(0.5f, 0.5f)
       .icon(BitmapDescriptorFactory.fromResource(R.drawable.arrow))
       .position(mPolylines.get(0)).rotate((float)getAngle(0));
      mMoveMarker=(Marker)mBaiduMap.addOverlay(markerOptions);}
   private double getAngle(int startIndex){  /***根据点获取图标转的角度*/
      if((startIndex+1)>= mPolyline.getPoints().size()){
         throw new RuntimeException("越界");}
      LatLng startPoint=mPolyline.getPoints().get(startIndex);
      LatLng endPoint=mPolyline.getPoints().get(startIndex+1);
      return getAngle(startPoint, endPoint);}
   /***根据两点算取图标转的角度*/
   private double getAngle(LatLng fromPoint,LatLng toPoint){
      double slope=getSlope(fromPoint,toPoint);float deltAngle=0;
      if(slope==Double.MAX_VALUE){
         if(toPoint.latitude >fromPoint.latitude){return 0;
         } else{return 180;}}
      if((toPoint.latitude-fromPoint.latitude)*slope<0){deltAngle=180;}
      double radio=Math.atan(slope);
      double angle=180*(radio/Math.PI)+deltAngle-90;return angle;}
   /***根据点和斜率算取截距*/
```

```
    private double getInterception(double slope, LatLng point){
        double interception=point.latitude - slope *point.longitude;
        return interception;}
    private double getSlope(LatLng fromPoint, LatLng toPoint){/***算斜率*/
        if(toPoint.longitude==fromPoint.longitude){return Double.MAX_VALUE;}
        double slope=((toPoint.latitude - fromPoint.latitude)/(toPoint.longitude-
fromPoint.longitude));return slope;}
    protected void onResume(){super.onResume();  mMapView.onResume();}
    protected void onPause(){super.onPause();  mMapView.onPause();}
    protected void onSaveInstanceState(Bundle outState){
        super.onSaveInstanceState(outState);
        mMapView.onSaveInstanceState(outState);}
    protected void onDestroy(){mMapView.onDestroy();  mBaiduMap.clear();
        mBaiduMap=null;  super.onDestroy();}
    private double getXMoveDistance(double slope){  /**计算x方向每次移动距离*/
        if(slope==Double.MAX_VALUE){return DISTANCE;}
        return Math.abs((DISTANCE *slope)/Math.sqrt(1+slope*slope));}
    public void setTraffic(View view){  /***设置是否显示交通图***/
        mBaiduMap.setTrafficEnabled(((CheckBox)view).isChecked());}
    public void setBaiduHeatMap(View view){  /***设置是否显示百度热力图***/
        mBaiduMap.setBaiduHeatMapEnabled(((CheckBox)view).isChecked());}
    public void setMapMode(View view){  /***设置底图显示模式***/
        boolean checked=((RadioButton)view).isChecked();
        switch(view.getId()){
        case R.id.normal:  if(checked){
            mBaiduMap.setMapType(BaiduMap.MAP_TYPE_NORMAL);}  break;
        case R.id.statellite:if(checked){
            mBaiduMap.setMapType(BaiduMap.MAP_TYPE_SATELLITE);}  break;}}
    /***定位SDK监听函数*/
    public class MyLocationListenner implements BDLocationListener{
        public void onReceiveLocation(BDLocation location){
            //地图视图销毁后不再处理新接收的位置
            if(location==null||mMapView==null){return;}
            mCurrentLat=location.getLatitude();
            mCurrentLon=location.getLongitude();
            mCurrentAccracy=location.getRadius();
            locData=new MyLocationData.Builder().accuracy(location.
            getRadius())//此处设置开发者获取到的方向信息，顺时针0~360。
            .direction(mCurrentDirection).latitude(location.getLatitude()).
            longitude(location.getLongitude()).build();
            mBaiduMap.setMyLocationData(locData);
            if(isFirstLoc){isFirstLoc=false;
        LatLng ll=new LatLng(location.getLatitude(), location.getLongitude());
```

```
        MapStatus.Builder builder=new MapStatus.Builder();
        builder.target(ll).zoom(18.0f);
    mBaiduMap.animateMapStatus(MapStatusUpdateFactory.newMapStatus(builder.
build()));}}}
```

本章小结

本章通过基于移动端GPS和传感器的运动打卡APP项目（足步天下App），让学生了解企业项目的整体开发流程：项目立项、需求分析、总体设计、详细设计、编码实现、系统调试。同时通过该项目让学生深入掌握软件界面的设计和实现，页面与页面之间数据交互应用技巧；掌握多媒体录音、拍照、音频播放和视频播放技术；掌握嵌入式SQLite数据库的设计和实现技术；掌握GPS系统服务访问、数据获取存储与地图应用技术。

强化练习

一、填空题

1. EditText的android:password可用来设置文本框中的内容是否显示为（　　）。
2. ImageButton的 android:src 属性用来设置ImageButton上显示的（　　）。
3. 状态开关按钮（ToggleButton）和开关（Switch）均派生于Button，它们的本质也是（　　）。
4. ImageView控件的（　　）属性可用来设置ImageView要显示的图像。
5. Handler.sendMessage（Message msg）用来（　　），handleMessage（Message msg）用来处理消息。
6. Simple API for XML解析器简称为（　　）。
7. Activity活动态与暂停态的转换需要调用的方法是（　　）。
8. 可扩展标记语言（EXtensible Markup Language）的缩写为（　　）。
9. 如果Service已经启动，当再次启动Service时，不会再执行onCreate()方法，而是直接执行（　　）方法。
10. PULL解析是属于（　　）驱动。

二、单选题

1. 活动（Activity）运行时首先执行的生命周期事件回调函数是（　　）。

    A. onCreate()　　B. onStart()　　C . onResume()　　D. onDestroy()

2. R.java类文件,在建立项目时由Android提供的工具自动生成,该工具名称是（　　）。

    A. 资源打包工具（Android Asset Packaging Tool，aapt）

    B. Dalvik虚拟机调试监控服务（Dalvik Debug Monitor Service,DDMS）

    C. 安卓开发工具（Android Development Tools, ADT）

D. 软件开发工具包（Software Development Kit, SDK）

3. 在 AndroidManifest.xml 中设置应用程序可调试的应用（Application）的属性是（　　）。

A. android:description="string resource"　　B. android:enabled=["true" | "false"]

C. android:debuggable=["true" | "false"]　　D. android:icon="drawable resource"

4. 资源目录res包含了Android应用项目的全部资源，命名规则正确的说法是（　　）。

A. 可以支持数字（0~9）、下画线(_)和大小写字符（a~z,A~Z）

B. 只能支持数字（0~9）和大小写字母（a~z,A~Z）

C. 只支持数字（0~9）、下画线(_)和小写字符（a~z），第一个可以是字母和下画线(_)

D. 只支持数字（0~9）、下画线(_)和小写字符（a~z），第一个必须是字母

5. Android应用项目启动最先加载的是AndroidManifest.xml，若项目中包含有多个活动（Activity），则决定最先加载活动的属性是（　　）。

A. android.intent.category.LAUNCHER　　B. android.intent.category.ICON

C. android.intent.action.MAIN　　D. android.intent.category.ACTIVITY

6. 若要向Android工程添加字符串资源，应将其添加到（　　）文件。

A. dimens.xml　　B. styles.xml　　C. strings.xml　　D. value.xml

7. 下面相对布局 LayoutParams属性中只能设置boolean值的属性为（　　）。

A. android:layout_centerHorizontal　　B. android:layout_toLeftOf

C. android:layout_above　　D. android:layout_alignLeft

8. 相对布局 LayoutParams属性中确定是否在父容器中位于中央的属性名为（　　）。

A. android:layout_centerHorizontal　　B. android:layout_centerVertical

C. android:layout_centerInParent　　D. android:layout_alignParentLeft

9. 在网格布局（GridLayout）中设置该网格列数的xml属性为（　　）。

A. android:alignmentMode　　B. android:columnCount

C. android:columnOrderPreserved　　D. android:rowCount

10. 网格布局（GridLayout）的LayoutParams属性中设置该子组件在GridLayout纵向横跨几行的xml属性名为（　　）。

A. android:layout_column　　B. android:layout_columnSpan

C. android:layout_row　　D. android:layout_rowSpan

第8章 Struggle 车牌识别系统 APP 项目

视频

Struggle车牌识别系统APP项目

学习目标

- 掌握车牌识别APP设计与实现技术。
- 掌握Android拍照、显示和存储技术。
- 掌握SQLite嵌入式数据库和MySQL数据库存储技术。
- 掌握文件读写操作、共享存储偏好数据存储和多线程编程技术。
- 掌握Android Studio综合应用APP项目应用编程技术。

8.1 项目概述

Struggle车牌识别系统APP项目是一款通过手机识别车牌的APP软件。该项目的主要特点如下：

（1）采用计算机视觉技术识别车牌，包括车辆图像采集、车牌定位、字符分割、光学字符识别、输出识别结果、识别速度快速高效。

（2）图库选择图片识别，包含车牌的图片，随时可识别车辆信息，能将车牌定位后自动分割用红色方框将其车牌圈出来，输出识别结果的同时还能用小字体将识别的结果打印在所圈出来的车的左上角，同时还能将分割出来的车牌探出窗口生成图片，窗口的命名包含车牌识别的结果。

（3）可以批量识别，大大提高识别效率，同时将识别的结构保存到后台数MySQL据库，生成相应文档。

（4）项目在满足车牌识别功能的同时，还进行相关界面优化的设计（如软件启动图，添加背景音乐、按键语音提示、视频拍照、查看记录车牌识别结果的后台MySQL数据等）。

8.2 项目设计

8.2.1　项目总体功能需求

项目总体功能性需求包括：

（1）针对不同角度下拍摄的车牌图片进行正确识别。

（2）针对不同的光线下对拍摄的车牌图片进行正确高效的识别，例如，烟雾、雨雪、日照不同角度的照射、车灯，以及大型广告灯等的影响能正确识别车牌。

（3）对识别的整个过程到识别出结果能够进行计时，同时显示所用的时间单位为毫秒。

项目的非功能性需求包括：车牌识别算法能够满足实时性要求，至少达到 5 帧/秒的要求。

8.2.2　项目总体设计

项目的总体功能框架如图 8-1 所示。

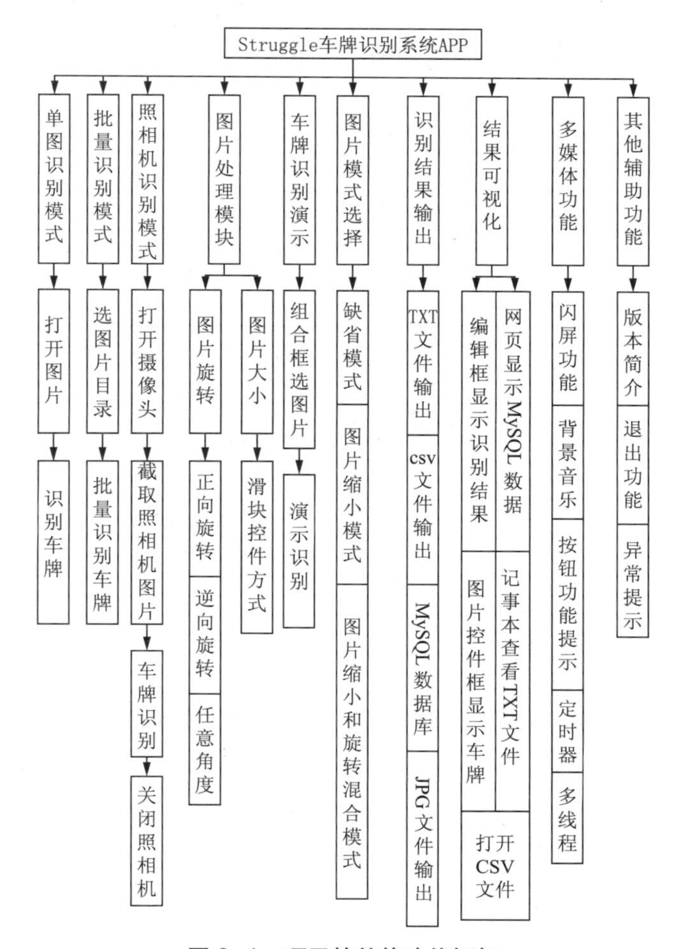

图 8-1　项目的总体功能框架

struggle《Android版本算法描述和软件设计说明书》

项目总共分成4个阶段：

第一阶段：拍照定位车牌位置。

第二阶段：车牌照片区域分割成单个字符。

第三阶段：光学字符识别输出识别结果。

第四阶段：识别结果写入SQLite嵌入式数据库和后台MySQL数据库。

车牌识别的处理流程如图8-2所示。

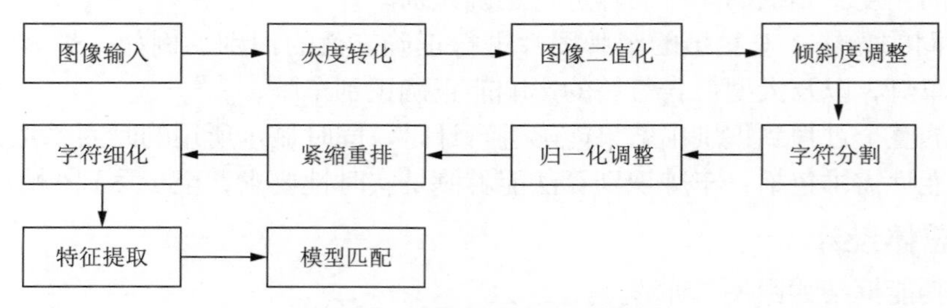

图8-2　车牌识别的处理流程

Struggle车牌识别系统APP核心功能图如图8-3所示。

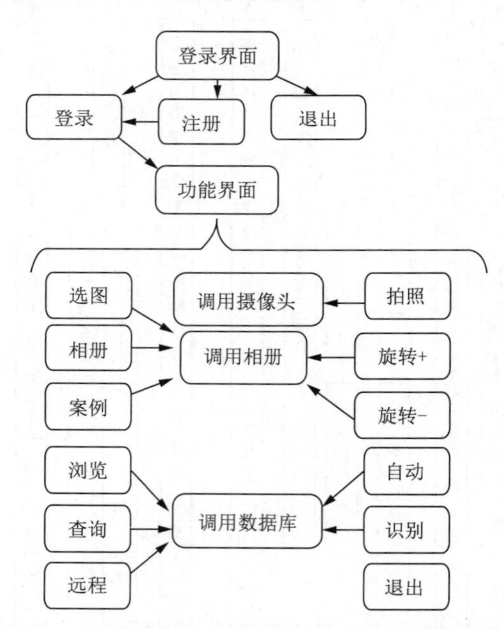

图8-3　Struggle车牌识别系统APP核心功能图

Struggle车牌识别系统APP核心功能模块描述如下：

（1）JavaWeb模块：主要负责登录界面、用户注册等。

（2）Open CV模块：主要负责机器视觉，调用手机摄像头进行初步的车牌识别，以及提供相应的opencv头文件等。

（3）数据库模块：主要负责存储识别结果的数据作为项目的组织、存储和管理数据的仓库，以及远程调用数据，查看识别过的数据。

（4）OCR-UI 模块：提供了一套默认的 UI。

（5）车牌倾斜矫正模块：提高不同角度识别的效率。

（6）分割字符模块：车牌分割成单个独立的字符，通常所拍摄的车牌图像的边缘信息都是高频信息，需要在水平、垂直方向上对车牌图像进行小波变换，对其高频信息进行重构，获得相应的高频信息方面的子图，在车牌垂直投影图像中找到每个车牌字符的边界所在位置，并记下边界位置的横坐标；同理，在水平投影图像中找到相应的边界的纵坐标，再根据相应的字符坐标值将字符分割，分割字符重组识别有利于提高识别中光线的影响。

Struggle 车牌识别系统 APP 软件的主要流程描述如下：

（1）通过 Open CV 机器视觉调用手机摄像头拍摄车牌或者拍完之后保存在手机相册中直接选择，读入图片，对图片进行灰度化处理，平滑化处理，进而再进行 sobel 运算，二值化处理，将图面转化为位数据，对边缘进行检车，定位车牌位置，进而再将车牌进行字符分割识别，与模板进行对比匹配。如果车牌定位不到或者匹配不到将进行重新读取，车牌正确识别后将识别结果写入 SQLite 嵌入式数据库和后台 MySQL 数据库。

（2）把数据存储到数据库作为项目的组织、存储和管理数据的仓库，SQLite 嵌入式数据库和从后台 MySQL 可以查看识别过的车牌信息、用时长短等。

（3）部署服务器，手机端通过向服务器发送请求获取以往车牌识别过的用时、车牌上次的识别时间等结果。

基于 Android Studio 的 Struggle 车牌识别系统 APP 项目组织结构如图 8-4 所示。

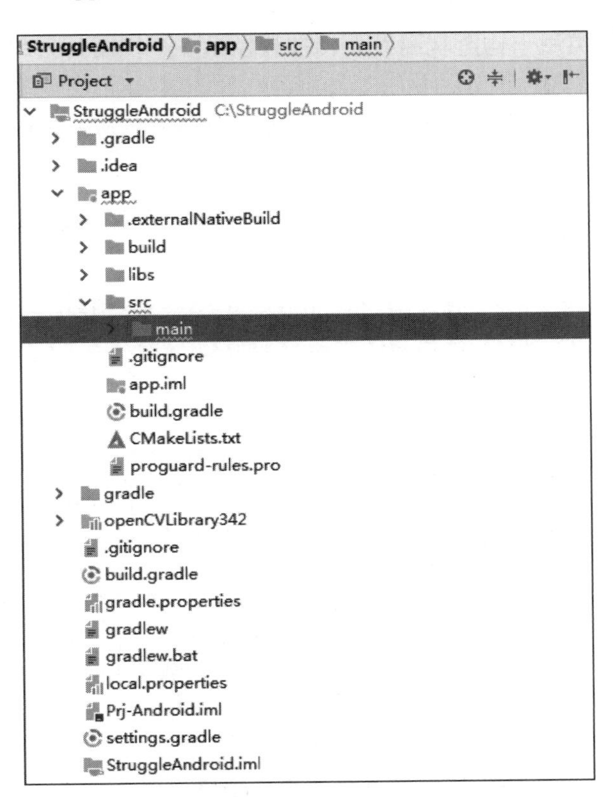

图 8-4　Struggle 车牌识别系统 APP 项目组织结构

8.3 必备的技术和知识点

本章必备的技术和知识点包括：

（1）MediaPlayer播放音频技术和方法。
（2）调用本地系统照相机应用实现拍照功能。
（3）嵌入式SQLite数据库和MySQL数据库基本操作。
（4）音乐池SoundPool应用编程技术。
（5）GPS定位和地图服务。
（6）意图(Intent)、服务（Service）、系统调用和数据存储技术。
（7）计算机视觉opencvlibrary342本地库系统调用技术。
（8）消息处理、单击事件监听和事件处理设计与实现方法。
（9）利用BitmapFactory创建、存储、旋转、平移和缩放操作Bitmap对象。

8.4 项目实施

8.4.1 欢迎页面

WelcomeSurfaceView类用于构造欢迎界面，加载显示多个位图，其源代码如下：

```
public class WelcomeSurfaceView extends SurfaceView
implements SurfaceHolder.Callback  //实现生命周期回调接口
{   //定义主活动开机闪屏对象
    pr.platerecognization.SplashActivity activity;
    private Paint paint; //画笔
    private int currentAlpha=0;  //当前的不透明值
    private DisplayMetrics dm=getResources().getDisplayMetrics();
    private int screenWidth=dm.widthPixels;  //屏幕宽度
    private int screenHeight=dm.heightPixels;  //屏幕高度
    private int sleepSpan=50;  //动画的时延ms
    private Bitmap[] logos=new Bitmap[2];  //logo图片数组
    private Bitmap currentLogo;  //当前logo图片引用
    private int currentX;private int currentY;
    private SurfaceHolder mholder;
    public WelcomeSurfaceView(SplashActivity activity){
        super(activity);this.activity=activity;
        this.getHolder().addCallback(this);  //设置生命周期回调接口的实现者
        paint=new Paint();  //创建画笔
        paint.setAntiAlias(true);  //打开抗锯齿
        logos[0]=BitmapFactory.decodeResource(activity.getResources(), R.drawable.
welcome1);  //加载图片
```

```
        logos[1]=BitmapFactory.decodeResource(activity.getResources(), R.drawable.
welcome2);  //加载图片}
    public void onDraw(Canvas canvas){  //绘制图形，进行平面贴图
      if(currentLogo==null)return;
      paint.setAlpha(currentAlpha);
      canvas.drawBitmap(currentLogo,currentX,currentY,paint);}
    public void surfaceChanged(SurfaceHolder arg0,int arg1,int arg2,int arg3){}
    public void surfaceCreated(SurfaceHolder holder){  //创建时被调用
        new Thread(){public void run(){
            for(Bitmap bm:logos)  {currentLogo=bm;
              currentX=screenWidth/2-bm.getWidth()/2;  //计算图片位置
              currentY=screenHeight/2-bm.getHeight()/2;
              for(int i=300;i>-10;i=i-10){  //动态更改图片的透明度值并不断重绘
                currentAlpha=i;
                if(currentAlpha<0){currentAlpha=0;}
                mholder=WelcomeSurfaceView.this.getHolder();
                Canvas canvas=mholder.lockCanvas();  //获取画布
                try{synchronized(mholder){onDraw(canvas);  //绘制}}
                catch(Exception e){e.printStackTrace();}
                finally{if(canvas!=null){
                    mholder.unlockCanvasAndPost(canvas);}}
                try{if(i==255){  //若是新图片，多等待一会
                    Thread.sleep(200);}Thread.sleep(sleepSpan);}
                catch(Exception e){e.printStackTrace();}}}
    activity.SAHandler.sendEmptyMessage(pr.platerecognization.SplashActivity.
MSG_SPLASH_ACTIVITY);}}.start();}
    public void surfaceDestroyed(SurfaceHolder arg0){  //销毁时被调用}}
```

从 WelcomeSurfaceView 切换到闪屏 SplashActivity 页面，其运行结果如图 8-5 所示。SplashActivity 类源代码如下：

图 8-5　欢迎页面运行结果

```
    public class SplashActivity extends Activity{
        private static final String TAG="SplashActivity";
      //向SQLITE嵌入式数据库插入数据
        public static final int MSG_SPLASH_ACTIVITY=666;
        //定义助手对象，处理消息循环(用于在欢迎界面到闪屏幕之间传递消息的助手)
        Handler SAHandler=new Handler(){
            public void handleMessage(Message msg){  //重写方法
                switch(msg.what){
                case SplashActivity.MSG_SPLASH_ACTIVITY:
                     gotoSplashActivity();break;  //主界面}}};
        protected void onCreate(Bundle savedInstanceState){
            super.onCreate(savedInstanceState);
            requestWindowFeature(Window.FEATURE_NO_TITLE);  //隐藏标题
            getWindow().setFlags(WindowManager.LayoutParams.FLAG_FULLSCREEN,
                 WindowManager.LayoutParams.FLAG_FULLSCREEN);
            gotoSplashActivity();}
        protected void onDestroy(){super.onDestroy();finish();}
        /***功能描述：欢迎界面*/
        public void gotoWelcomeSurfaceView(){
            WelcomeSurfaceView mView=new WelcomeSurfaceView(this);
            setContentView(mView);}
        /***功能描述：切换到闪屏活动页面*/
        private void gotoSplashActivity(){
            setContentView(R.layout.activity_splash);  //设置闪屏活动页面布局
            //程序已经启动，直接跳转到运行界面
            if( MainActivity.GetRunningStatus()==true ){
                //创建一个新的Intent，指定当前应用程序上下文和要启动的StepActivity类
                Intent intent=new Intent(SplashActivity.this, pr.platerecognization.
 LoginActivity.class);
                startActivity(intent);  //传递这个intent给startActivity
                this.finish();} else{new CountDownTimer(200L,200L){
                     public void onFinish(){  //启动界面淡入淡出效果
                         Intent intent=new Intent(SplashActivity.this,
pr.platerecognization.LoginActivity.class);
                try{Thread.sleep(1000);  //延迟1000
                }catch(InterruptedException e){e.printStackTrace();}
                     startActivity(intent);
                overridePendingTransition(R.anim.fade_in, R.anim.fade_out);
                finish();}
                public void onTick(long paramLong){}}.start();}}}}
```

8.4.2 登录界面

登录界面布局设计效果如图8-6所示，其activity_login.xml实现代码如下：

图 8-6　登录界面布局设计效果

```
<LinearLayout xmlns:android="http://schemas.android.com/apk/res/android"
    android:layout_width="match_parent"
    android:layout_height="match_parent"
    android:orientation="vertical">
    <LinearLayout
        android:layout_width="match_parent"
        android:layout_height="wrap_content"
        android:orientation="horizontal">
        <TextView
            android:layout_width="match_parent"
            android:layout_height="wrap_content"
            android:layout_gravity="center_horizontal"
            android:gravity="center_horizontal"
            android:layout_marginLeft="2dip"
            android:background="@drawable/btn_bg"
            android:padding="0dip"
            android:textSize="30sp"
            android:text="Struggle用户登录界面"
            android:textColor="@android:color/white"/>
    </LinearLayout>
<LinearLayout
    android:layout_width="fill_parent"
    android:layout_height="fill_parent"
    android:orientation="vertical"
    android:background="@drawable/background">
    <RelativeLayout
```

```
        android:layout_width="fill_parent"
        android:layout_height="wrap_content"
        android:layout_marginTop="@dimen/activity_vertical_margin">
        <TextView
            android:id="@+id/tel_tv"
            android:layout_width="wrap_content"
            android:layout_height="wrap_content"
            android:text="@string/username"
            android:textColor="@android:color/black"
            android:textSize="20sp"/>
        <EditText
            android:id="@+id/usernameEditText"
            android:layout_width="fill_parent"
            android:layout_height="wrap_content"
            android:layout_below="@+id/tel_tv"
            android:hint="@string/username_prompt"
            android:textColor="#787878"/>
    </RelativeLayout>
    <RelativeLayout
        android:layout_width="fill_parent"
        android:layout_height="wrap_content"
        android:layout_marginTop="@dimen/activity_vertical_margin">
        <TextView
            android:id="@+id/pswd_tv"
            android:layout_width="wrap_content"
            android:layout_height="wrap_content"
            android:text="@string/password"
            android:textColor="@android:color/black"
            android:textSize="20sp"/>
        <EditText
            android:id="@+id/pswdEditText"
            android:layout_width="fill_parent"
            android:layout_height="wrap_content"
            android:layout_below="@+id/pswd_tv"
            android:hint="@string/password_prompt"
            android:inputType="textPassword"
            android:password="true"
            android:textColor="#787878"/>
    </RelativeLayout>
    <RelativeLayout
        android:layout_width="fill_parent"
        android:layout_height="wrap_content"
        android:layout_marginTop="@dimen/activity_vertical_margin">
```

```
    <CheckBox
        android:id="@+id/remeber_pswdCheckBox"
        android:layout_width="wrap_content"
        android:layout_height="wrap_content"
        android:text="@string/rember_pswd"
         android:textColor="@android:color/black"
        android:textSize="20sp"/>
    </RelativeLayout>
    <LinearLayout
        android:layout_width="match_parent"
        android:layout_height="match_parent"
        android:orientation="horizontal">
        <Button
            android:id="@+id/loginButton"
            android:layout_width="wrap_content"
            android:layout_height="wrap_content"
            android:layout_weight="1"
            android:text="@string/login"
            android:textColor="@android:color/white"/>
        <Button
            android:id="@+id/registerButton"
            android:layout_width="wrap_content"
            android:layout_height="wrap_content"
            android:layout_weight="1"
            android:text="@string/register"
            android:textColor="@android:color/white"/>
        <Button
            android:id="@+id/exitButton"
            android:layout_width="wrap_content"
            android:layout_height="wrap_content"
            android:layout_weight="1"
            android:text="@string/quit"
            android:textColor="@android:color/white"/>
        <ImageView
            android:id="@+id/ivpausesong"
            android:layout_width="wrap_content"
            android:layout_height="wrap_content"
            android:layout_weight="1"
            android:src="@drawable/pause1">
        </ImageView>
    </LinearLayout>
</LinearLayout>
</LinearLayout>
```

Login Activity类实现登录界面，其派生于活动Activity类，实现了单击侦听器接口（On Click Listener），运行结果如图8-7所示，其源代码如下所示：

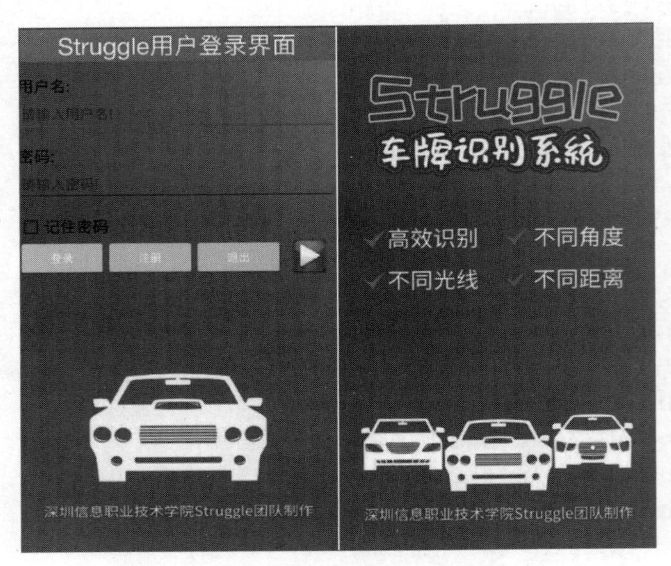

图8-7 登录界面LoginActivity运行结果

```
public class LoginActivity extends Activity implements OnClickListener{
    private static final String TAG="LoginActivity";
    private EditText usernameEditText;  //用户名输入编辑控件
    private EditText pswdEditText;  //密码输入编辑控件
    private Button loginButton;  //登录按钮
    private Button exitButton;  //退出按钮
    private CheckBox remeber_pswdCheckBox;  //是否记住密码复选框控件
    private SharedPreferences spref;  //共享偏好存储对象
    private SharedPreferences.Editor editor;  //编辑对象
    private String userName, password;  //用户名和密码
    private static final int SOUNDPOOL_login0=1;  //登录系统
    private static final int SOUNDPOOL_exit10=2;  //退出系统
    private static final int SOUNDPOOL_registeruser0=3;  //退出系统
    private String sutusername="";  //从数据库查询的用户名
    private String sutpassword="";  //从数据库查询的密码
    private MediaPlayer mMediaPlayer;  //播放音乐的对象(背景音乐)
    private ImageView ivpausesong;  //暂停图片和注册图片
    private boolean imagebflag=true;
    SoundPool soundPool;  //定义一个SoundPool用于播放功能提示声音
    HashMap<Integer,Integer>soundMap=new HashMap<Integer,Integer>();
    //用于验证用户名和密码(登录功能)
    private DBLpr mDBLpr=new DBLpr(this);private SQLiteDatabase mDB=null;
    private Cursor mCursor=null;private suserdata supdata=null;
```

```
    /***功能描述：重写onCreate()方法*/
    protected void onCreate(Bundle savedInstanceState){
      Log.d(TAG,"onCreate()");super.onCreate(savedInstanceState);
      requestWindowFeature(Window.FEATURE_NO_TITLE);
      setContentView(R.layout.activity_login);  initModle();initView();
      try{mMediaPlayer=MediaPlayer.create(this, R.raw.higher);
        StartOrPauseMediaPlayer();
    }catch(IllegalArgumentException e){e.printStackTrace();
    }catch(IllegalStateException e){e.printStackTrace();}
      //设置最多可容纳10个音频流，音频的品质为5
      soundPool=new SoundPool(10,AudioManager.STREAM_SYSTEM,5);
      soundMap.put(SOUNDPOOL_exit10,soundPool.load(this,R.raw.exit,1));
       initSUTUserPwdDB();}
  /***功能描述：初始化共享偏好存储类对象*/
  private void initModle(){Log.d(TAG, "initModle()");
      spref=PreferenceManager.getDefaultSharedPreferences(this);}
  /***功能描述：初始化视图*/
  private void initView(){
      Button registerButton=(Button)findViewById(R.id.registerButton);
      registerButton.setOnClickListener(this);
      loginButton=(Button)findViewById(R.id.loginButton);
      loginButton.setOnClickListener(this);
      exitButton=(Button)findViewById(R.id.exitButton);
      exitButton.setOnClickListener(this);
      usernameEditText=(EditText)findViewById(R.id.usernameEditText);
      pswdEditText=(EditText)findViewById(R.id.pswdEditText);remeber_
pswdCheckBox=(CheckBox)findViewById(R.id.remeber_pswdCheckBox);
      boolean isRemeber=spref.getBoolean("remeberPswd",false);
      if(isRemeber){userName=spref.getString("userName","");  //用户名
          password=spref.getString("password","");  //密码
          usernameEditText.setText(userName);
          pswdEditText.setText(password);
          remeber_pswdCheckBox.setChecked(true);}
      ivpausesong=(ImageView)findViewById(R.id.ivpausesong);
      ivpausesong.setOnClickListener(new OnClickListener(){
      public void onClick(View v){StartOrPauseMediaPlayer();}});}
      void initSUTUserPwdDB(){  /***初始化SQLite嵌入式数据库*/
      try{mDBLpr=new DBLpr(this);mDB=mDBLpr.openDB();
          supdata=new suserdata();setDefaultUserPassword();
      }catch(SQLiteException se){se.printStackTrace();
      }catch(Exception e){e.printStackTrace();}}
  /***功能：为方便专家测试APP功能，特设置用户登录默认密码*/
  private void setDefaultUserPassword(){  //设置两个默认用户名和密码
```

```
    supdata=new suserdata();supdata.sut_name="root";
      supdata.sut_password="123456";
      mDBLpr.insert_SUTUserPwd(supdata);supdata.sut_name="0";
      supdata.sut_password="0";mDBLpr.insert_SUTUserPwd(supdata);}
   public void onClick(View v){/***功能描述：实现单击事件回调处理。*/
      switch(v.getId()){  //根据按钮控制ID号做不同功能处理
      case R.id.loginButton:doLoginAction();break;  //处理登录按钮
      case R.id.registerButton:
      soundPool.play(soundMap.get(SOUNDPOOL_registeruser0),1,1,0,0,1);
      Intent intent=new Intent(LoginActivity.this,RegisterActivity.class);
      startActivity(intent);LoginActivity.this.finish();break;
      case R.id.exitButton:  //处理退出按钮
           soundPool.play(soundMap.get(SOUNDPOOL_exit10),1,1,0,0,1);
           showExitAlert();break;}}
  /***功能：执行用户登录*/
  private void doLoginAction(){
     soundPool.play(soundMap.get(SOUNDPOOL_login0),1,1,0,0,1);
     String username=usernameEditText.getText().toString().trim();
     String password=pswdEditText.getText().toString().trim();  //密码
       if( !username.isEmpty()&&!password.isEmpty()){
          try{if(mDB!=null){
               mCursor=mDB.rawQuery("Select *from  user_table where sut_name =?
and sut_password =?", new String[]{username, password});
               if(mCursor!=null){mCursor.moveToFirst();
               sutusername=mCursor.getString(mCursor.getColumnIndex("sut_name"));
               sutpassword=mCursor.getString(mCursor.getColumnIndex("sut_password"));
               }else{}}}catch(SQLiteException se){
                   se.printStackTrace();}catch(Exception e){
                   e.printStackTrace();}}
         if(username.equals(sutusername)&&password.equals(sutpassword)){
            editor=spref.edit();  //获得编辑器对象
            if(remeber_pswdCheckBox.isChecked()){  //如果选择记住密码
                editor.putBoolean("remeberPswd",true);  //保存记住密码项
                editor.putString("userName",username);  //存储用户名
                editor.putString("password",password);  //存储密码}else{
                editor.clear();  //清空编辑器}
                editor.commit();  //提交存储
       //创建意图(或意向)对象，从当前登录页面向主界面MainActivity切换
       Intent intent=new Intent(LoginActivity.this,MainActivity.class);
       startActivity(intent);  //启动意图
       LoginActivity.this.finish();  //关闭当前登录界面
    }else{Log.d(TAG,getString(R.string.pls_input_username));}}
    /***销毁用户登录活动页面，释放所有全部资源*/
```

```
    protected void onDestroy(){
        if(mMediaPlayer.isPlaying())mMediaPlayer.stop();
        if(mMediaPlayer!=null)mMediaPlayer.release();
        super.onDestroy();}
    /*** 功能: 停止或播放动画帧 */
    private void StartOrPauseMediaPlayer(){
        if(imagebflag){
            ivpausesong.setImageResource(R.drawable.start1);
            imagebflag=false;}else{
            ivpausesong.setImageResource(R.drawable.pause1);
            imagebflag=true;}
        if(mMediaPlayer.isPlaying())mMediaPlayer.pause();
        else  mMediaPlayer.start();}}
```

8.4.3　程序主界面

程序主界面布局效果如图 8-8 所示，其 activity_main.xml 实现代码如下：

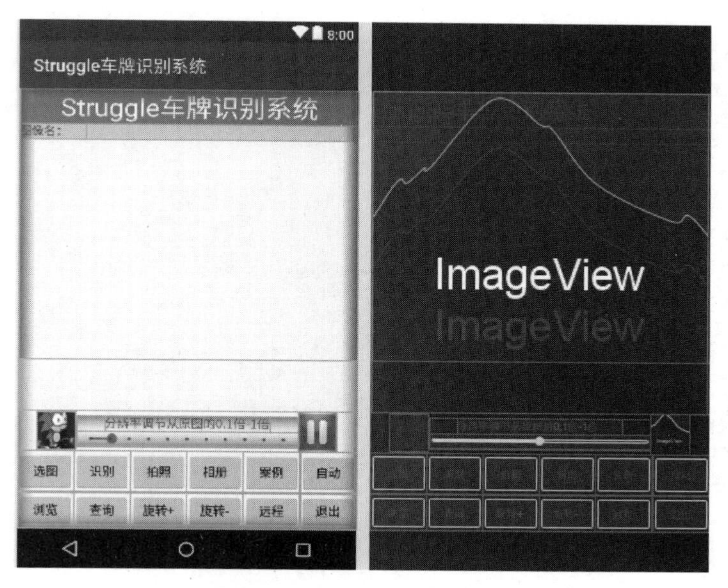

图 8-8　程序主界面布局效果

```
    <?xml version="1.0"encoding="utf-8"?>
    <android.support.constraint.ConstraintLayout xmlns:android="http://schemas.
android.com/apk/res/android"
        xmlns:app="http://schemas.android.com/apk/res-auto"
        xmlns:tools="http://schemas.android.com/tools"
        android:layout_width="match_parent"
        android:layout_height="match_parent"
        tools:context="pr.platerecognization.MainActivity">
        <LinearLayout
```

```
        android:layout_width="match_parent"
        android:layout_height="match_parent"
        android:orientation="vertical"
        tools:layout_editor_absoluteX="0dp"
        tools:layout_editor_absoluteY="61dp">
    <LinearLayout
        android:layout_width="match_parent"
        android:layout_height="wrap_content"
        android:orientation="horizontal">
        <TextView
            android:layout_width="match_parent"
            android:layout_height="wrap_content"
            android:layout_gravity="center_horizontal"
            android:gravity="center_horizontal"
            android:layout_marginLeft="2dip"
            android:background="@drawable/btn_bg"
            android:padding="0dip"
            android:textSize="30sp"
            android:text="Struggle车牌识别系统"
            android:textColor="@android:color/white"/>
    </LinearLayout>
    <LinearLayout
        android:layout_width="match_parent"
        android:layout_height="wrap_content"
        android:orientation="horizontal">
        <TextView
            android:id="@+id/textViewPictureNameLabel"
            android:layout_width="wrap_content"
            android:layout_height="wrap_content"
            android:layout_weight="1"
            android:background="#00FF00"
            android:text="@string/picture_name_label"
            android:textAppearance="@style/TextAppearance.AppCompat.Small"/>
        <TextView
            android:id="@+id/textViewPicturePathName"
            android:layout_width="wrap_content"
            android:layout_height="wrap_content"
            android:layout_weight="15"
            android:background="#00FF00"
            android:text=""android:textAppearance="@style/TextAppearance.
AppCompat.Small"/>
    </LinearLayout>
    <LinearLayout
```

```
        android:layout_width="match_parent"
        android:layout_height="wrap_content"
        android:layout_gravity="center"
        android:gravity="center_horizontal"
        android:orientation="horizontal">
        <ImageView
            android:id="@+id/imageViewCarPicture"
            android:layout_width="match_parent"
            android:layout_height="316dp"
            android:layout_marginLeft="2dp"
            android:layout_marginRight="2dp"/>
    </LinearLayout>
    <LinearLayout
        android:layout_width="match_parent"
        android:layout_height="wrap_content"
        android:layout_gravity="center"
        android:gravity="center_horizontal"
        android:orientation="horizontal">
        <LinearLayout
            android:layout_width="wrap_content"
            android:layout_height="wrap_content"
            android:layout_gravity="center"
            android:gravity="center_horizontal"
            android:orientation="horizontal">
            <FrameLayout
                android:id="@+id/runningPicFrame"
                android:layout_width="45sp"
                android:layout_height="45sp"
                android:background="@color/display_background"
                android:foreground="@drawable/s1"
                android:foregroundGravity="center_vertical|center_horizontal"
                android:gravity="center_vertical|center_horizontal">
            </FrameLayout>
        </LinearLayout>
        <LinearLayout
            android:layout_width="wrap_content"
            android:layout_height="wrap_content"
            android:layout_gravity="center"
            android:gravity="center_horizontal"
            android:orientation="vertical">
            <TextView
                android:id="@+id/textViewResolution"
                android:layout_width="wrap_content"
```

```
                android:layout_height="wrap_content"
                android:text="@string/resolution_range"
android:textAppearance="@style/TextAppearance.AppCompat.Small"/>
            <SeekBar
                android:id="@+id/seekBarResolution"
                style="@style/Widget.AppCompat.SeekBar.Discrete"
                android:layout_width="250dp"
                android:layout_height="wrap_content"
                android:max="10"
                android:min="1"
                android:progress="2"/>
        </LinearLayout>
        <LinearLayout
            android:layout_width="wrap_content"
            android:layout_height="wrap_content"
            android:layout_gravity="center"
            android:gravity="center_horizontal"
            android:orientation="horizontal">
            <ImageView
                android:id="@+id/ivpausesong"
                android:layout_width="wrap_content"
                android:layout_height="wrap_content"
                android:src="@drawable/pause1">
            </ImageView>
        </LinearLayout>
        </LinearLayout>
        <LinearLayout
            android:layout_width="match_parent"
            android:layout_height="wrap_content"
            android:layout_gravity="center"
            android:gravity="center_horizontal"
            android:orientation="horizontal">
            <Button
                android:id="@+id/buttonSelectImg"
                android:layout_width="wrap_content"
                android:layout_height="wrap_content"
                android:layout_weight="1"
                android:text="@string/select_img"/>
            <Button
                android:id="@+id/buttonRecogPlate"
                android:layout_width="wrap_content"
                android:layout_height="wrap_content"
                android:layout_weight="1"
```

```
            android:text="@string/recog_plate"/>
        <Button
            android:id="@+id/buttonTakePhoto"
            android:layout_width="wrap_content"
            android:layout_height="wrap_content"
            android:layout_weight="1"
            android:text="@string/take_photo"/>
        <Button
            android:id="@+id/buttonSelectFromGALLERY"
            android:layout_width="wrap_content"
            android:layout_height="wrap_content"
            android:layout_weight="1"
            android:text="@string/select_from_gallery"/>
        <Button
            android:id="@+id/buttonDemo"
            android:layout_width="wrap_content"
            android:layout_height="wrap_content"
            android:layout_weight="1"
            android:text="@string/recog_demo"/>
        <Button
            android:id="@+id/buttonAutoRotate"
            android:layout_width="wrap_content"
            android:layout_height="wrap_content"
            android:layout_weight="1"
            android:text="@string/recog_auto_rotate"/>
    </LinearLayout>
    <LinearLayout
        android:layout_width="match_parent"
        android:layout_height="wrap_content"
        android:layout_gravity="center"
        android:gravity="center_horizontal"
        android:orientation="horizontal">
        <Button
            android:id="@+id/buttonPlateBrowser"
            android:layout_width="wrap_content"
            android:layout_height="wrap_content"
            android:layout_weight="1"
            android:text="浏览"/>
        <Button
            android:id="@+id/buttonPlateQuery"
            android:layout_width="wrap_content"
            android:layout_height="wrap_content"
            android:layout_weight="1"
```

```
                android:text="查询"/>
            <Button
                android:id="@+id/buttonRotateBmpForward"
                android:layout_width="wrap_content"
                android:layout_height="wrap_content"
                android:layout_weight="1"
                android:text="旋转+"/>
            <Button
                android:id="@+id/buttonRotateBmpBack"
                android:layout_width="wrap_content"
                android:layout_height="wrap_content"
                android:layout_weight="1"
                android:text="旋转-"/>
            <Button
                android:id="@+id/buttonMySQL"
                android:layout_width="wrap_content"
                android:layout_height="wrap_content"
                android:layout_weight="1"
                android:text="远程"/>
            <Button
                android:id="@+id/buttonExit"
                android:layout_width="wrap_content"
                android:layout_height="wrap_content"
                android:layout_weight="1"
                android:text="退出"/>
        </LinearLayout>
        <LinearLayout
            android:layout_width="match_parent"
            android:layout_height="wrap_content"
            android:orientation="horizontal">
            <TextView
                android:id="@+id/textViewRecogResultLabel"
                android:layout_width="wrap_content"
                android:layout_height="wrap_content"
                android:text="@string/recog_result_label"
                android:background="#0000FF"
                android:textColor="#FFFFFF"
                android:textAppearance="@style/TextAppearance.AppCompat.Medium"/>
            <TextView
                android:id="@+id/textViewRecogResult"
                android:layout_width="wrap_content"
                android:layout_height="wrap_content"
                android:layout_weight="3"
```

```
            android:background="#0000FF"
            android:textColor="#FFFFFF"
            android:text="@string/recog_result_prompt"
            android:textAppearance="@style/TextAppearance.AppCompat.Medium"/>
    </LinearLayout>
    <LinearLayout
        android:layout_width="match_parent"
        android:layout_height="wrap_content"
        android:orientation="horizontal">
        <TextView
            android:id="@+id/textViewRuntimeLabel"
            android:layout_width="wrap_content"
            android:layout_height="wrap_content"
            android:text="@string/runtime_label"
            android:layout_weight="1"
            android:background="#0000FF"
            android:textColor="#FFFFFF"
            android:textAppearance="@style/TextAppearance.AppCompat.Medium"/>
        <TextView
            android:id="@+id/textViewRuntime"
            android:layout_width="wrap_content"
            android:layout_height="wrap_content"
            android:background="#0000FF"
            android:textColor="#FFFFFF"
            android:layout_weight="1"
            android:text=""
            android:textAppearance="@style/TextAppearance.AppCompat.Medium"/>
        <TextView
            android:id="@+id/tvDemoFileNaleLable"
            android:layout_width="wrap_content"
            android:layout_height="wrap_content"
            android:text="文件："
            android:layout_weight="1"
            android:background="#0000FF"
            android:textColor="#FFFFFF"
            android:textAppearance="@style/TextAppearance.AppCompat.Medium"/>
        <Spinner
            android:id="@+id/demoSpinner"
            android:layout_weight="2"
            android:layout_width="wrap_content"
            android:layout_height="wrap_content">
        </Spinner>
    </LinearLayout>
```

```
        <LinearLayout
            android:layout_width="match_parent"
            android:layout_height="wrap_content"
            android:orientation="horizontal">
            <TextView
                android:id="@+id/textViewPictureWidthLabel"
                android:layout_width="wrap_content"
                android:layout_height="wrap_content"
                android:text="图像宽: "
                android:textAppearance="@style/TextAppearance.AppCompat.Small"/>
            <TextView
                android:id="@+id/textViewPictureWidth"
                android:layout_width="wrap_content"
                android:layout_height="wrap_content"
                android:layout_weight="2"
                android:text=""
                android:textAppearance="@style/TextAppearance.AppCompat.Small"/>
            <TextView
                android:id="@+id/textViewPictureHeightLabel"
                android:layout_width="wrap_content"
                android:layout_height="wrap_content"
                android:text="图像高: "
                android:textAppearance="@style/TextAppearance.AppCompat.Small"/>
            <TextView
                android:id="@+id/textViewPictureHeight"
                android:layout_width="wrap_content"
                android:layout_height="wrap_content"
                android:layout_weight="2"
                android:text=""
                android:textAppearance="@style/TextAppearance.AppCompat.Small"/>
            <TextView
                android:id="@+id/textViewBmpDegreesLabel"
                android:layout_width="wrap_content"
                android:layout_height="wrap_content"
                android:text="角度: "
                android:textAppearance="@style/TextAppearance.AppCompat.Small"/>
            <TextView
                android:id="@+id/textViewBmpDegrees"
                android:layout_width="wrap_content"
                android:layout_height="wrap_content"
                android:layout_weight="2"
                android:text=""
                android:textAppearance="@style/TextAppearance.AppCompat.Small"/>
```

```
        </LinearLayout>
    </LinearLayout>
    <LinearLayout
        android:layout_width="match_parent"
        android:layout_height="wrap_content"
        android:layout_gravity="center"
        android:gravity="center_horizontal"
        android:orientation="horizontal">
        <ImageView
            android:id="@+id/imageViewCarPicture1"
            android:layout_width="match_parent"
            android:visibility="invisible"
            android:layout_height="316dp"
            android:layout_marginLeft="2dp"
            android:layout_marginRight="2dp"/>
    </LinearLayout>
</android.support.constraint.ConstraintLayout>
```

车牌识别主界面MainActivity类，从活动Activity基类派生，实现报警对话框AlertDialog，单击侦听器接口OnClickListener，运行结果如图8-9所示，其源代码如下：

视频

Struggle《Android测试用例与测试报告》

图 8-9　程序主界面运行结果

```
    public class MainActivity extends Activity implements
AlertDialog.OnClickListener,View.OnClickListener{
        //用于在应用程序启动时加载"本机lib"库。
        static{if(OpenCVLoader.initDebug()){
```

```
    Log.d("OpenCV","OpenCV加载成功！ ");}else{
    Log.d("OpenCV","OpenCV无法加载！");}}
  //定义类私有数据成员
  private cardata data;  //写入SQLite和MySQL数据库的车牌识别数据类
  private Button buttonSelectImg;  //定义按钮对象
  private Button buttonRecogPlate;  //定义识别按钮
  private Button buttonTakePhoto;  //定义拍照按钮
  private Button buttonSelectFromGALLERY;  //定义拍照按钮
  private Button buttonPlateBrowser;  //定义浏览SQLITE车牌识别信息数据库
  private Button buttonPlateQuery;  //定义查询车牌识别信息
  private Button buttonRotateBmpForward;  //图像正向旋转
  private Button buttonRotateBmpBack;  //图像反向旋转
  private Button buttonExit;  //退出系统按钮
  private Button buttonMySQL;  //访问远程MySQL或本地MySQL服务器
  private Button buttonAutoRotate;  //自动旋转和调整图像分辨率
  private Button buttonDemo;  //用于演示比赛照片车牌识别效果的按钮
  private Spinner demoSpinner;  //用于演示比赛照片车牌识别效果的列表控件
  private String [] m_demoFileName;  //保存演示比赛照片名称
  private ArrayAdapter<String>m_demoAdapter;
  private TextView textViewRecogResult;  //定义文本视图按钮
  private TextView textViewRuntime;  //定义文本视图按钮：显示运行时间
  private ImageView imageViewCarPicture;  //定义图像视图
  private ImageView imageViewCarPicture1;  //定义图像视图
  private SeekBar seekBarResolution;  //定义SeekBar控件对象
  private TextView textViewPicturePathName;  //显示图片路径和名称
  private TextView textViewPictureWidth;  //用于显示数据库中图像最大宽度
  private TextView textViewPictureHeight;  //用于显示数据库中图像最大高度
  private TextView textViewBmpDegrees;  //用于显示数据库中图像旋转角度
  //定义onActivityResult回调事件处理常量
  private static final int REQUEST_CODE_IMAGE_CAMERA=1;
  private static final int REQUEST_CODE_IMAGE_OP=2;
  private static final int REQUEST_CODE_OP=3;
  private static final int PHOTO_REQUEST_CAREMA=4;  //拍照
  private static final int PHOTO_REQUEST_GALLERY=5;  //从相册中选择
  private static final int PHOTO_REQUEST_CUT=6;  //照相机图像裁剪
  private static final int PHOTO_REQUEST_CUT_GALLERY=7;  //相册图像裁剪
  //定义助手Handler处理的消息常量
  private static final int MSG_DISPLAY_BITMAP=111;  //显示位图消息
  private static final int MSG_INSERT_SQLITE=222;  //向SQLITE插入数据
  private static final int MSG_MYSQL_TEST_0=240;
  private static final int MSG_MYSQL_TEST_1=241;
  private static final int MSQ_MYSQL_INSERT_SUCESS =88881;
  private static final int MSQ_MYSQL_INSERT_FAIL =88880;
```

```
private int g_spinner_positon =0;  //初始化列表选项位置
private Uri mPath;  //定义Uri路径
private Bitmap latestBitmap;  //定义最新位图对象（已被加载图像）
private Bitmap g_bmp;  //演示位
private float g_degrees=(float)0.0;  //全局旋转角度
private static final int MAX_DEGREES=8;  //定义最大允许车牌图像旋转角度
private static final int MAX_WIDTH=1700;  //定义最大允许车牌图像宽度
private static final int MAX_HEIGHT=2300;  //定义最大允许车牌图像高度
private static boolean isRunning =false;  //播放动画帧标记
//用于标记声音ID
private static final int SOUNDPOOL_SelectImg=1;  //选择图像
private static final int SOUNDPOOL_RecogPlate=2;  //识别车牌
private static final int SOUNDPOOL_TakePhoto=3;  //拍照识别
private static final int SOUNDPOOL_SelectFromGALLERY =4;  //相册选图
private static final int SOUNDPOOL_PlateBrowser=5;  //浏览车牌识别数据
private static final int SOUNDPOOL_PlateQuery=6;  //定制查询车牌识别数据
private static final int SOUNDPOOL_RotateBmpForward=7;  //正向旋转图像
private static final int SOUNDPOOL_RotateBmpBack =8;  //逆向旋转图像
private static final int SOUNDPOOL_seekBarResolution=9;  //调整图像比例
private static final int SOUNDPOOL_Exit=10;  //退出系统
private static final int SOUNDPOOL_Demo=8000;  //案例提升声音
private static final int SOUNDPOOL_MySQL=8001;  //访问远程MySQL提升声音
private static final int SOUNDPOOL_AutoRotate=8002;  //自动提升声音
private static final int SOUNDPOOL_Spinnerdemo=8003;  //从列表选择车牌
/*拍摄图片名称 */
private static final String PHOTO_FILE_NAME="temp_photo.jpg";
private File tempFile;  //临时文件对象
//初始化嵌入式SQLite数据库对象
private DBLpr mDBLpr=new DBLpr(this);  //车牌识别嵌入式数据库对象
private SQLiteDatabase mDB=null;  //定义SQLiteDatabase对象
private Cursor mCursor=null;  //定义访问嵌入式SQLite数据库的游标对象
private cardata m_cardata=null;  //定义保存车牌识别数据对象（类似JavaBean）
public long handlePR;  //定义公有PlateRecognition句柄对象
private final String TAG=this.getClass().toString();
private MediaPlayer mMediaPlayer;  //播放音乐的对象（背景音乐）
private ImageView ivpausesong; //暂停图片和注册图片
private boolean imagebflag=true;
private final int MSG_CTRL_RUNNING=0x123456;  //处理帧动画
FrameLayout runningPicFrame=null;  //初始化变量，帧布局（用于显示动画）
final int[] names;  //在Activity中创建 Drawable组件数组
private Timer runningTimer=new Timer();  //定时器（用于控制帧动画）
TimerTask runningTimerTask=CreateRunningTimerTask();
private boolean isRun=false;
```

```
    //处理MySQL数据库查询、插入和数据删除等功能的助手
    Handler mysql_handler=new Handler(new Handler.Callback(){
        public boolean handleMessage(Message message){
            String str="MySQL插入不存在";
            if(message.what==MSQ_MYSQL_INSERT_SUCESS){
                str="MySQL插入成功!";
            }else if( message.what==MSQ_MYSQL_INSERT_FAIL){
                str="MySQL插入失败! ";} return false;}});
    SoundPool soundPool;  //声音池，用于播放按键功能提示信息
    //保存已经加载mp3声音的哈希映射表
    HashMap<Integer,Integer>soundMap=new HashMap<Integer, Integer>();
    /***从资源拷贝文件*/
    public void copyFilesFromAssets(Context context, String oldPath,
String newPath){
       try{  //获得资源文件清单
            String[] fileNames=context.getAssets().list(oldPath);
            if(fileNames.length >0){  //如果文件名长度大于0
                File file=new File(newPath);  //目录
                if(!file.exists()){
                    if(!file.mkdir()){Log.d("mkdir","无法创建文件夹!");}
                    else{}}else{}
                for(String fileName : fileNames){//递归复制
                    //从原始目录将文件拷贝到新建目录
                    copyFilesFromAssets(context,oldPath+"/"+fileName,
                            newPath+"/"+fileName);}}else{
                InputStream is=context.getAssets().open(oldPath);
                FileOutputStream fos=new FileOutputStream(new File(newPath));
                byte[] buffer=new byte[1024];int byteCount;
                while((byteCount=is.read(buffer))!= -1){
                    fos.write(buffer, 0, byteCount);}
                fos.flush();is.close();fos.close();}
        }catch(Exception e){e.printStackTrace();}}
    public void initRecognizer(){/***初始化车牌识别对象*/
        String assetPath="pr";//定义assets资源文件夹下"pr"目录
        String state;String path;
        state=Environment.getExternalStorageState();
        if(state.equals(Environment.MEDIA_MOUNTED)){
          path=Environment.getExternalStorageDirectory().getAbsolutePath();}
    else
          {Log.d("initRecognizer", "sdcard不存在!");return;}
        //获得SD卡上资源文件路径: mnt/sdcard/pr
        String sdcardPath=Environment.getExternalStorageDirectory()
                +File.separator+assetPath;
```

```
        //将assets/pr资源复制到mnt/sdcard/pr
        copyFilesFromAssets(this,assetPath,sdcardPath);
        //定义"cascade.xml"路径:mnt/sdcard/pr/cascade.xml
        String cascade_filename=sdcardPath
            +File.separator+"cascade.xml";
        String finemapping_prototxt=sdcardPath
              +File.separator+"HorizonalFinemapping.prototxt";
        String finemapping_caffemodel=sdcardPath
              +File.separator+"HorizonalFinemapping.caffemodel";
        String segmentation_prototxt=sdcardPath
              +File.separator+"Segmentation.prototxt";
        String segmentation_caffemodel=sdcardPath
              +File.separator+"Segmentation.caffemodel";
        String character_prototxt=sdcardPath
              +File.separator+"CharacterRecognization.prototxt";
        String character_caffemodel=sdcardPath
              +File.separator+"CharacterRecognization.caffemodel";
        //初始化车牌识别句柄(要将这7个文件提前放到assets/pr目录)
        handlePR=PlateRecognition.InitPlateRecognizer(
                cascade_filename,
                finemapping_prototxt,finemapping_caffemodel,
                segmentation_prototxt,segmentation_caffemodel,
                character_prototxt,character_caffemodel);}
    void initLPRDB(){  /***初始化SQLite嵌入式数据库*/
        mDBLpr=new DBLpr(this);mDB=mDBLpr.openDB();
        m_cardata=new cardata();g_degrees =(float)0.0;
        isRunning =true;}
    /***返回播放动画帧的状态*/
    public static boolean GetRunningStatus(){return isRunning;}
    public static String getDataColumn(Context context,Uri uri,String
selection,String[]selectionArgs){Cursor cursor=null;
        final String column="_data";
        final String[] projection={column};
        try{cursor=context.getContentResolver().query(uri, projection,
selection, selectionArgs,null);
            if(cursor!=null && cursor.moveToFirst()){
                final int index=cursor.getColumnIndexOrThrow(column);
                return cursor.getString(index);}
        }finally{if(cursor!=null)cursor.close();}return null;}
    private String getPath(Uri uri){  /***获得Uri路径/
        if(Build.VERSION.SDK_INT>=Build.VERSION_CODES.KITKAT){
            if(DocumentsContract.isDocumentUri(this, uri)){
                final String docId=DocumentsContract.getDocumentId(uri);
```

```
                final String[] split=docId.split(":");
                final String type=split[0];Uri contentUri=null;
                if("image".equals(type)){  //存放图像Uri地址
                contentUri=MediaStore.Images.Media.EXTERNAL_CONTENT_URI;
                }else if("video".equals(type)){  //存放视频Uri地址
                   contentUri=MediaStore.Video.Media.EXTERNAL_CONTENT_URI;
                       return null;}else if("audio".equals(type)){
                       //存放音频Uri地址
                   contentUri=MediaStore.Audio.Media.EXTERNAL_CONTENT_URI;
                       return null;}
                final String selection="_id=?";
                final String[]selectionArgs=new String[]{split[1]};
                //获得数据列表（类似SQL查询语句）
          return getDataColumn(this,contentUri,selection,selectionArgs);}}
          String[] proj={MediaStore.Images.Media.DATA};
          Cursor actualimagecursor=managedQuery(uri, proj,null,null,null);
          int actual_image_column_index=actualimagecursor.getColumnIndexOrThrow
(MediaStore.Images.Media.DATA);
          actualimagecursor.moveToFirst();
          String img_path=actualimagecursor.getString(actual_image_column_index);
          String end=img_path.substring(img_path.length()-4);
          //查询".jpg"或".png"文件
          if(0!=end.compareToIgnoreCase(".jpg")&&0!=end.
          compareToIgnoreCase(".png")){return null;}return img_path;}
      //解码给定路径图像为位图
      public static Bitmap decodeImage(String path){
          try{Bitmap res;  //定义位图对象
             ExifInterface exif=new ExifInterface(path);  //ExIF接口
          int orientation=exif.getAttributeInt(ExifInterface.TAG_ORIENTATION,
    ExifInterface.ORIENTATION_NORMAL);  //获得方向参数
             //获得位图选填
             BitmapFactory.Options op=new BitmapFactory.Options();
             op.inSampleSize=1;op.inJustDecodeBounds=false;
             res=BitmapFactory.decodeFile(path, op);  //解码图像
             Matrix matrix=new Matrix();
             if(orientation==ExifInterface.ORIENTATION_ROTATE_90){
                 matrix.postRotate(90);  //旋转90°
             }else if(orientation==ExifInterface.ORIENTATION_ROTATE_180){
                 matrix.postRotate(180);  //旋转180°
             }else if(orientation==ExifInterface.ORIENTATION_ROTATE_270){
                 matrix.postRotate(270);  //旋转270° }
             //创建旋转和缩放后位图图像
             Bitmap temp=Bitmap.createBitmap(res,0,0,res.getWidth(),
```

```
res.getHeight(), matrix, true);
        if(!temp.equals(res)){res.recycle(); //释放原始位图图像}
    return temp;}catch(Exception e){e.printStackTrace();return null;}}
    /***简单车牌识别*/
    public void SimpleRecog(Bitmap bmp,int dp){
        if(bmp==null){return;}
        if(dp<=1)dp=1; //限制最小分辨率为0.1倍
        float dp_asp=dp/10.f;
        Mat mat_src=new Mat(bmp.getWidth(), bmp.getHeight(), CvType.CV_8UC4);
        float new_w=bmp.getWidth()*dp_asp;
        float new_h=bmp.getHeight()*dp_asp;
        if(new_w>=MAX_WIDTH){new_w=MAX_WIDTH;
            dp_asp=new_w/bmp.getWidth();new_h=bmp.getHeight()*dp_asp;}
        if(new_h>=MAX_HEIGHT){new_h=MAX_HEIGHT;
            dp_asp=new_h/bmp.getHeight();new_w=bmp.getWidth()*dp_asp;}
        Size sz=new Size(new_w,new_h); //构建新图像的长和宽
        //构造方法的字符格式这里如果小数不足1位,会以0补足
        DecimalFormat decimalFormat=new DecimalFormat(".0");
        textViewPictureWidth.setText(decimalFormat.format(new_w));
        textViewPictureHeight.setText(decimalFormat.format(new_h));
        textViewBmpDegrees.setText(decimalFormat.format(g_degrees));
        Utils.bitmapToMat(bmp, mat_src); //根据指定文件,将位图转换成图像矩阵
        Imgproc.resize(mat_src,mat_src,sz); //重新调整图像尺寸
        new Thread(){ //构造进程
            public void run(){ //实现进程执行体
                try{Message msg=new Message();
                    msg.what=MainActivity.MSG_DISPLAY_BITMAP;
                    msg.obj=latestBitmap;m_handler.sendMessage(msg);
                }catch(Exception e){e.printStackTrace();}} //处理异常
        }.start(); //启动进程
        long currentTime=System.currentTimeMillis(); //获得系统当前时间
        String res=PlateRecognition.SimpleRecognization
(mat_src.getNativeObjAddr(),handlePR);
        long diff=System.currentTimeMillis()-currentTime; //车牌识别时间
        if(res.isEmpty()){
            textViewRecogResult.setText("请调整比例、角度和位置! ");
            m_cardata.recog_result="";return;}else{
            textViewRecogResult.setText(res); //显示车牌识别结果
            textViewRuntime.setText(String.valueOf(diff)); //显示车牌识别时间
            m_cardata.recog_result=res;m_cardata.recog_runtime=(int)diff;
            m_cardata.img_width=new_w;m_cardata.img_height=new_h;
            m_cardata.cur_time=DateTools.getCurDateStr();
            new Thread(){ //构造进程
```

```
                public void run(){  //实现进程执行体
                    try{Message msg=new Message();
                        msg.what=MainActivity.MSG_INSERT_SQLITE;
                    msg.obj=(Object)m_cardata;m_handler.sendMessage(msg);
                    }catch(Exception e){e.printStackTrace();}}  //处理异常
            }.start();//启动进程}}
    /***返回结果*/
    protected void onActivityResult(int requestCode,int resultCode,Intent data)
{if(requestCode==REQUEST_CODE_IMAGE_OP&&resultCode==RESULT_OK){
        //请求图像操作
        mPath=data.getData();  //返回数据：文件路径
        String file=getPath(mPath);  //根据路径，获得文件
        textViewPicturePathName.setText(file.toString());
        m_cardata.car_name=file;m_cardata.img_degrees=(float)0.0;
        m_cardata.src_type="选择图片(SELECT_PHOTO)";
        Bitmap bmp=decodeImage(file);  //图像解码，获得位图文件
        if(bmp==null||bmp.getWidth()<=0||bmp.getHeight()<=0){
            Log.e(TAG, "错误！");}else{
            Log.i(TAG,"bmp["+bmp.getWidth()+","+bmp.getHeight());
            latestBitmap=bmp;  //获得最新位图
            //传递图像，进行车牌识别
            SimpleRecog(bmp,seekBarResolution.getProgress());}
        } else if(requestCode==REQUEST_CODE_OP){  //请求结果操作
            Log.i(TAG, "RESULT ="+resultCode);  //获得车牌识别结果
            if(data==null){return;}
            Bundle bundle=data.getExtras();  //获得绑定数据
            String path=bundle.getString("imagePath");  //获得图像路径
        }else if(requestCode==REQUEST_CODE_IMAGE_CAMERA&&resultCode
         ==RESULT_OK){  //请求照相机图像
            String file=getPath(mPath);  //根据路径获得图像文件
            Bitmap bmp=decodeImage(file);  //从文件中解码图像获得位图图像
            latestBitmap=bmp;  //最新图像
            //根据图像，进行车牌识别
            SimpleRecog(bmp,seekBarResolution.getProgress());
        }else if(requestCode==PHOTO_REQUEST_GALLERY){  //从相册返回的数据
            if(data!=null){Uri uri=data.getData();  //得到图片的全路径
                crop_photo(uri);}
        }else if(requestCode==PHOTO_REQUEST_CAREMA){  //从照相机返回的数据
            if(hasSdcard()){crop(Uri.fromFile(tempFile));
                textViewPicturePathName.setText(Uri.fromFile(tempFile).toString());
                m_cardata.car_name=Uri.fromFile(tempFile).toString();
                m_cardata.img_degrees=(float)0.0;
                m_cardata.src_type="拍照识别(CAMERA)";}else{}
```

```
        }else if(requestCode==PHOTO_REQUEST_CUT){  //从剪切图片返回的数据
            if(data!=null){Bitmap bitmap=data.getParcelableExtra("data");
                new Thread(){  //构造进程
                    public void run(){  //实现进程执行体
                        try{  //把位图bitmap以msg形式发送至主线程
                            Message msg=new Message();
                            msg.what=MainActivity.MSG_DISPLAY_BITMAP;
                            msg.obj=latestBitmap;
                            m_handler.sendMessage(msg);}
                        catch(Exception e){e.printStackTrace();}}  //处理异常
                }.start();  //启动进程
            latestBitmap=bitmap;  //最新图像
            //根据图像，进行车牌识别
            SimpleRecog(latestBitmap,seekBarResolution.getProgress());}
    }else if(requestCode==PHOTO_REQUEST_CUT_GALLERY){  //从剪切图片返回数据
            if(data!=null){Bitmap bitmap=data.getParcelableExtra("data");
                new Thread(){  //构造进程
                    public void run(){  //实现进程执行体
                        try{  //把位图bitmap以msg形式发送至主线程
                            Message msg=new Message();
                            msg.what=MainActivity.MSG_DISPLAY_BITMAP;
                            msg.obj=latestBitmap;m_handler.sendMessage(msg);
                        }catch(Exception e){e.printStackTrace();}}
                }.start();  //启动进程
                latestBitmap=bitmap;  //最新图像
                try{
                    tempFile=FileTools.getOutputMediaFile(FileTools.FILE_TYPE_GALLERY);
                    saveToLocal(bitmap, tempFile);
                    textViewPicturePathName.setText(tempFile.toString());
                    m_cardata.car_name=tempFile.toString();
                    m_cardata.img_degrees=(float)0.0;
                    m_cardata.src_type="相册选图(GALLERY)";
                }catch(FileNotFoundException e){e.printStackTrace();
                }catch(IOException e){e.printStackTrace();}
                //根据图像，进行车牌识别
            SimpleRecog(latestBitmap,seekBarResolution.getProgress());}}
        super.onActivityResult(requestCode, resultCode, data);}
    /***功能：识别演示图片的车牌信息*/
    private void SimpleRecognizationMatDemo(Bitmap bmp){
        Mat mat_src=new Mat(bmp.getWidth(),bmp.getHeight(),
CvType.CV_8UC4);
        Size sz=new Size(720,682);
        Utils.bitmapToMat(bmp, mat_src);  //将位图数据转换成MAT数据
```

```
            Imgproc.resize(mat_src,mat_src,sz);
            String res=PlateRecognition.SimpleRecognization
    (mat_src.getNativeObjAddr(),handlePR);
            if(!res.isEmpty()){textViewRecogResult.setText(res);
                m_cardata.recog_result=res;}
            else{
                textViewRecogResult.setText(R.string.recog_result_prompt);
                m_cardata.recog_result="";}}
        Handler m_handler=new Handler(){  //助手更新的作用
            public void handleMessage(Message msg){
                switch( msg.what)
               {   case MainActivity.MSG_DISPLAY_BITMAP:
                        Bitmap bmp=(Bitmap)msg.obj; //显示
                        imageViewCarPicture.setImageBitmap(bmp);  //显示车牌图像
                        imageViewCarPicture1.setImageBitmap(bmp);  //显示车牌图像
                        break;
                    case MainActivity.MSG_INSERT_SQLITE:data=(cardata)msg.obj;
                        if(!data.car_name.isEmpty()&&!data.src_type.isEmpty()&&
                            !data.recog_result.isEmpty()&&!data.cur_time.
isEmpty()&&data.img_height>0&& data.img_width>0&&data.recog_runtime>0){
                            mDBLpr.insert_LprData(data);
                            insertMySQLLprData();  //将数据写入MySQL数据库
                            uploadImage(data.car_name);  //上传图片到服务器}break;
                    case MainActivity.MSG_MYSQL_TEST_1:String str1="查询成功!";
                        textViewPicturePathName.setText(str1);break;
                    case MainActivity.MSG_MYSQL_TEST_0:String str0="查询失败!";
                         textViewPicturePathName.setText(str0);break;}}};};
        /***活动初始化回调对象：一旦活动页面被创建时调用此方法*/
        protected void onCreate(Bundle savedInstanceState){
            super.onCreate(savedInstanceState);  //调用父类或基类方法
            requestWindowFeature(Window.FEATURE_NO_TITLE);  //取消标题栏
            getWindow().setFlags(WindowManager.LayoutParams.FLAG_FULLSCREEN,
WindowManager.LayoutParams.FLAG_FULLSCREEN);
//全屏getWindow().addFlags(WindowManager.LayoutParams.FLAG_KEEP_SCREEN_ON);
            setContentView(R.layout.activity_main);  //加载(活动)屏幕布局文件
            //初始化外部存储器(SD卡)写权限
            String permission=Manifest.permission.WRITE_EXTERNAL_STORAGE;
            //检查外部存储器(SD卡)写权限
            if(ActivityCompat.checkSelfPermission(MainActivity.this,permission)
                    != PackageManager.PERMISSION_GRANTED){
                //如果没有被授权，请求授权
                ActivityCompat.requestPermissions(MainActivity.this,
    new String[]{permission},123);}
```

```
        AddView();  //添加视图控件，用于将代码功能与活动布局中控件进行绑定
        //初始化车牌识别函数
        initRecognizer();  //初始化车牌识别动态库
        initLPRDB();  //用户初始化车牌识别嵌入式SQlite数据库和MySQL数据库
        InitMediaPlayer();  //初始化多媒体播放器或音乐池对象
        setAllClickListerner();  //为所有需要的控件添加单击事件侦听器回调处理函数
        //设置最多可容纳10个音频流，音频的品质为5
        soundPool=new SoundPool(10, AudioManager.STREAM_SYSTEM, 5);
    /***功能：绑定GUI控件*/
    private void AddView(){
        buttonSelectImg=(Button)findViewById(R.id.buttonSelectImg);
        buttonRecogPlate=(Button)findViewById(R.id.buttonRecogPlate);
        buttonTakePhoto=(Button)findViewById(R.id.buttonTakePhoto);
        //绑定查询车牌识别信息按钮
        buttonPlateQuery=(Button)findViewById(R.id.buttonPlateQuery);}
        //绑定反向旋转图片按钮
    /***设置所需按钮、文本视图、列表对象等单击侦听器*/
    private void setAllClickListerner(){
        buttonSelectImg.setOnClickListener(this);  //设置按钮单击侦听器}
    /***初始化播放背景音乐和按键提示声音的多媒体播放器对象和音乐池对象*/
    private void InitMediaPlayer(){
        try{mMediaPlayer=MediaPlayer.create(MainActivity.this,
R.raw.background_muisc);StartOrPauseMediaPlayer();
        }catch(IllegalArgumentException e){e.printStackTrace();
        }catch(IllegalStateException e){e.printStackTrace();}}
    /***选择按钮单击事件侦听回调处理*/
    private void buttonSelectImgClickRecog(){
        buttonSelectImg.setOnClickListener(new View.OnClickListener(){
            public void onClick(View view){RecogDemoPlate();}});}
    /***单击单击事件处理函数*/
    @Override
    public void onClick(DialogInterface dialogInterface, int which){
        switch(which){
            case 1:  //通过照相机捕获图像意图对象
                Intent getImageByCamera=new Intent(
  "android.media.action.IMAGE_CAPTURE");
                ContentValues values=new ContentValues(1);
                values.put(MediaStore.Images.Media.MIME_TYPE,"image/jpeg");
                mPath=getContentResolver().insert(MediaStore.Images.
  Media.EXTERNAL_CONTENT_URI, values);
                getImageByCamera.putExtra(MediaStore.EXTRA_OUTPUT, mPath);
                //请求获得拍照图像
                startActivityForResult(getImageByCamera,REQUEST_CODE_IMAGE_CAMERA);
```

```
                break;
            case 0: //获得内容意图
                Intent getImageByalbum=new Intent(Intent.ACTION_GET_CONTENT);
                //通过相册获得图像
                getImageByalbum.addCategory(Intent.CATEGORY_OPENABLE);
                getImageByalbum.setType("image/jpeg");  //相册图像类型
         //请求获得相册图像
         startActivityForResult(getImageByalbum,REQUEST_CODE_IMAGE_OP);}}
    public void onClick(View view){
        switch(view.getId()){
            case R.id.buttonSelectImg: //打开图像(选择图像)
                g_degrees=(float)0.0;
                new AlertDialog.Builder(this).setTitle("打开方式")
                .setItems(new String[]{"打开图片"},this).show();break;
            case R.id.buttonRecogPlate:
              if(latestBitmap!=null){buttonRecogPlate.setEnabled(false);
  SimpleRecog(latestBitmap,seekBarResolution.getProgress());  //进行车牌识别
                  buttonRecogPlate.setEnabled(true);}else{
                  buttonRecogPlate.setEnabled(true);}break;
            case R.id.buttonTakePhoto:  //拍照,取得汽车图像
                g_degrees=(float)0.0;getPlatePhotoFromCamera( );break;
            case R.id.buttonSelectFromGALLERY: //从相册选择汽车图片
                g_degrees=(float)0.0;getPlatePhotoFromGallery();break;
            case R.id.buttonPlateBrowser:doPlateBrowser();break;
            case R.id.buttonPlateQuery:doCustomPlateQuery();break;
            case R.id.buttonRotateBmpForward:doRotateBmpForward();break;
            case R.id.buttonMySQL:doMySQLQuery();break;
            case R.id.buttonDemo:doRecogPlateDemo();break;
            case R.id.buttonRotateBmpBack:doRotateBmpBack();break;
            case  R.id.buttonExit:showExitAlert();break;}}
    private void doCustomPlateQuery(){  /*执行定制查询SQlite数据库功能*/
        Intent intent=new Intent(MainActivity.this, pr.platerecognization.
CustomizedQueryPlateActivity.class);
        startActivity(intent);} //启动意图,切换到定制查询车牌识别界面
    private void doPlateBrowser(){  /***执行查询SQLite数据库全部数据*/
        Intent intent=new Intent(MainActivity.this,
  pr.platerecognization.PlateBrowserActivity.class);
    startActivity(intent);}  //启动意图,切换到查询浏览全部车牌识别数据活动页面
    private void doMySQLQuery(){  /***执行查询、删除和更新MySQL数据库数据*/
        Intent intent=new Intent(MainActivity.this,
  MySQLQueryPlateActivity.class);
        startActivity(intent);}
    private void doRotateBmpForward(){  /***正向旋转车牌图像*/
```

```
        if(latestBitmap!=null){buttonRotateBmpForward.setEnabled(false);
            g_degrees +=(float)1.0;  //正向旋转1°
            if( g_degrees>=MAX_DEGREES)g_degrees=MAX_DEGREES;
            textViewBmpDegrees.setText(Float.toString(g_degrees));
            try{latestBitmap=BitmapTools.createRotateBitmap
 (latestBitmap, g_degrees);
                m_cardata.img_degrees=g_degrees;
            }catch(RuntimeException re){re.printStackTrace();
                buttonRotateBmpForward.setEnabled(true);return ;
            } catch(Exception ex){ex.printStackTrace();
                buttonRotateBmpForward.setEnabled(true);return;}
            if(latestBitmap!=null){  //进行车牌识别
             SimpleRecog(latestBitmap, seekBarResolution.getProgress());}
            buttonRotateBmpForward.setEnabled(true);}else{
            buttonRotateBmpForward.setEnabled(true);}}
    private  void doRecogPlateDemo(){  /***演示车牌识别功能*/
        Intent intent=new Intent(MainActivity.this,
pr.platerecognization.DemoGridViewActivity.class);
        startActivity(intent);}  //启动意图，切换到车牌识别功能演示活动页面
    public void getPlatePhotoFromGallery( ){  /***从相册获取车牌图像*/
        //激活系统图库，选择一张图片
        Intent intent=new Intent(Intent.ACTION_PICK);
        intent.setType("image/*");
        //开启一个带有返回值的Activity，请求码为PHOTO_REQUEST_GALLERY
        startActivityForResult(intent, PHOTO_REQUEST_GALLERY);
    public void getPlatePhotoFromCamera( ){  /***从照相机获取车牌图像*/
         Intent intent=new Intent("android.media.action.IMAGE_CAPTURE");
         //判断存储卡是否可以用，可用进行存储
         if(hasSdcard()){
             tempFile=FileTools.getOutputMediaFile(FileTools.FILE_TYPE_CAMERA);
             Uri uri=Uri.fromFile(tempFile); //从文件中创建uri
             textViewPicturePathName.setText(uri.toString());
             m_cardata.car_name=uri.toString();
             intent.putExtra(MediaStore.EXTRA_OUTPUT, uri);}
         else{return ;}
         //开启一个带有返回值的Activity，请求码为PHOTO_REQUEST_CAREMA
         startActivityForResult(intent, PHOTO_REQUEST_CAREMA);}
    private void crop(Uri uri){/***功能：剪切拍照图像**/
       Intent intent=new Intent("com.android.camera.action.CROP");
       intent.setDataAndType(uri,"image/*");intent.putExtra("crop","true");
       intent.putExtra("aspectX", 4);  //裁剪框的比例，1:1
       intent.putExtra("aspectY", 4);
       intent.putExtra("outputX", 250);  //裁剪后输出图片的尺寸大小
```

```
        intent.putExtra("outputY", 250);
        intent.putExtra("outputFormat", "JPEG");  //图片格式
        intent.putExtra("noFaceDetection", true);  //取消人脸识别
        intent.putExtra("return-data", true);
        //开启一个带有返回值的Activity，请求码为PHOTO_REQUEST_CUT
         startActivityForResult(intent, PHOTO_REQUEST_CUT);}
    /***功能：将图片保存到本地指定路径*/
    private void saveToLocal(Bitmap bitmap, String fileName)throws
IOException{File file=new File(fileName);
        if(file.exists()){file.delete();}
        FileOutputStream out;
        try{out=new FileOutputStream(file);
            if(bitmap.compress(Bitmap.CompressFormat.PNG, 90, out)){
                out.flush();out.close();
                //保存图片后发送广播通知更新数据库
                Intent intent=new Intent(Intent.ACTION_MEDIA_SCANNER_SCAN_FILE);
                Uri uri=Uri.fromFile(file);intent.setData(uri);
                this.sendBroadcast(intent);}
        } catch(FileNotFoundException e){e.printStackTrace();
        } catch(IOException e){e.printStackTrace();}}
    /***功能：将图片保存到本地指定路径*/
    private void saveToLocal(Bitmap bitmap,File file)throws IOException{
        if(file.exists()){file.delete();}FileOutputStream out;
        try{out=new FileOutputStream(file);
            if(bitmap.compress(Bitmap.CompressFormat.PNG,90,out)){
                out.flush();out.close();
                //保存图片后发送广播通知更新数据库
                Intent intent=new Intent(Intent.ACTION_MEDIA_SCANNER_SCAN_FILE);
                Uri uri=Uri.fromFile(file);intent.setData(uri);
                this.sendBroadcast(intent);}
        }catch(FileNotFoundException e){e.printStackTrace();
        } catch(IOException e){e.printStackTrace();}}
    void insertMySQLLprData( ){  /***将车牌识别结果写入MySQL服务器*/
        new Thread(new Runnable(){public void run(){
         try{pr.platerecognization.DBService m_DBService=getDbService();
              String result=m_DBService.insertLprData(data);
              Message msg=new Message();msg.what=MSQ_MYSQL_INSERT_SUCESS;
              msg.obj =result ;mysql_handler.sendMessage(msg);
            }catch(Exception e){Message msg=new Message();
              msg.what=MSQ_MYSQL_INSERT_FAIL;msg.obj="失败:"+e.toString();
              mysql_handler.sendMessage(msg);e.printStackTrace();}}
        }).start();}
    private void uploadImage(String imagePath){  /***功能：上传图片*/
```

```
        new NetworkTask().execute(imagePath);}
    /***访问网络AsyncTask,访问网络在子线程进行并返回主线程通知访问的结果*/
    class NetworkTask extends AsyncTask<String,Integer,String>{
        protected void onPreExecute(){super.onPreExecute();}
        protected String doInBackground(String... params){
            return doPost(params[0]);}
        protected void onPostExecute(String result){
            if(!"error".equals(result)){
        Glide.with(MainActivity.this).load(Constant.SERVER_URL+result)
           .into(imageViewCarPicture1);}}}
    private String doPost(String imagePath){
        OkHttpClient mOkHttpClient=new OkHttpClient();
        String result="error";
        MultipartBody.Builder builder=new MultipartBody.Builder();
        //这里添加用户ID
        builder.addFormDataPart("image",imagePath,RequestBody.create
(MediaType.parse("image/jpeg"),new File(imagePath)));
        RequestBody requestBody=builder.build();
        Request.Builder reqBuilder=new Request.Builder();
        Request request=reqBuilder.url(Constant.SERVER_URL+"/uploadimage")
        .post(requestBody).build();
        try{Response response=mOkHttpClient.newCall(request).execute();
            if(response.isSuccessful()){
                String resultValue=response.body().string();
                Log.d(TAG,"响应体 "+resultValue);return resultValue;}
        }catch(Exception e){e.printStackTrace();}return result;}
```

8.4.4 SQLite 嵌入式数据库 DBLpr 类

SQLite嵌入式数据库DBLpr类代码如下：

```
public class DBLpr{
    private static final String TAG="DBLpr";  //定义日志标记字符串常量
    //定义车牌识别APP的SQlite嵌入式数据库数据库名
    private static final String DATABASENAME="LPRCar.db";
    private static final int VERSION=1; //版本
    //定义选择图片车牌识别数据
    public static final String PLATE_TABLENAME="app_plate_table";
    public static final String APT_ID="id";  //编码ID
    public static final String APT_CARNAME="car_name"; //汽车图像名
    public static final String SPT_RECOG_RESULT="recog_result";
    public static final String SPT_RUNTIME="recog_runtime";  //车牌识别时间
    public static final String SPT_CUR_TIME="cur_time";  //识别系统当前时间
    public static final String SPT_WIDTH="img_width";  //图像宽度
    public static final String SPT_HEIGHT="img_height";  //图像高度
```

```
    public static final String SPT_TYPE="src_type";
    public static final String SPT_DEGREES="img_degrees";  //图像角度
    //定义用户和密码管理数据表
    public static final String USER_TABLENAME="user_table";  //用户表
    public static final String SUT_ID="sut_id"; //用户ID
    public static final String SUT_NAME="sut_name";  //用户名
    public static final String SUT_PASSWORD="sut_password";  //登录密码
    private final Context ct;private SQLiteDatabase db;
    private SQLiteDB sdb;
    /**构造嵌入式数据库SQLiteOpenHelper对象，并创建数据库LPRCar.db*/
    private static class SQLiteDB extends SQLiteOpenHelper{
        public SQLiteDB(Context context){super(context,DATABASENAME,null,1);}
            /**: 初始化或创建嵌入式数据SQLite中车牌识别数据表"app_plate_table"
            *和用户登录数据表"user_table"*@param sdb*/
        public void onCreate(SQLiteDatabase sdb){
            //建数据表(PLATE_TABLENAME)
            String lpr_car_sql="CREATE TABLE "+PLATE_TABLENAME+"("+APT_ID
                +"INTEGER PRIMARY KEY  AUTOINCREMENT,"+APT_CARNAME
                +"TEXT,"+SPT_RECOG_RESULT+"TEXT,"+SPT_RUNTIME
                +"INTEGER,"+SPT_CUR_TIME+"DATE,"+SPT_WIDTH
                +"DOUBLE,"+SPT_HEIGHT+"DOUBLE,"+SPT_TYPE
                +"TEXT,"+SPT_DEGREES+"DOUBLE);";
            sdb.execSQL(lpr_car_sql);
            //建立用户和密码管理数据表
            String user_sql="CREATE TABLE "+USER_TABLENAME+"("+SUT_ID
                +"INTEGER PRIMARY KEY  AUTOINCREMENT,"+SUT_NAME
            +"VARCHAR(255),"+SUT_PASSWORD+"VARCHAR(255));";
        sdb.execSQL(user_sql);try{
          String StrSql=String.format("INSERT INTO "+USER_TABLENAME+"("
                  +SUT_NAME+","+SUT_PASSWORD +")"
                  +"values( '%s' ,' %s' )", "root","123456");
          sdb.execSQL(StrSql);
          StrSql=String.format("INSERT INTO "+USER_TABLENAME+"("
                  +SUT_NAME+","+SUT_PASSWORD +")"
                  +"values( '%s' ,' %s' )", "0","0");
          sdb.execSQL(StrSql);
                  StrSql=String.format("INSERT INTO "+USER_TABLENAME+"("
                        +SUT_NAME+","+SUT_PASSWORD +")"
                        +"values( '%s' ,' %s' )", "r","r");
                  sdb.execSQL(StrSql);Log.d(TAG,StrSql);
        }catch(SQLiteException se){se.printStackTrace();
        }catch(Exception e){e.printStackTrace();}}
      /***功能: 更新嵌入式数据库SQlite*/
```

```
    public void onUpgrade(SQLiteDatabase sdb,int oldVersion,int newVersion)
{//删除表PLATE_TABLENAME
        String lpr_car_sql="DROP TABLE IF EXISTS "+PLATE_TABLENAME;
        sdb.execSQL(lpr_car_sql);
        //删除SQLITE_USER_TABLENAME
        String user_sql="DROP TABLE IF EXISTS "+USER_TABLENAME;
        sdb.execSQL(user_sql);}}
  public DBLpr(Context context){  //初始化数据库
      ct=context;sdb=new SQLiteDB(ct);}
  public SQLiteDatabase openDB(){  //打开数据库
      return db=sdb.getWritableDatabase();}
  public void closeDB(){sdb.close();}  //关闭数据库
  public Cursor select(String table){  //查询数据表
      SQLiteDatabase db=sdb.getReadableDatabase();
      Cursor cursor=db.query(table, null, null, null, null, null, null);
      return cursor;}
  public Cursor select_plate(){  //查询车牌识别数据
      SQLiteDatabase db=sdb.getReadableDatabase();
      Cursor cursor=db.query(PLATE_TABLENAME, null, null, null, null,
          null, null);return cursor;}
  /***功能：将车牌识别结果cdata写入到嵌入式数据库SQlite中*/
  public long insert_LprData(cardata cdata){
      SQLiteDatabase db=sdb.getReadableDatabase();
      ContentValues cv=new ContentValues();
      cv.put(APT_CARNAME, cdata.car_name);
      cv.put(SPT_RECOG_RESULT, cdata.recog_result);
      cv.put(SPT_RUNTIME, cdata.recog_runtime);
      cv.put(SPT_CUR_TIME, cdata.cur_time);
      cv.put(SPT_WIDTH, cdata.img_width);
      cv.put(SPT_HEIGHT, cdata.img_height);
      cv.put(SPT_TYPE, cdata.src_type);
      cv.put(SPT_DEGREES, cdata.img_degrees);
      try{long row=db.insert(PLATE_TABLENAME,null,cv);
          return row;}catch(Exception e){return -1;}}}
```

8.4.5 实现访问MySQL数据库操作接口

实现访问MySQL数据库操作接口MySQLDBOpenHelper的源代码如下：

```
  public class MySQLDBOpenHelper{
      private static final String TAG="MySQLDBOpenHelper";
      private static String m_ip="120.78.81.93";//"127.0.0.1";10.0.2.2"
      private static String m_driver="com.mysql.jdbc.Driver";  //MySQL 驱动
      private static String m_dbName="lprcar";
  private static String m_url="jdbc:mysql://"+m_ip+":3306/"+m_dbName+"?
```

```
useUnicode=true&characterEncoding=UTF-8";  //MySQL数据库连接Url
    private static String m_user="root";  //用户名
    private static String m_password="123456";  //密码
    private static Connection getConnection(String dbName){  //连接数据库
        Connection conn=null;try{Class.forName(m_driver);  //加载驱动
            conn=DriverManager.getConnection("jdbc:mysql://"+m_ip+
 ":3306/"+dbName,m_user,m_password);
        }catch(SQLException ex){ex.printStackTrace();return null;
        }catch(ClassNotFoundException ex){ex.printStackTrace();
            return null;}return conn;}
    public static Connection getConn(){  /***连接数据库**/
        Connection conn=null;try{Class.forName(m_driver);  //获取MySQL驱动
 conn=(Connection)DriverManager.getConnection(m_url,m_user,m_password);
        }catch(ClassNotFoundException e){e.printStackTrace();
        }catch(SQLException e){e.printStackTrace();}return conn;}
    /***关闭数据库**/
    public static void closeAll(Connection conn,PreparedStatement ps){
        if(conn!=null){try{conn.close();
            }catch(SQLException e){e.printStackTrace();}}
        if(ps!=null){try{ps.close();
            }catch(SQLException e){e.printStackTrace();}}}
    /***关闭数据库**/
    public static void closeAll(Connection conn,PreparedStatement ps,
ResultSet rs){if(conn!=null){try{conn.close();
        } catch(SQLException e){e.printStackTrace();}}
        if(ps!=null){try{ps.close();
            }catch(SQLException e){e.printStackTrace();}}
        if(rs!=null){try{rs.close();}catch(SQLException e){
                e.printStackTrace();}}}
    public static HashMap<String,String>getUserInfoByName(String name){
        HashMap<String, String>map=new HashMap<>();
        Connection conn=getConnection("lprcar");
        if(conn==null ){return null;}
        try{Statement st=conn.createStatement();
            String sql="select *from app_plate_table where car_name='"+name+"' ";
            ResultSet res=st.executeQuery(sql);
            if(res==null){return null;} else{
                int cnt=res.getMetaData().getColumnCount();res.next();
                for(int i=1; i<=cnt; ++i){
                    String field=res.getMetaData().getColumnName(i);
                    map.put(field, res.getString(field));}
                conn.close();st.close();res.close();return map;}
        }catch(Exception e){e.printStackTrace();return null;}}}
```

8.4.6 PlateBrowserActivity 显示查询 SQlite 数据库功能

PlateBrowserActivity 类派生于 Activity，实现 OnClickListener 接口，其源代码如下：

```
   public class PlateBrowserActivity extends Activity implements
OnClickListener{private DBLpr mDBLpr=new DBLpr(this);
      private SQLiteDatabase mDB=null ;private Cursor mCursor=null;
      private pr.platerecognization.cardata m_cardata=new cardata();
      private String EXPORT_PLATE_FILE_NAME="/myexportplate.txt";
      private String myimportPlate="/myimportplate.txt";
      //定义助手Handler处理的消息长量
      private static final int MSG_DISPLAY_EXPORT_FILE=333;  //显示位图消息
      Handler m_handler=new Handler(){  //助手更新的作用
          public void handleMessage(Message msg){
            switch( msg.what){
              case PlateBrowserActivity.MSG_DISPLAY_EXPORT_FILE:
                tvExportFileName.setText(EXPORT_PLATE_FILE_NAME);break;}};};
      /**当活动被第一次创建时调用 .*/
      public void onCreate(Bundle savedInstanceState){
          super.onCreate(savedInstanceState);
          requestWindowFeature(Window.FEATURE_NO_TITLE);  //取消标题栏
          getWindow().setFlags(WindowManager.LayoutParams.FLAG_FULLSCREEN,
WindowManager.LayoutParams.FLAG_FULLSCREEN);
       //全屏getWindow().addFlags(WindowManager.LayoutParams.FLAG_KEEP_SCREEN_ON);
       setContentView(R.layout.activity_query_lprdata);
       AddView();InitData();}
    /***功能: 查询车牌识别全部数据*/
    private Cursor QueryLprPlateData(){try{
    mCursor=mDB.rawQuery("select *from "+DBLpr.PLATE_TABLENAME,null);
       return mCursor;}catch(SQLiteException se){return null;}}
    private void ShowPlatePhoto(String fileName){  /***显示车牌图片*/
      if(!(fileName.trim().equals(""))){
        Intent intent=new Intent(PlateBrowserActivity.this, PhotoActivity.class);
        mID =qlprID.getText().toString();
        mCarName=qlprCarName.getText().toString();
        mRecogResult=qlprRecogResult.getText().toString();
        mRecogRuntime=qlprRecogRuntime.getText().toString();
        mCurTime=qlprCurTime.getText().toString();
        mImgWidth=qlprImgWidth.getText().toString();
        mImgHeight=qlprImgHeight.getText().toString();
        mSrcType=qlprSrcType.getText().toString();
        mImgDegrees=qlprImgDegrees.getText().toString();
        intent.putExtra("photo", fileName);intent.putExtra("mID",mID);
        intent.putExtra("mCarName",mCarName);
```

```
        intent.putExtra("mRecogResult",mRecogResult);
        intent.putExtra("mRecogRuntime",mRecogRuntime);
        intent.putExtra("mCurTime",mCurTime);
        intent.putExtra("mImgWidth",mImgWidth);
        intent.putExtra("mImgHeight",mImgHeight);
        intent.putExtra("mSrcType",mSrcType);
        intent.putExtra("mImgDegrees",mImgDegrees);
        startActivity(intent);}}
    public void onClick(View view){  /***功能: 鼠标单击事件回调函数*/
        switch(view.getId()){
        case R.id.qFirestQuery:DisplayFirstPlate();break;
        case R.id.qNextQuery:DisplayNextPlate();  break;
        case R.id.qPrevQuery:DisplayPrevPlate();  break;
        case R.id.qLastQuery:DisplayLastPlate();  break;
        case R.id.qDeleteCurRecord:showDeleteCurRecordAlert();break;
        case R.id.qUpdateCurRecord:showUpdateCurRecordAlert();break;
        case R.id.qBtnExport:DoExportPlate();  break;
        case R.id.qBtnImport:DoImportPlate();  break;
        case R.id.qBtnClear:  showClearAllAlert();  break;
        case R.id.qlprCarName:
          if(!(qlprCarName.getText().toString().trim().equals(""))){
             ShowPlatePhoto(qlprCarName.getText().toString().trim());
        }break;
        case R.id.qReturnExit:mCursor.close();mDBLpr.closeDB();
           mDB.close();finish();break;}}}
```

PlateBrowserActivity运行结果如图8-10所示

图 8-10　PlateBrowserActivity 运行结果

8.4.7　实现显示选定车牌图像的车牌识别信息

PhotoActivity 类派生于 Activity，实现 OnClickListener 接口，运行结果如图 8-11 所示。其原代码如下：

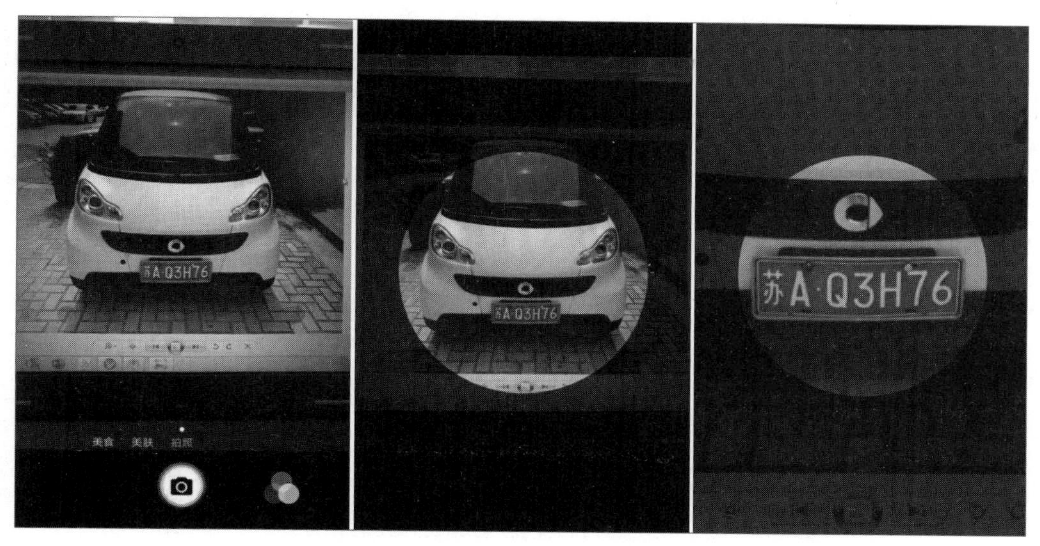

图 8-11　PhotoActivity 运行结果

```
public class PhotoActivity extends Activity implements OnClickListener{
    protected void onCreate(Bundle savedInstanceState){
        super.onCreate(savedInstanceState);
        requestWindowFeature(Window.FEATURE_NO_TITLE);
        setContentView(R.layout.activity_photo);  AddView();
        InitData();getSelectedData();
        imageViewSP=(ImageView)findViewById(R.id.imageViewSP);}
    /** 功能：从Intent获得传递的数据并显示在控件上 */
    private void getSelectedData(){Intent intent=getIntent();
      mID=intent.getStringExtra("mID");
      mCarName=intent.getStringExtra("mCarName");
      mRecogResult=intent.getStringExtra("mRecogResult");
      mRecogRuntime=intent.getStringExtra("mRecogRuntime");
      mCurTime=intent.getStringExtra("mCurTime");
      mImgWidth=intent.getStringExtra("mImgWidth");
      mImgHeight=intent.getStringExtra("mImgHeight");
      mSrcType=intent.getStringExtra("mSrcType");
      mImgDegrees=intent.getStringExtra("mImgDegrees");
      spID.setText(mID);
      spCarName.setText(mCarName);
      spRecogResult.setText(mRecogResult);
      spRecogRuntime.setText(mRecogRuntime);
```

```
        spCurTime.setText(mCurTime);
        spImgWidth.setText(mImgWidth);
        spImgHeight.setText(mImgHeight);
        spSrcType.setText(mSrcType);
        spImgDegrees.setText(mImgDegrees);
        //取出图片路径，并解析成Bitmap对象，然后在ZoomImageView中显示
        try{mBitmap=BitmapFactory.decodeFile
(intent.getStringExtra("photo"));
         if(mBitmap!=null){imageViewSP.setImageBitmap(mBitmap);}
        }catch(RuntimeException re){re.printStackTrace();}}
        /***功能：初始化SQLITE数据库和显示数据*/
    private void InitData(){mDBLpr=new DBLpr(this);
         mDB=mDBLpr.openDB();m_cardata=new cardata();
         mCursor=mDBLpr.select(DBLpr.PLATE_TABLENAME);}
    public void onClick(View view){  /**功能：鼠标单击事件回调函数*/
         switch(view.getId()){
               case R.id.spDeleteCurRecord:showDeleteSPCurRecordAlert();break;
               case R.id.spUpdateCurRecord:showUpdateSPCurRecordAlert();  break;
               case R.id.spBtnExport:DoExportSPPlate();  break;
               case R.id.spReturnExit:mCursor.close();mDBLpr.closeDB();
               mDB.close();finish();break;}}
    /***功能：显示车牌识别数据库前一条记录*/
    private void DisplayPrevSPPlate(){
         try{if(mCursor==null)return ;
            if(mCursor.moveToPrevious()){DisplaySPPlateData(mCursor);}
         }catch(SQLiteException se){}}
    /***功能：显示SQLITE数据库中某条车牌识别记录*/
    void DisplaySPPlateData(Cursor cursor);
    /***功能：更新当前记录*/
    void DoUpdateSPCurRecord();
    /***功能：从数据库导出车牌识别数据到文件*/
    private void DoExportSPPlate();
    /***功能：导出车牌识别数据*/
    private void exportSPPlate();
    /***功能描述：遍历车牌识别数据*/
    protected void writeSPPlateToFile(Cursor cursor,String filename);
```

8.4.8 CustomizedQueryPlateActivity 定制查询车牌页面

CustomizedQueryPlateActivity 实现定制访问嵌入式数据库SQLite中数据：插入、更新、查询、删除等，运行结果如图8-12所示。其源代码如下：

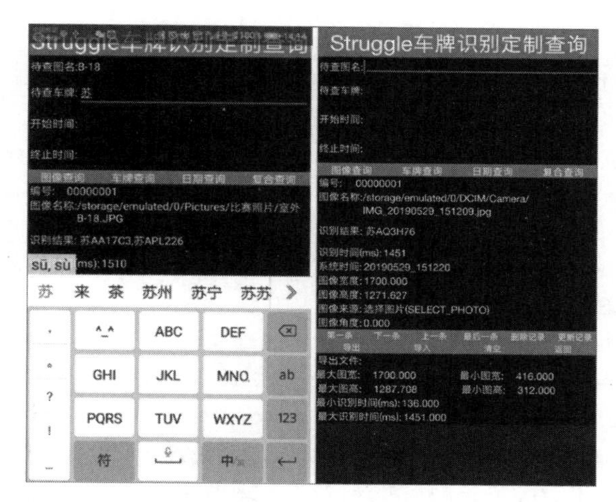

图 8-12　CustomizedQueryPlateActivity 运行结果

```
   public class CustomizedQueryPlateActivity extends Activity implements
OnClickListener{
      /**功能:当活动被第一次创建时调用*/
      public void onCreate(Bundle savedInstanceState){
         super.onCreate(savedInstanceState);
         requestWindowFeature(Window.FEATURE_NO_TITLE);  //取消标题栏
         setContentView(R.layout.activity_customized_query);
         AddCqView();InitCqData();}
      /***功能:初始化SQLITE数据库和显示数据*/
      private void InitCqData(){
          try{mDBLpr=new DBLpr(this);mDB=mDBLpr.openDB();
             m_cardata=new cardata();
             mCursor=mDBLpr.select(DBLpr.PLATE_TABLENAME);
             if(mCursor!=null){mCursor.moveToFirst();}
             if(mCursor!=null && mCursor.moveToFirst()){
                DisplayCqLprPlateData(mCursor);}
             mCursor=mDB.rawQuery("select *from app_plate_table where img_width=(select
max(img_width)from app_plate_table)", null);
          if(mCursor!=null && mCursor.moveToFirst()){
             cqMaxWidth.setText(String.format("%.3f", mCursor.getDouble(mCursor.
getColumnIndex(DBLpr.SPT_WIDTH))));}
          mCursor=mDB.rawQuery("select *from app_plate_table where img_width=(select
min(img_width)from app_plate_table)", null);
          if(mCursor!=null && mCursor.moveToFirst()){
             cqMinWidth.setText(String.format("%.3f", mCursor.getDouble(mCursor.
getColumnIndex(DBLpr.SPT_WIDTH))));}
          mCursor=mDB.rawQuery("select *from app_plate_table where img_
height=(select max(img_height)from app_plate_table)", null);
```

```
            if(mCursor!=null && mCursor.moveToFirst()){
                cqMaxHeight.setText(String.format("%.3f", mCursor.getDouble(mCursor.
getColumnIndex(DBLpr.SPT_HEIGHT))));}
            mCursor=mDB.rawQuery("select *from app_plate_table where img_
height=(select min(img_height)from app_plate_table)", null);
            if(mCursor!=null && mCursor.moveToFirst()){
                cqMinHeight.setText(String.format("%.3f", mCursor.getDouble(mCursor.
getColumnIndex(DBLpr.SPT_HEIGHT))));}
            mCursor=mDB.rawQuery("select *from app_plate_table where recog_
runtime=(select max(recog_runtime)from app_plate_table)", null);
            if(mCursor!=null && mCursor.moveToFirst()){
                cqMaxRecogRuntime.setText(String.format("%.3f", mCursor.
getDouble(mCursor.getColumnIndex(DBLpr.SPT_RUNTIME))));}
            mCursor=mDB.rawQuery("select *from app_plate_table where recog_
runtime=(select min(recog_runtime)from app_plate_table)", null);
            if(mCursor!=null && mCursor.moveToFirst()){
                cqMinRecogRuntime.setText(String.format("%.3f", mCursor.
getDouble(mCursor.getColumnIndex(DBLpr.SPT_RUNTIME))));}
        }catch( SQLiteException se){se.printStackTrace();
        }catch(Exception e){e.printStackTrace();}}
        /***功能：显示SQLITE数据库中某条车牌识别记录*/
        void DisplayCqLprPlateData(Cursor cursor){if(cursor==null)return;
            cqID.setText(String.format("%08d", cursor.getInt(cursor.
getColumnIndex(DBLpr.APT_ID))));
            String car_name_file =String.valueOf(cursor.getString(cursor.
            getColumnIndex(DBLpr.APT_CARNAME)));
            cqCarName.setText(car_name_file);
            cqRecogResult.setText(String.valueOf(cursor.getString
            (cursor.getColumnIndex(DBLpr.SPT_RECOG_RESULT))));
            cqRecogRuntime.setText(String.valueOf(cursor.getInt
            (cursor.getColumnIndex(DBLpr.SPT_RUNTIME))));
            cqCurTime.setText(String.valueOf(cursor.getString
            (cursor.getColumnIndex(DBLpr.SPT_CUR_TIME))));
            cqImgWidth.setText(String.format("%.3f",
            cursor.getDouble(cursor.getColumnIndex(DBLpr.SPT_WIDTH))));
            cqImgHeight.setText(String.format("%.3f",
            cursor.getDouble(cursor.getColumnIndex(DBLpr.SPT_HEIGHT))));
            cqSrcType.setText(String.valueOf(cursor.getString
            (cursor.getColumnIndex(DBLpr.SPT_TYPE))));
            cqImgDegrees.setText(String.format("%.3f",
            cursor.getDouble(cursor.getColumnIndex(DBLpr.SPT_DEGREES))));}
            /***功能：查询车牌识别全部数据*/
```

```
private Cursor QueryLprPlateData(){try{mCursor=mDB.rawQuery
("select *from "+DBLpr.PLATE_TABLENAME, null);return mCursor;
} catch(SQLiteException se){return null;}}
private void ShowPlatePhoto(String fileName){  /***功能:显示车牌图片*/
  if(!(fileName.trim().equals(""))){
    Intent intent=new Intent(CustomizedQueryPlateActivity.this,
PhotoActivity.class);
    mID =cqID.getText().toString();
    mCarName=cqCarName.getText().toString();
    mRecogResult=cqRecogResult.getText().toString();
    mRecogRuntime=cqRecogRuntime.getText().toString();
    mCurTime=cqCurTime.getText().toString();
    mImgWidth=cqImgWidth.getText().toString();
    mImgHeight=cqImgHeight.getText().toString();
    mSrcType=cqSrcType.getText().toString();
    mImgDegrees=cqImgDegrees.getText().toString();
    intent.putExtra("photo", fileName);
    intent.putExtra("mID",mID);
    intent.putExtra("mCarName",mCarName);
    intent.putExtra("mRecogResult",mRecogResult);
    intent.putExtra("mRecogRuntime",mRecogRuntime);
    intent.putExtra("mCurTime",mCurTime);
    intent.putExtra("mImgWidth",mImgWidth);
    intent.putExtra("mImgHeight",mImgHeight);
    intent.putExtra("mSrcType",mSrcType);
    intent.putExtra("mImgDegrees",mImgDegrees);
    startActivity(intent);}}
public void onClick(View view){/***功能:鼠标单击事件回调函数*/
  switch(view.getId()){
  case R.id.cqFirestQuery:DisplayFirstPlate();break;
  case R.id.cqNextQuery:DisplayNextPlate();break;
  case R.id.cqPrevQuery:DisplayPrevPlate();break;
  case R.id.cqLastQuery:DisplayLastPlate();break;
  case R.id.cqDeleteCurRecord:showDeleteCurRecordAlert();break;
  case R.id.cqUpdateCurRecord:showUpdateCurRecordAlert();break;
  case R.id.cqBtnExport:DoExportPlate();break;
  case R.id.cqBtnImport:DoImportPlate();break;
  case R.id.cqBtnClear:showClearAllAlert();break;
  case R.id.customQueryImgButton:doCustomQueryImg();break;
  case R.id.customQueryPlateButton:
    doCustomQueryPlateResult(cqEditTextRecogResult, "WHERE recog_result LIKE
'%", "请输入待查询的车牌名称!");break;
```

```
        case R.id.cqCarName:
            if(!(cqCarName.getText().toString().trim().equals(""))){
                ShowPlatePhoto(cqCarName.getText().toString().trim());}
            break;
        case R.id.cqReturnExit:
            if(mCursor!=null){mCursor.close();}
            if(mDBLpr!=null){mDBLpr.closeDB();}
            if(mDB!=null){mDB.close();}finish();break;}}
    /***功能:执行定制查询车牌识别结果，即根据输入车牌信息，进行相关模糊查询/
    private void doCustomQueryPlateResult(EditText cqEditTextRecogResult,
String s, String s2);
    /***功能:执行车牌图像查询功能。即查询给定车牌图像的相关模糊查询结果。*/
    private void doCustomQueryImg();
    /***功能: 清除所有数据*/
    private void showClearAllAlert();
    private void showDeleteCurRecordAlert();  /***功能: 清除当前数据记录*/
    private void ShowPhoto(String fileName);  /***功能: 显示给定名称车牌图像*/
    /***功能: 从文件导入车牌识别数据到数据库*/
    private void DoImportPlate();
    void DoDeleteCurRecord();  /***功能: 删除当前记录*/
    void DoUpdateCurRecord();  /***功能: 更新当前记录*/
    /***功能: 从数据库导出车牌识别数据到文件*/
    private void DoExportPlate();
    /***功能: 显示车牌识别数据库最后一条记录*/
    private void DisplayLastPlate();
    /***功能: 显示车牌识别数据库前一条记录*/
    private void DisplayPrevPlate();
    /***功能: 显示车牌识别数据库下一条记录*/
    private void DisplayNextPlate();
    /***功能: 显示车牌识别数据库第一条记录*/
    private void DisplayFirstPlate();
    private void importPlate();  /***功能: 导入车牌数据到数据库*/
    public void read(String filename); /***功能: 将数据从文件导入数据库*/
    private void exportPlate();  /***功能: 导出车牌识别数据*/
    /***功能描述: 遍历车牌识别数据*/
    protected void writePlateToFile(Cursor cursor,String filename);
    /***清除SQLITE数据库中所有车牌识别记录*/
    private void clearAllPlateData();
```

8.4.9 DemoGridViewActivity 页面

DemoGridViewActivity类通过GridView视图显示演示车牌信息，运行结果如图8-13所示。其源代码如下：

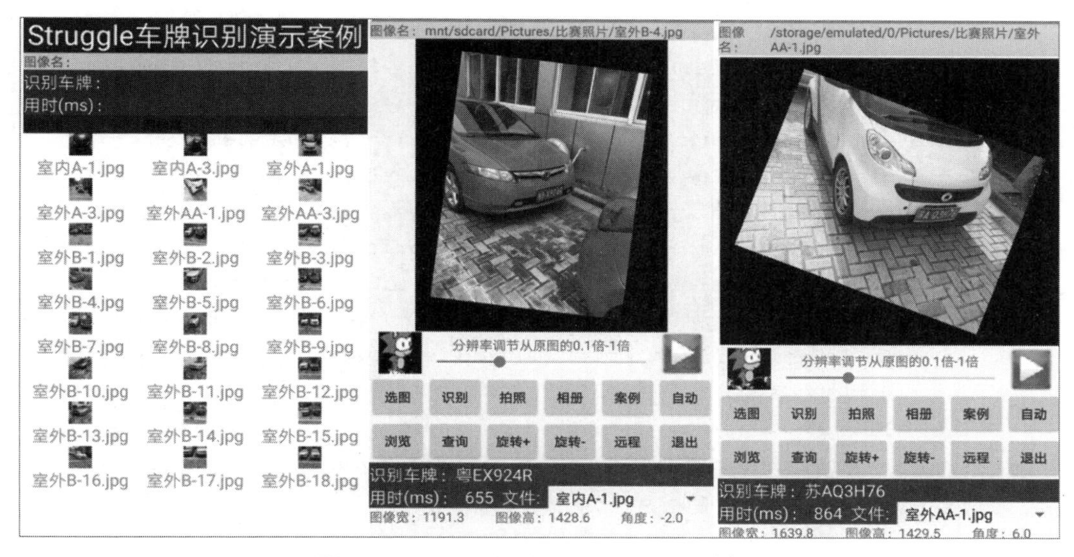

图 8-13　DemoGridViewActivity 运行结果

```
public class DemoGridViewActivity extends Activity{
    private  GridView m_gridView;  //声明GridView对象
    private TextView dmPicturePathName;
    private TextView dmRecogResult;
    private TextView dmTextViewRuntime;
    private TextView dmPictureWidth;
    private TextView dmPictureHeight;
    private TextView dmBmpDegrees;
    private Bitmap latestBitmap;  //定义最新位图对象
   private ImageView dmCarPicture;
    /**初始化DemoGridViewActivity活动页面*/
    protected void onCreate(Bundle savedInstanceState){
        super.onCreate(savedInstanceState);  //基类调用 onCreate()方法
        setContentView(R.layout.activity_demo);  //设置活动界面布局
        dmPicturePathName=(TextView)findViewById(R.id.dmPicturePathName);
        dmRecogResult=(TextView)findViewById(R.id.dmRecogResult);
        dmTextViewRuntime=(TextView)findViewById(R.id.dmTextViewRuntime);
        dmPictureWidth=(TextView)findViewById(R.id.dmPictureWidth);
        dmPictureHeight=(TextView)findViewById(R.id.dmPictureHeight);
        dmBmpDegrees=(TextView)findViewById(R.id.dmBmpDegrees);
             dmCarPicture=(ImageView)findViewById(R.id.dmCarPicture);
        //创建一个List对象，List对象的元素是Map
List<Map<String,Object>>listItems=new ArrayList<Map<String,Object>>();
      for(int i=0;i<24;i++){  //向listItems添加数据成员
          Map<String,Object>listItem=new HashMap<String,Object>();
          listItem.put("itemImage",mThumbIds[i]);  //添加图标
```

```
            listItem.put("itemText", mCarName[i]);  //添加文本信息
            listItems.add(listItem);}  //将列表项添加到listItems中
        //创建一个SimpleAdapter
        SimpleAdapter adapter=new SimpleAdapter(this,listItems,
            R.layout.cell,new String[]{"itemImage","itemText"},
            new int[]{R.id.itemImage,R.id.itemText});
        m_gridView=(GridView)findViewById(R.id.gridView);
        m_gridView.setAdapter(adapter);  //为GridView设置Adapter
        m_gridView.setOnItemClickListener(new OnItemClickListener(){
            public void onItemClick(AdapterView<?>arg0,View arg1,int arg2,
                long arg3){
                HashMap<String, Object>item=(HashMap<String, Object>)arg0
                .getItemAtPosition(arg2);  //访问被单击AdapterView对象中数据
                dmPicturePathName.setText((String)item.get("itemText"));
                Intent aIntent=new Intent(DemoGridViewActivity.this,
    pr.platerecognization.MainActivity.class);
            aIntent.putExtra("demo_id",arg2);
            startActivity(aIntent);}});}
    /***功能：对给定车牌图像进行解码获得相应图像*/
    public static Bitmap decodeImage(String path){
        try{Bitmap res;  //定义位图对象
            ExifInterface exif=new ExifInterface(path);  //ExIF接口
            int orientation=exif.getAttributeInt(ExifInterface.
    TAG_ORIENTATION,ExifInterface.ORIENTATION_NORMAL);  //获得方向参数
            //获得位图选填
            BitmapFactory.Options op=new BitmapFactory.Options();
            op.inSampleSize=1;op.inJustDecodeBounds=false;
            res=BitmapFactory.decodeFile(path,op);  //解码图像
            Matrix matrix=new Matrix();  //旋转和缩放矩阵
            if(orientation==ExifInterface.ORIENTATION_ROTATE_90){
              matrix.postRotate(90); //旋转90°
          } else if(orientation==ExifInterface.ORIENTATION_ROTATE_180){
              matrix.postRotate(180);//旋转180°
          } else if(orientation==ExifInterface.ORIENTATION_ROTATE_270){
              matrix.postRotate(270);//旋转270° }
            //创建旋转和缩放后位图图像
            Bitmap temp=Bitmap.createBitmap(res, 0, 0, res.getWidth(), res.
getHeight(), matrix, true);
            if(!temp.equals(res)){res.recycle();}  //释放原始位图图像
            return  temp;
        } catch(Exception e){e.printStackTrace();return null;}}
```

8.4.10　清单文件 AndroidManifest

清单文件 AndroidManifest.xml 如下：

```
<?xml version="1.0"encoding="utf-8"?>
<manifest xmlns:android="http://schemas.android.com/apk/res/android"
     package="pr.platerecognization">
     <uses-feature android:name="android.hardware.camera"/>
     <uses-feature android:name="android.hardware.camera.autofocus"/>
     <uses-permission android:name="android.permission.INTERACT_ACROSS_USERS_
FULL"/>
     <uses-permission android:name="android.permission.RESTART_PACKAGES"/>
     <!-- 访问INTERNET的权限 -->
     <uses-permission android:name="android.permission.INTERNET"/>
     <uses-permission android:name="android.permission.ACCESS_NETWORK_STATE"/>
     <uses-permission android:name="android.permission.CHANGE_WIFI_STATE"/>
     <uses-permission android:name="android.permission.ACCESS_WIFI_STATE"/>
     <!-- 手机信息 -->
     <uses-permission android:name="android.permission.READ_PHONE_STATE"/>
     <uses-permission android:name="android.permission.WRITE_EXTERNAL_STORAGE"/>
     <!-- 在SD卡中创建文件与删除文件权限 -->
     <uses-permission android:name="android.permission.MOUNT_UNMOUNT_
FILESYSTEMS"/>
     <!-- 传感器 -->
     <uses-permission android:name="android.permission.VIBRATE"/>
     <!-- 摄像头权限 -->
     <uses-permission android:name="android.permission.CAMERA">
     </uses-permission>
     <uses-permission android:name="android.permission.FLASHLIGHT"/>
     <uses-permission android:name="com.meilapp.meila.permission.MIPUSH_
RECEIVE"/>
   <uses-permission android:name="android.permission.READ_EXTERNAL_STORAGE"/>
       <!-- 开启闪光灯权限 -->
       <uses-permission android:name="android.permission.FLASHLIGHT"/>
       <application
           android:allowBackup="true"
           android:icon="@mipmap/struggle"
           android:label="@string/app_name"
           android:roundIcon="@mipmap/struggle"
           android:theme="@style/AppTheme">
           <activity android:name=".SplashActivity1">
               <intent-filter>
                   <action android:name="android.intent.action.MAIN"/>
                   <category android:name="android.intent.category.LAUNCHER"/>
```

```
            </intent-filter>
        </activity>
        <activity  android:name=".SplashActivity"></activity>
        <activity  android:name=".MySQLQueryPlateActivity"></activity>
        <activity
            android:name="pr.platerecognization.MainActivity"
            android:label="@string/app_name">
        </activity>
        <activity
            android:name="pr.platerecognization.PhotoActivity"
            android:label="@string/photoactivity_name">
        </activity>
        <activity
            android:name="pr.platerecognization.PlateBrowserActivity"
            android:label="@string/platebrowser_activity_name">
        </activity>
        <activity
            android:name="pr.platerecognization.RegisterActivity"
            android:label="Struggle注册界面">
        </activity>
        <activity
            android:name="pr.platerecognization.LoginActivity"
            android:label="Struggle用户登录界面">
        </activity>
        <activity
            android:name="pr.platerecognization.DemoGridViewActivity"
            android:label="Struggle案例演示界面">
        </activity>
        <activity  android:name=".CustomizedQueryPlateActivity"
            android:label="Struggle定制查询界面">
        </activity>
    </application>
</manifest>
```

本章小结

本章通过Struggle车牌识别系统APP项目，让学生了解企业项目的整体开发流程：项目立项、需求分析、总体设计、详细设计、编码实现、系统调试。通过该项目让学生扎实掌握车牌识别APP项目中所涉及的Android拍照、显示和存储技术、SQLite嵌入式数据库和MySQL数据库存储技术、文件读写操作、共享存储偏好数据存储和多线程编程技术。

强化练习

一、填空题

1. 活动（Activity）销毁时最后执行的生命周期事件回调函数为（　　）。
2. 安卓开发工具（Android Development Tools）的简写为（　　）。
3. Java 开发工具（Java Development Kit）的简写为（　　）。
4. Android 虚拟设备（Android Virtual Device）的简写为（　　）。
5. 在字符串资源源文件 strings.xml 定义字符串的标签名为（　　）。
6. 在 android 的 dom 解析中 Element element = document.getDocumentElement() 方法可以得到 xml 文档中标签（　　）。
7. 在播放音乐期间调用 stop () 方法实现功能为（　　）播放音乐。
8. 在 Android 编程过程中，可采用（　　）方法根据控件的 ID 号获取对控件的引用。
9. Adapter 配置好以后，需要用（　　）函数将 ListView 和 Adapter 绑定。
10. TextView 的 android:textColor 属性用来设置 TextView 的（　　）。

二、单选题

1. 在 Android 中常用来显示程序执行进度的控件为________。

A. 文本视图（TextView）　　B. 编辑框（EditText）
C. 按钮（Button）　　D. 进度条（ProgressBar）

2. 在 Android 中用于显示图片的控件为（　　）。

A. 按钮（Button）　　B. 文本视图（TextView）
C. ImageView　　D. EditText

3. 在 Android 中能简单有效地处理单击事件响应 OnClickListener 的控件为（　　）。

A. 按钮（Button）　　B. 文本视图（TextView）
C. ImageView　　D. EditText

4. 控制虚拟键盘输入类型的属性为________。

A. android:text　　B. android:src　　C. android: inputType　　D. android:id

5. 在 Android 中，若要向工程添加字符串资源，应将其添加到（　　）文件。

A. dimens.xml　　B. styles.xml　　C. strings.xml　　D. value.xml

6. ListView 与数组或 List 集合的多个值进行数据绑定时使用（　　）。

A. ArrayAdapter　　B. BaseAdapter
C. SimpleAdapter　　D. SimpleCursorAdapter

7. Android 中包含多种基本 UI 控件和高级 UI 控件，这些 UI 控件均派生于（　　）类。

A. 视图组（ViewGroup）　　B. 控件（Control）
C. 文本视图（TextView）　　D. 视图（View）

8. Android 线性布局分为两种方式：纵向和横向，设置该方式的属性为（　　）。

A. android: orientation　　B. android:layout_gravity

C. android:layout_width　　D. android:layout_height

9. Android开发中用户经常用到打印日志的方式进行调试，其中日志类型不包括下面的（　　）。

A. Log.v　　B.Log.c　　C.Log.e　　D. Log.d

10. 在调用对话框Dialog时，需要最后调用（　　）方法来显示对话框。

A. onLongClick()　　B. onClick()　　C. onTouch（ ）　　D. show()

第 9 章

基于 Android 智能仓储系统项目

学习目标

- 界面的设计和实现。
- 页面与页面之间数据的交互。
- 数据库的设计和实现。
- 视频数据处理。

视频

基于Android智能仓储系统项目

9.1 项目概述

物联网智能仓储系统是基于RFID技术、ZigBee技术、Wi-Fi技术等可用于真实项目的系统。整个系统具有五大子系统：环境监控系统、安防报警系统、入库系统、出库系统、查询系统。

无论何时何地，都可以通过手持设备或者计算机控制仓储环境状况、货物入库记录、货物出库记录等功能，同时还支持防盗报警、火灾报警等功能。集成了目前智能仓储中众多声光电因素，能够很好地展示现代仓储管理的安全化和智能化。

智能仓储实训台高度结合了物联网工程技术与行业体系架构，还原了行业真实环境，将感知层、网络层和应用层3个区域分开、区域清晰；预留扩展接口，方便用户二次开发调试；烧写接口集中管控方便烧写；智能断电系统，过电流可自动切断电源。

9.2 项目设计

9.2.1 项目总体功能需求

1. 环境监控系统

在手持设备或平板计算机上能查看到：温度、湿度、光照度、空气质量等传感器的状况。

2. 安防报警系统

(1) 防盗报警：当仓库里没有人，系统检测到有人入侵时，启动报警器，同时，手持设备或平板计算机能显示报警信息。

(2) 防灾报警：当仓库里有烟雾异常时，启动报警器，打开通风设备，同时，手持设备或平板计算机能显示报警信息。

3. 入库系统

(1) 通过RFID设备对货物进行自动入库，货物的入库信息经智能网关定时上传到系统服务器中。

(2) 在入库系统中通过选择不同的时间段可以直观地查看每天、每周、每月货物入库信息的曲线图。

(3) 在入库系统中通过选择具体的货物，可以查看该货物的入库详情：货物名称、货物产地、货物入库人员名称、货物入库时间。

4. 出库系统

(1) 通过RFID设备对货物进行自动出库，货物的出库信息经智能网关定时上传到系统服务器中。

(2) 在出库系统中通过选择不同的时间段可以直观地查看每天、每周、每月货物出库信息的曲线图。

(3) 在出库系统中通过选择具体的货物，可以查看该货物的出库详情：货物名称、货物销售地、货物出库人员名称、货物出库时间。

5. 查询系统

在查询系统中通过输入具体的货物，可以查看该货物的总库存数据、入库总数量、入库价格、出库总数量、出库价格。

9.2.2 项目总体设计

现在全球在进行信息化建设，而信息化建设最终能否落地，移动终端上的应用开发将起到决定性作用。Android是目前市场占有率最高的移动设备操作系统，同时它的操作系统是开源的，这样有利于其上应用程序的开发。因此，本系统的应用程序选择在Android操作系统上进行开发。本项目整体设计框架如图9-1所示。

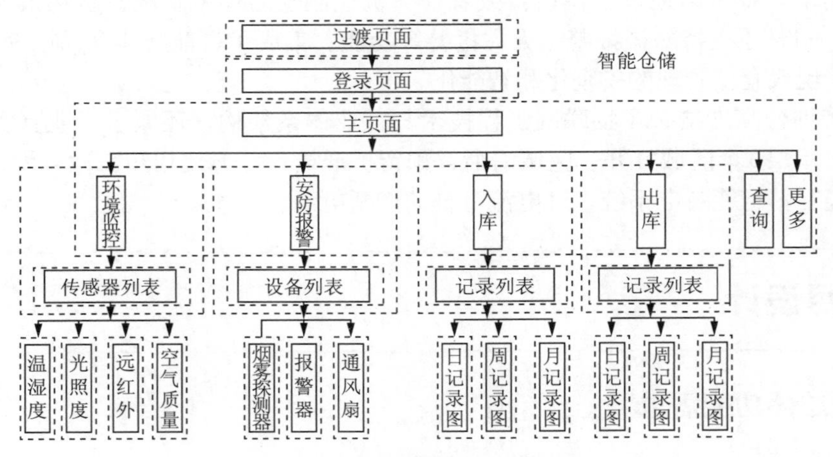

图 9-1 项目整体设计框架图

从安全性和便利性方面考虑，本系统将采用云与端的方式进行开发。

本系统采用常规的用户与密码的方式，登录进入系统，在系统主页面中有6个采用扁平化设计的功能选项：环境监控、安防报警、物品入库、物品出库、物品查询设置和更多。

（1）“环境监控"中包括：温湿度、光照度、空气质量、远红外和“+”；“+”的功能主要是考虑功能的扩展，用户可以根据仓库的情况添加新的探测器，同时也可以对探测器进行删除操作。在界面中查看到简单的探测器状况信息。

（2）“安防报警"中包括：烟雾探测器、报警器、通风扇和“+”；“+”的功能主要是考虑功能的扩展，用户可以根据实际情况添加新的设备，同时也可以对设备进行删除操作。在界面中查看到简单的设备状况信息。对于可控制的设备，点击具体的设备后，可以查看设备具体信息，并可以对设备进行控制。

（3）“入库”中包括：日记录图、周记录图、月记录图和“+”；“+”的功能主要是考虑功能的扩展。

（4）“出库”中包括：日记录图、周记录图、月记录图和“+”；“+”的功能主要是考虑功能的扩展。

（5）“查询”中主要用于货物的查询。

（6）“更多”选项用于后期功能的扩展。

9.3 必备的技术和知识点

本章必备的技术和知识点包括：

（1）界面编程与视图组件。

（2）布局管理器。

（3）TextView 及其子类。

（4）ImageView 及其子类。

（5）AdapterView 及其子类。

（6）ViewAnimator 及其子类。

（7）对话框和菜单。

（8）基于监听的事件处理。

（9）基于回调的事件处理。

（10）响应系统设置的事件。

（11）Handle 消息传递机制。

（12）基于 TCP 的网络通信。

（13）使用 URL 访问网络资源。

（14）使用 WebServer 进行网络编程。

（15）数据存储与访问。

（16）嵌入式关系型 SQLite 数据库存储数据。

（17）视频采集。

9.4 项目实施

9.4.1 登录页面

实现账号和密码输入文本框、实现登录按键功能、实现保存密码和自动登录功能。页面布局如图 9-2 所示。

图 9-2 登录页面图

当用户输入账号和密码，点击“登录”按键时，系统读取账号和密码文本框信息，并把账号和密码信息经网络传输给智能网关，与智能网关数据库中的账号和密码数据信息进行比对，如果账号和密码同时存在于智能网关中的数据库中，则允许用户登录系统。如果账号或密码不正确，则根据返回标记码的不同，提示“账号不存在”或“密码不正确”。

当用户选中“保存密码”时，系统会把账号和密码保存在系统相关的配置文件中，当下次再登录系统时，则不需要再次输入账号和密码。如果用户只选中“自动登录”时，则系统会保存账号和密码，下次自动登录系统。

程序清单：CH09\WareHousing\src\com\warehousing \smarthome activity\ LoginActivity.java。

```
   package com.zigcloud.warehousing.activity;
   public class LoginActivity extends BaseActivity{
   /***用户登录**/
       public void onCreate(Bundle savedInstanceState){
           super.onCreate(savedInstanceState);
           setContentView(R.layout.activity_login);
           MainApplication.getInstance().addActivity(this);
           findViewById(R.id.btn_login).setOnClickListener
(new View.OnClickListener(){
             public void onClick(View v){
                startActivity(new Intent(LoginActivity.this,MainActivity.class));
           }});}
       public void onBackPressed(){super.onBackPressed();}
       public boolean onKeyDown(int keyCode, KeyEvent event){
           if(keyCode==KeyEvent.KEYCODE_BACK)  {exitBy2Click();}
           return false;}
       private static Boolean isExit=false;
       private void exitBy2Click(){  /*双击“后退”按钮退出*/
           Timer tExit=null;
           if(isExit==false){isExit=true;
               Toast.makeText(this, getResources().getString(R.string.exit_dialog_
message),Toast.LENGTH_SHORT).show();
               tExit=new Timer();
```

```
            tExit.schedule(new TimerTask(){
                public void run(){
                    isExit=false; }}, 2000);
        } else{
            MainApplication.getInstance().exit();}}}
```

9.4.2 主页面

实现“返回”和6个示图功能。页面布局如图9-3所示。

“返回"功能的作用返回上一级页面；“环境监控”功能的作用是查看当前仓库中的传感器设备信息。其他的五张示图功能类似。

程序的界面布局代码清单：CH09\WareHousing\res\layout\activity_main.xml。

图 9-3　主页面图

```
    <?xml version="1.0"encoding="utf-8"?>
    <RelativeLayout xmlns:android="http://schemas. android.com/
apk/res/android"
        android:layout_width="fill_parent"
        android:layout_height="fill_parent"
        android:background="@drawable/main_default_ bg">
        <include android:id="@id/bt_createtask_title_ layout"
            layout="@layout/activity_common_title_ bar"/>
        <RelativeLayout
            android:id="@+id/RelativeLayout1"
            android:layout_width="match_parent"
            android:layout_height="match_parent"
            android:layout_below="@id/bt_createtask_title_layout"
            android:layout_centerHorizontal="true"
            android:layout_marginLeft="15dp"
            android:layout_marginRight="15dp"
            android:orientation="vertical"
            android:visibility="visible">
            <LinearLayout
                android:id="@+id/LinearLayout1"
                android:layout_width="match_parent"
                android:layout_height="match_parent"
                android:layout_alignParentTop="true"
                android:orientation="vertical">
                <GridView
                    android:id="@+id/gdv_main"
                    android:layout_width="match_parent"
                    android:layout_height="wrap_content"
                    android:listSelector="@drawable/selector_list"
                    android:horizontalSpacing="10dp"
```

```
            android:numColumns="2"
            android:verticalSpacing="10dp">
        </GridView>
        <include
            android:id="@+id/view_loading_error"
            android:layout_width="match_parent"
            android:layout_height="match_parent"
            layout="@layout/view_loading_error"
            android:visibility="gone"/>
        <include
            android:id="@+id/view_loading"
            android:layout_width="match_parent"
            android:layout_height="match_parent"
            layout="@layout/view_loading"
            android:visibility="gone"/>
    </LinearLayout>
  </RelativeLayout>
</RelativeLayout>
```

程序清单：CH09\ WareHousing\src\com\zigcloud\warehousing \activity\MainActivity.java。

```
package com.zigcloud.warehousing.activity;
public class MainActivity extends BaseActivity{/***主界面**/
    private TextView titlebar_left;
    private TextView titlebar_title;
    public void onCreate(Bundle savedInstanceState){
        super.onCreate(savedInstanceState);
        setContentView(R.layout.activity_main);
        MainApplication.getInstance().addActivity(this);
        titlebar_left=(TextView)findViewById(R.id.titlebar_left);
        titlebar_left.setOnClickListener(new View.OnClickListener(){
            public void onClick(View v){
                onBackPressed();}});
        titlebar_title=(TextView)findViewById(R.id.titlebar_title);
        titlebar_title.setText(R.string.app_name);
        GridView gridView=(GridView)findViewById(R.id.gdv_main);
        ArrayList<HashMap<String,Object>>lstImageItem=new ArrayList
<HashMap<String, Object>>();
        HashMap<String, Object>map=new HashMap<String,Object>();
         map.put("ItemImage", R.drawable.icon_environment);
         map.put("ItemText", "环境监控");
         lstImageItem.add(map);
         map=new HashMap<String, Object>();
         map.put("ItemImage", R.drawable.icon_alarm);
```

```
            map.put("ItemText", "安防报警");
            lstImageItem.add(map);
            map=new HashMap<String, Object>();
            map.put("ItemImage", R.drawable.icon_import);
            map.put("ItemText", "物品入库");
            lstImageItem.add(map);
            map=new HashMap<String, Object>();
            map.put("ItemImage", R.drawable.icon_outport);
            map.put("ItemText", "物品出库");
            lstImageItem.add(map);
            map=new HashMap<String, Object>();
            map.put("ItemImage", R.drawable.icon_search);
            map.put("ItemText", "物品查询");
            lstImageItem.add(map);
            map=new HashMap<String, Object>();
            map.put("ItemImage", R.drawable.icon_setting);
            map.put("ItemText", "设置");
            lstImageItem.add(map);
            map=new HashMap<String, Object>();
            map.put("ItemImage", R.drawable.icon_add);
            map.put("ItemText", "预留");
            lstImageItem.add(map);
            map=new HashMap<String, Object>();
            map.put("ItemImage", R.drawable.icon_add);
            map.put("ItemText", "预留");
            lstImageItem.add(map);
        SimpleAdapter saImageItems=new SimpleAdapter(this, lstImageItem, R.
layout.view_squared_item, new String[]{"ItemImage","ItemText"}, new int[]{R.
id.itemImage, R.id.itemText});
            gridView.setAdapter(saImageItems);
            gridView.setOnItemClickListener(new OnItemClickListener(){
            public void onItemClick(AdapterView<?>arg0, View arg1, int arg2,long arg3){
              switch(arg2){
                case 0:
                    startActivity(new Intent ntent(MainActivity.this,Environment
MonitorActivity.class)); break;
                case 2:
                    startActivity(new ntent(MainActivity.
this,GoodsImportActivity.class));break;
                case 3:
                    startActivity(new Intent(MainActivity.this,
GoodsOutportActivity.class));break;
                case 4:
```

```
                    startActivity(new Intent(MainActivity.this,
GoodsSearchActivity.class));break;}}});
            gridView.setSelector(new ColorDrawable(Color.TRANSPARENT));}
       public boolean onKeyDown(int keyCode, KeyEvent event){
         if(keyCode==KeyEvent.KEYCODE_BACK){
            exitBy2Click();}
         return false;}
       private static Boolean isExit=false;
       private void exitBy2Click(){Timer tExit=null;
         if(isExit==false){isExit=true;
            Toast.makeText(this,getResources().getString(R.string.exit_dialog_
message), Toast.LENGTH_SHORT).show();
            tExit=new Timer();
            tExit.schedule(new TimerTask(){
               public void run(){
                   isExit=false;}}, 2000);
       } else{
         MainApplication.getInstance().exit();}}}
```

9.4.3 环境监控页面

实现查看仓库中各个区域的传感器信息，如火焰传感器、温度传感器、湿度传感器等环境信息。环境监控页面如图 9-4 所示。程序界面布局代码：CH09\WareHousing\res\layout\ activity_environmentmonitor.xml。

```
   <?xml version="1.0"encoding="utf-8"?>
   <RelativeLayout xmlns:android="http://schemas.android.com/
apk/res/android"
       android:layout_width="fill_parent"
       android:layout_height="fill_parent"
       android:background="@drawable/main_default_bg">
       <include
           android:id="@id/bt_createtask_title_layout"
           layout="@layout/activity_common_title_bar"/>
       <RelativeLayout
           android:id="@+id/RelativeLayout1"
           android:layout_width="match_parent"
           android:layout_height="match_parent"
           android:layout_below="@id/bt_createtask_title_layout"
           android:layout_centerHorizontal="true"
           android:layout_marginLeft="15dp"
           android:layout_marginRight="15dp"
           android:background="@drawable/content_default_bg"
           android:orientation="vertical"
```

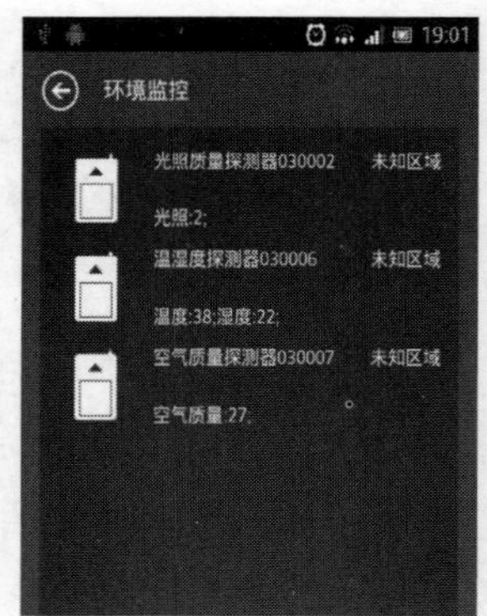

图 9-4　环境监控页面图

```
        android:visibility="visible">
        <com.zigcloud.warehousing.widget.Gallery3D
            android:id="@+id/gal_rooms"
            android:layout_width="fill_parent"
            android:layout_height="wrap_content"
            android:layout_alignParentBottom="true"
            android:layout_alignParentLeft="true"
            android:spacing="30dp"
            android:unselectedAlpha="128"/>
        <LinearLayout
            android:id="@+id/LinearLayout1"
            android:layout_width="match_parent"
            android:layout_height="match_parent"
            android:layout_above="@+id/gal_rooms"
            android:layout_alignParentTop="true"
            android:orientation="vertical">
            <include
                android:id="@+id/view_loading_error"
                android:layout_width="match_parent"
                android:layout_height="match_parent"
                layout="@layout/view_loading_error"
                android:visibility="gone"/>
            <include
                android:id="@+id/view_loading"
                android:layout_width="match_parent"
                android:layout_height="match_parent"
                layout="@layout/view_loading"
                android:visibility="gone"/>
            <GridView
                android:id="@+id/gdv_equipments"
                android:layout_width="match_parent"
                android:layout_height="wrap_content"
                android:numColumns="1"
                android:padding="10dp"
                android:verticalSpacing="10dp"
                android:visibility="visible">
            </GridView>
        </LinearLayout>
    </RelativeLayout>
</RelativeLayout>
```

程序清单：CH09\WareHousing\src\com\zigcloud\warehousing\activity\ EnvironmentMonitor Activity.java。

```
package com.zigcloud.warehousing.activity;
@SuppressLint("HandlerLeak")
public class EnvironmentMonitorActivity extends BaseActivity{/***环境监控**/
    private TextView titlebar_left;  private TextView titlebar_title;
    private GridView equipmentsGridView;
    private ArrayList<HashMap<String, Object>>equipmentsArrayList=new
ArrayList<HashMap<String, Object>>();
    private HashMap<String,Object>equipmentsHashMap=new HashMap<String,Object>();
    private SimpleAdapter equipmentsAdapter;
    /***获取设备列表**/
    private EquipmentListHttpRequestTask equipmentListHttpRequestTask=new
EquipmentListHttpRequestTask();
    public void onCreate(Bundle savedInstanceState){
        super.onCreate(savedInstanceState);
        setContentView(R.layout.activity_environmentmonitor);
        MainApplication.getInstance().addActivity(this);
        titlebar_left=(TextView)findViewById(R.id.titlebar_left);
        titlebar_left.setOnClickListener(new View.OnClickListener(){
            public void onClick(View v){
                onBackPressed();}});
        titlebar_title=(TextView)findViewById(R.id.titlebar_title);
        titlebar_title.setText(R.string.title_environment_monitor);
        initialRoomsGallery();
        initialEquipmentsGridView();}
    private void initialEquipmentsGridView(){/***初始化设备信息列表**/
        equipmentsGridView=(GridView)findViewById(R.id.gdv_equipments);
        equipmentsAdapter=new SimpleAdapter(getApplicationContext(),
equipmentsArrayList,R.layout.activity_equipment_list_item,new String[]{"ItemImage",
"ItemNodeName","ItemDataValue","ItemArea"},new int[]{R.id.img_image,R.id.tv_name,
R.id.tv_datavalue,R.id. tv_area});
        equipmentsGridView.setAdapter(equipmentsAdapter);
        equipmentsGridView.setOnItemClickListener(new OnItemClickListener(){
          @SuppressWarnings("unchecked")
          public void onItemClick(AdapterView<?>arg0, View arg1, int arg2,
long arg3){
                HashMap<String, Object>item=(HashMap<String, Object>)arg0.
getItemAtPosition(arg2);
                String nodeId=item.get("ItemNodeId")==null ?null:item.get("ItemNodeId").
toString();
                String nodeName=item.get("ItemNodeName")==null ?null:item.
get("ItemNodeName").toString();
                String nodeTypeId=item.get("ItemNodeTypeId")==null ?null:item.get
```

```
("ItemNodeTypeId").toString();
            String nodeTypeName=item.get("ItemNodeTypeName")==null ?null:item.get
("ItemNodeTypeName").toString();
            String dataValueString=item.get("ItemDataValueString")==null ?null:item.
get("ItemDataValueString").toString();
            String updateTimeString=item.get("ItemUpdateTimeString")==null
?null:item.get("ItemUpdateTimeString").toString();
            Intent intent=new Intent(getApplicationContext(),Equipment Activity.class);
            intent.putExtra("nodeId", nodeId);
            intent.putExtra("nodeName", nodeName);
            intent.putExtra("nodeTypeId", nodeTypeId);
            intent.putExtra("nodeTypeName", nodeTypeName);
            intent.putExtra("dataValueString", dataValueString);
            intent.putExtra("updateTimeString",updateTimeString);
            startActivity(intent);}});
        equipmentListHttpRequestTask.execute();}
    private EquipmentDAO mEquipmentDAO=new EquipmentDAO();/***设备业务控制类**/
    private Handler mHandler=new Handler(){
        public void handleMessage(Message msg){
            super.handleMessage(msg);
            switch(msg.what){
            case 10001:              //正在加载
                findViewById(R.id.gdv_equipments).setVisibility(8);
                findViewById(R.id.view_loading).setVisibility(0);
                findViewById(R.id.view_loading_error).setVisibility(8);
                break;
            case 10002:              //加载成功
                findViewById(R.id.gdv_equipments).setVisibility(0);
                findViewById(R.id.view_loading).setVisibility(8);
                findViewById(R.id.view_loading_error).setVisibility(8);
                break;
            case 10003:             //加载失败
                findViewById(R.id.gdv_equipments).setVisibility(8);
                findViewById(R.id.view_loading).setVisibility(8);
                findViewById(R.id.view_loading_error).setVisibility(0);
                break;}}};
    /***获取equipment列表**/
    private class EquipmentListHttpRequestTask extends AsyncTask<String, Integer,
List<EquipmentJson>>{protected void onPreExecute(){
        super.onPreExecute();
        if(!(equipmentsAdapter!=null&&equipmentsAdapter.getCount()>0)){
        mHandler.sendEmptyMessage(10001);}}
    protected List<EquipmentJson>doInBackground(String... params){
```

```
            return mEquipmentDAO.getAll();}
        protected void onProgressUpdate(Integer... values){
            super.onProgressUpdate(values);}
        protected void onPostExecute(List<EquipmentJson>result){
            if(result!=null){
                EquipmentJson equipmentEntity=null;
                BaseEquipmentEntity equipment=null;
                for(int i=0;i<result.size();i++){
                    equipmentEntity= result.get(i);
                    equipment= BaseEquipmentEntity.parse(equipmentEntity);
                    if(equipment!=null){
                        equipmentsHashMap=new HashMap<String, Object>();
                        equipmentsHashMap.put("ItemImage", equipment.getIconRes());
                        equipmentsHashMap.put("ItemNodeName", equipment.getName());
                        equipmentsHashMap.put("ItemDataValue", equipment.getData
ValueString());
                        equipmentsHashMap.put("ItemArea", "未知区域");
                        equipmentsHashMap.put("ItemNodeId",equipment.getNodeId());
                        equipmentsHashMap.put("ItemNodeTypeId",equipment. getNodeTypeId());
                        equipmentsHashMap.put("ItemNodeTypeName",equipment. getTypeName());
                        equipmentsHashMap.put("ItemDataValueString", equipment.
getDataValueString());
                        equipmentsHashMap.put("ItemUpdateTimeString",equipment.
getUpdateTimeString());
                        equipmentsArrayList.add(equipmentsHashMap);}}
                equipmentsAdapter.notifyDataSetChanged();
                mHandler.sendEmptyMessage(10002);}
            else{
                mHandler.sendEmptyMessage(10003);}}};
    private void initialRoomsGallery(){
        Gallery3D   gallery=(Gallery3D)findViewById(R.id.gal_rooms);
        ArrayList<HashMap<String, Object>>lstImageItem=new ArrayList
<HashMap<String, Object>>();
         HashMap<String, Object>map=new HashMap<String, Object>();
         map.put("ItemImage", R.drawable.scene_gallery_0);
         map.put("ItemText", "仓库1");
         map.put("ItemContent", "仓库1");
         lstImageItem.add(map);
         map=new HashMap<String, Object>();
         map.put("ItemImage", R.drawable.scene_gallery_1);
         map.put("ItemText", "仓库2");
         map.put("ItemContent", "仓库2");
         lstImageItem.add(map);
```

```
        map=new HashMap<String, Object>();
        map.put("ItemImage", R.drawable.scene_gallery_2);
        map.put("ItemText", "仓库3");
        map.put("ItemContent", "仓库3");
        lstImageItem.add(map);
        map=new HashMap<String, Object>();
        map.put("ItemImage", R.drawable.scene_gallery_3);
        map.put("ItemText", "仓库4");
        map.put("ItemContent", "仓库4");
        lstImageItem.add(map);
        SimpleAdapter saImageItems=new SimpleAdapter(this, lstImageItem,R.
layout.view_grallery3d_item,new String[]{"ItemImage","ItemText","ItemContent"},
new int[]{R.id.itemImage,R.id.itemText,R.id.itemContent});
        gallery.setFadingEdgeLength(0);
        gallery.setAdapter(saImageItems);
        gallery.setOnItemSelectedListener(new OnItemSelectedListener(){
        public void onItemSelected(AdapterView<?>parent, View view, int position,
long id){
          Toast.makeText(RoomListActivity.this, "img "+(position+1)+"selected",
Toast.LENGTH_SHORT).show();}
        public void onNothingSelected(AdapterView<?>parent){}});
        gallery.setOnItemClickListener(new OnItemClickListener(){
        public void onItemClick(AdapterView<?>parent, View view, int position,
long id){
           Toast.makeText(RoomListActivity.this,
"img "+(position+1)+"selected",Toast.LENGTH_SHORT).show();}});}}
```

9.4.4　物品入库页面

实现货物入库的功能。页面布局如图 9-5 所示。

程序的界面布局代码如下：CH09\WareHousing\res\layout\ activity_goods_import.xml。

```
  <?xml version="1.0"encoding="utf-8"?>
  <RelativeLayout xmlns:android=
"http://schemas.android.com/apk/res/android"
      android:layout_width="fill_parent"
      android:layout_height="fill_parent"
      android:background="@drawable/main_default_bg">
      <include
          android:id="@+id/bt_createtask_title_layout"
          layout="@layout/activity_common_title_bar"/>
      <RelativeLayout
          android:id="@+id/RelativeLayout1"
          android:layout_width="match_parent"
```

图 9-5　物品入库页面

```
    android:layout_height="match_parent"
    android:layout_below="@id/bt_createtask_title_layout"
    android:layout_centerHorizontal="true"
    android:layout_marginLeft="15dp"
    android:layout_marginRight="15dp"
    android:background="@drawable/content_default_bg"
    android:orientation="vertical"
    android:visibility="visible">
    <TableLayout
        android:layout_width="wrap_content"
        android:layout_height="wrap_content"
        android:layout_alignParentBottom="true"
        android:layout_alignParentLeft="true"
        android:layout_alignParentRight="true"
        android:layout_alignParentTop="true">
        <TableRow
            android:id="@+id/tableRow1"
            android:layout_width="wrap_content"
            android:layout_height="wrap_content">
            <TextView
                android:id="@+id/textView1"
                android:layout_width="wrap_content"
                android:layout_height="wrap_content"
                android:text="@string/goods_cardid"/>
            <EditText
                android:id="@+id/edt_goods_cardId"
                android:layout_width="wrap_content"
                android:layout_height="wrap_content"
                android:ems="10">
                <requestFocus />
            </EditText>
        </TableRow>
        <TableRow
            android:id="@+id/tableRow2"
            android:layout_width="wrap_content"
            android:layout_height="wrap_content">
            <TextView
                android:id="@+id/textView2"
                android:layout_width="wrap_content"
                android:layout_height="wrap_content"
                android:text="@string/goods_name"/>
            <EditText
                android:id="@+id/edt_goods_name"
```

```
                android:layout_width="wrap_content"
                android:layout_height="wrap_content"
                android:ems="10"/>
        </TableRow>
        <TableRow
            android:id="@+id/tableRow3"
            android:layout_width="wrap_content"
            android:layout_height="wrap_content">
            <TextView
                android:id="@+id/textView3"
                android:layout_width="wrap_content"
                android:layout_height="wrap_content"
                android:text="@string/goods_address"/>
            <EditText
                android:id="@+id/edt_goods_address"
                android:layout_width="wrap_content"
                android:layout_height="wrap_content"
                android:ems="10"/>
        </TableRow>
        <TableRow
            android:id="@+id/tableRow4"
            android:layout_width="wrap_content"
            android:layout_height="wrap_content">
            <TextView
                android:id="@+id/textView4"
                android:layout_width="wrap_content"
                android:layout_height="wrap_content"
                android:text="@string/operate_option"/>
            <Button
                android:id="@+id/btn_ok"
                android:layout_width="wrap_content"
                android:layout_height="wrap_content"
                android:text="@string/ok"/>
        </TableRow>
    </TableLayout>
  </RelativeLayout>
</RelativeLayout>
```

程序清单：CH09\ WareHousing\src\com\zigcloud\warehousing\activity\GoodsImportActivity.java。

```
package com.zigcloud.warehousing.activity;
public class GoodsImportActivity extends BaseActivity{/***物品入库**/
    private TextView titlebar_left;  private TextView titlebar_title;
    private GoodsDAO mGoodsDAO=new GoodsDAO();/***物品操作业务类**/
```

```
    private GoodsImportTask mGoodsImportTask=new GoodsImportTask();
    /***物品入库任务类**/
    protected void onCreate(Bundle savedInstanceState){
        super.onCreate(savedInstanceState);
        setContentView(R.layout.activity_goods_import);
        MainApplication.getInstance().addActivity(this);
        titlebar_left=(TextView)findViewById(R.id.titlebar_left);
        titlebar_left.setOnClickListener(new View.OnClickListener(){
          public void onClick(View v){
            onBackPressed(); }});
        titlebar_title=(TextView)findViewById(R.id.titlebar_title);
        titlebar_title.setText(R.string.title_goods_import);
        /***商品入库按钮事件**/
        findViewById(R.id.btn_ok).setOnClickListener(new OnClickListener(){
           public void onClick(View v){
                EditText edt_goods_cardId=(EditText)findViewById(R.id.edt_goods_
cardId);
                EditText edt_goods_name=(EditText)findViewById(R.id.edt_goods_ name);
                EditText edt_goods_address=(EditText)findViewById(R.id.edt_goods_
address);
                mGoodsImportTask=new GoodsImportTask();
                mGoodsImportTask.execute(edt_goods_cardId.getText().toString(),edt_
goods_name.getText().toString(),edt_goods_address.getText().toString(),"admin");
        }});}
    /***物品入库任务类**/
    private class GoodsImportTask extends AsyncTask<String,Integer,ControlResultJ
son>{
       protected ControlResultJson doInBackground(String... params){
           if(params!=null&&params.length>3){
                String cardNum=params[0];
                String dataName=params[1];
                String originPlace=params[2];
                String staffName=params[3];
                if(cardNum!=null&&dataName!=null&&originPlace!=null&&staff Name!=null)
                    return mGoodsDAO.goodsImport(cardNum, dataName, originPlace,
staffName);}
                return null;}
      protected void onPostExecute(ControlResultJson result){
          super.onPostExecute(result);  String resStr=null;
          if(result!=null){resStr=result.flag==0?"添加成功! ":"添加失败";}
          else{resStr="添加失败";}
        Toast.makeText(GoodsImportActivity.this,resStr,Toast.LENGTH_ SHORT).
show();}
```

```
        protected void onPreExecute(){super.onPreExecute();}
        protected void onProgressUpdate(Integer... values){
          super.onProgressUpdate(values);}}}
```

9.4.5 具体设备页面

实现“返回”、设备名称显示、设备图片呈现、“打开”、“关闭”功能。页面布局如图9-6所示。“返回"功能的作用是返回上一级页面；在“返回”后面显示当前设备的名称。设备图片采用设备实物图；“打开”按键和“关闭”按键分别实现对设备的开关控制。程序的界面布局代码清单：CH09\WareHousing\res\ layout\activity_equipment.xml。

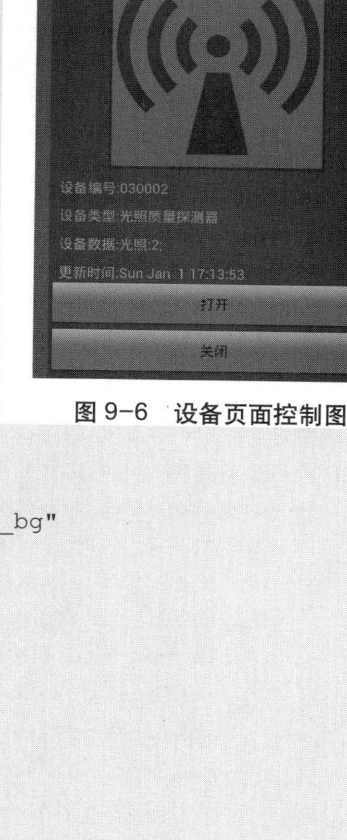

图 9-6 设备页面控制图

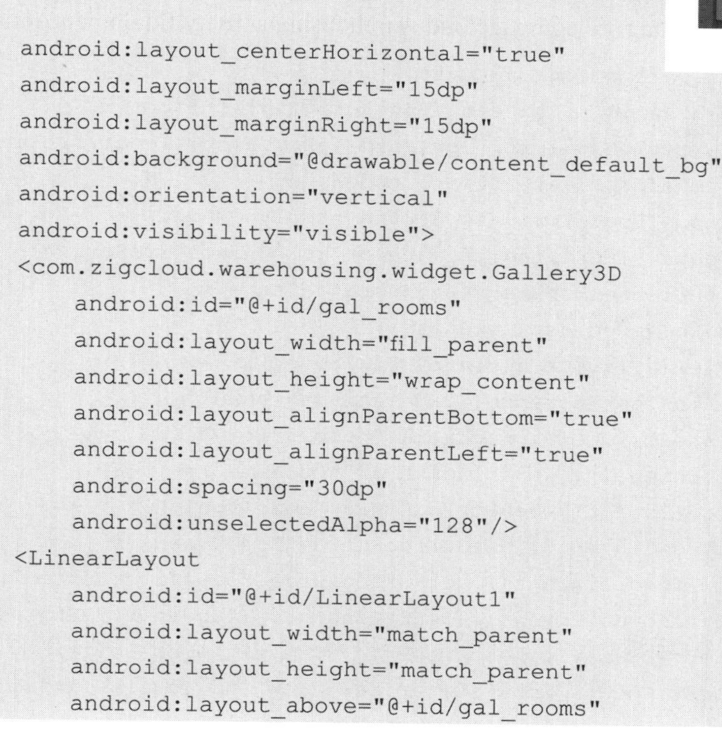

```
<?xml version="1.0"encoding="utf-8"?>
<RelativeLayout xmlns:android="http://schemas.
android.com/apk/res/android"
    android:layout_width="fill_parent"
    android:layout_height="fill_parent"
    android:background="@drawable/main_ default_bg">
    <include android:id="@id/bt_createtask_title_
layout"
  layout="@layout/activity_common_title_ bar"/>
    <RelativeLayout
        android:id="@+id/RelativeLayout1"
        android:layout_width="match_parent"
        android:layout_height="match_parent"
        android:layout_below="@id/bt_createtask_title_
layout"
        android:layout_centerHorizontal="true"
        android:layout_marginLeft="15dp"
        android:layout_marginRight="15dp"
        android:background="@drawable/content_default_bg"
        android:orientation="vertical"
        android:visibility="visible">
        <com.zigcloud.warehousing.widget.Gallery3D
            android:id="@+id/gal_rooms"
            android:layout_width="fill_parent"
            android:layout_height="wrap_content"
            android:layout_alignParentBottom="true"
            android:layout_alignParentLeft="true"
            android:spacing="30dp"
            android:unselectedAlpha="128"/>
        <LinearLayout
            android:id="@+id/LinearLayout1"
            android:layout_width="match_parent"
            android:layout_height="match_parent"
            android:layout_above="@+id/gal_rooms"
```

```
            android:layout_alignParentTop="true"
            android:orientation="vertical">
            <include
                android:id="@+id/view_loading_error"
                android:layout_width="match_parent"
                android:layout_height="match_parent"
                layout="@layout/view_loading_error"
                android:visibility="gone"/>
            <include
                android:id="@+id/view_loading"
                android:layout_width="match_parent"
                android:layout_height="match_parent"
                layout="@layout/view_loading"
                android:visibility="gone"/>
            <LinearLayout
            android:id="@+id/ll_equipment"
            android:layout_width="match_parent"
            android:layout_height="match_parent"
            android:orientation="vertical">
            </LinearLayout>
        </LinearLayout>
    </RelativeLayout>
</RelativeLayout>
```

程序清单：CH09\WareHousing\src\com\zigcloud\warehousing\activity\EquipmentActivity.java。

```
package com.zigcloud.warehousing.activity;
public class EquipmentActivity extends BaseActivity{/***设备**/
    private TextView titlebar_left;  private TextView titlebar_title;
    private String nodeId;private String nodeName;
    private String nodeTypeId;private String nodeTypeName;
    private String dataValueString;private String updateTimeString;
    protected void onCreate(Bundle savedInstanceState){
        super.onCreate(savedInstanceState);
        setContentView(R.layout.activity_equipment);
        MainApplication.getInstance().addActivity(this);
        nodeId=getIntent().getStringExtra("nodeId");
        nodeName=getIntent().getStringExtra("nodeName");
        nodeTypeId=getIntent().getStringExtra("nodeTypeId");
        nodeTypeName=getIntent().getStringExtra("nodeTypeName");
        dataValueString=getIntent().getStringExtra("dataValueString");
        updateTimeString=getIntent().getStringExtra("updateTimeString");
         titlebar_left=(TextView)findViewById(R.id.titlebar_left);
         titlebar_left.setOnClickListener(new View.OnClickListener(){
```

```
            public void onClick(View v){onBackPressed();}});
        titlebar_title=(TextView)findViewById(R.id.titlebar_title);
        titlebar_title.setText(nodeName);initalEquipment();}
    private void initalEquipment(){  /***初始化设备信息**/
        LinearLayout ll_equipment=(LinearLayout)findViewById(R.id.ll_ equipment);
        if(ll_equipment!=null){
            if(nodeTypeId!=null){
                BaseEquipmentWidget equipmentWidget=null;
                if(nodeTypeId.equals("20")){
                  equipmentWidget=new CurtainWidget(EquipmentActivity.this);}
                else if(nodeTypeId.equals("24")){
                  equipmentWidget=new  LampWidget(EquipmentActivity.this);}
                else if(nodeTypeId.equals("25")){
                  equipmentWidget=new BedLampWidget(EquipmentActivity.this);}
                else if(nodeTypeId.equals("26")){
                equipmentWidget=new WallLampWidget(EquipmentActivity.this);}
        else{
                equipmentWidget=new BaseEquipmentWidget(EquipmentActivity.this);}
                equipmentWidget.setNodeId(nodeId);
                equipmentWidget.setNodeTypeName(nodeTypeName);
                equipmentWidget.setNodeDataString(dataValueString);
                equipmentWidget.setUpdateTimeString(updateTimeString);
                equipmentWidget.addSendCmdListener(new BaseEquipmentWidget.
  Listener(){
                public void sendCmd(String nodeId, String stateFlag,String stateValue){
                    if(mSendCmdHttpRequestTask.getStatus()!=Status.RUNNING){
                        mSendCmdHttpRequestTask=new SendCmdHttpRequestTask();
                mSendCmdHttpRequestTask.execute(nodeId, stateFlag, stateValue);}}});
              ll_equipment.addView(equipmentWidget,new LayoutParams(LayoutParams.
MATCH_PARENT,LayoutParams.MATCH_PARENT));}}
      /***设备控制业务类**/
    protected EquipmentDAO mEquipmentDAO=new EquipmentDAO();
    /***异步发送命令任务类**/
    protected SendCmdHttpRequestTask mSendCmdHttpRequestTask=new Send
CmdHttpRequestTask();/***切换当前模式**/
    public class SendCmdHttpRequestTask extends  AsyncTask<String, Integer,
ControlResultJson>{
        protected ControlResultJson doInBackground(String... params){
            if(params!=null&&params.length>2){
                mEquipmentDAO.sendCmd(params[0], params[1], params[2]);}
            return null;} }
    protected EquipmentHttpRequestTask mEquipmentHttpRequestTask=new Equipment
HttpRequestTask();
```

```
    /***获取equipment**/
    protected class EquipmentHttpRequestTask extends AsyncTask<String, Integer, 
EquipmentJson>{
        protected void onPreExecute(){super.onPreExecute();}
        protected void onProgressUpdate(Integer... values){
            super.onProgressUpdate(values);}
        protected EquipmentJson doInBackground(String... arg0){
            String nodeId=arg0!=null&&arg0.length>0?arg0[0]:null;
            return mEquipmentDAO.getByNodeId(nodeId);}
        protected void onPostExecute(EquipmentJson result){
            super.onPostExecute(result);
            BaseEquipmentEntity baseEquipment=BaseEquipmentEntity.parse(result);
            if(baseEquipment!=null){}}}}
```

9.4.6 物品出库页面

实现物品出库功能，页面布局如图9-7所示。

程序的界面布局代码如：CH09\WareHousing \res\layout\activity_goods_outport.xml。

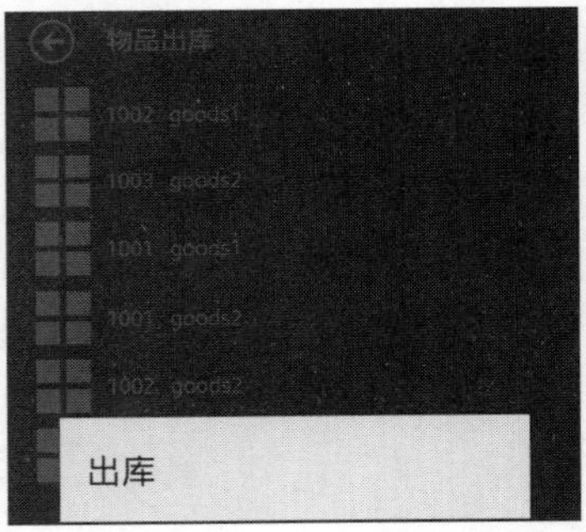

图 9-7 物品出库页面列表图

```
<?xml version="1.0"encoding="utf-8"?>
<RelativeLayout xmlns:android="http://schemas.
android.com/apk/res/android"
    android:layout_width="fill_parent"
    android:layout_height="fill_parent"
    android:background="@drawable/main_ default_bg">
    <include
        android:id="@+id/bt_createtask_title_layout"
        layout="@layout/activity_common_title_bar"/>
    <RelativeLayout
        android:id="@+id/RelativeLayout1"
        android:layout_width="match_parent"
        android:layout_height="match_parent"
        android:layout_below="@id/bt_createtask_title_layout"
        android:layout_centerHorizontal="true"
        android:layout_marginLeft="15dp"
        android:layout_marginRight="15dp"
        android:background="@drawable/content_default_bg"
        android:orientation="vertical"
        android:visibility="visible">
        <GridView
            android:id="@+id/gridView1"
            android:layout_width="match_parent"
            android:layout_height="wrap_content"
            android:layout_alignParentLeft="true"
```

```
            android:layout_alignParentTop="true"
            android:horizontalSpacing="10dp"
            android:numColumns="1"
            android:verticalSpacing="10dp"
            android:visibility="visible">
        </GridView>
        <include
            android:id="@+id/view_loading_error"
            android:layout_width="match_parent"
            android:layout_height="match_parent"
            android:layout_alignParentLeft="true"
            android:layout_below="@+id/gdv_models"
            layout="@layout/view_loading_error"
            android:visibility="gone"/>
    </RelativeLayout>
</RelativeLayout>
```

源代码清单：CH09\WareHousing\src\com\zigcloud\warehousing\activity\GoodsOutport Activity.java。

```
package com.zigcloud.warehousing.activity;
public class GoodsOutportActivity extends BaseActivity{  /***物品出库**/
    private TextView titlebar_left;
    private TextView titlebar_title;
    private GridView goodsGridView;
    private ArrayList<HashMap<String,Object>>goodsArrayList=new ArrayList
<HashMap<String, Object>>();
    private HashMap<String,Object>goodsHashMap=new HashMap<String, Object>();
    private SimpleAdapter goodsAdapter;
    private GoodsDAO mGoodsDAO=new GoodsDAO();  /***物品业务处理类**/
     /***获取物品信息任务类**/
    private GoodsRequestTask mGoodsRequestTask=new GoodsRequestTask();
     /***获取物品出库**/
    private GoodsOutportTask mGoodsOutportTask=new GoodsOutportTask();
    protected void onCreate(Bundle savedInstanceState){
        super.onCreate(savedInstanceState);
        setContentView(R.layout.activity_goods_outport);
        MainApplication.getInstance().addActivity(this);
        titlebar_left=(TextView)findViewById(R.id.titlebar_left);
        titlebar_left.setOnClickListener(new View.OnClickListener(){
          public void onClick(View v){onBackPressed();}});
        titlebar_title=(TextView)findViewById(R.id.titlebar_title);
        titlebar_title.setText(R.string.title_goods_outport);
        initialGoodsGridView();}
    private void initialGoodsGridView(){/***初始化情景模式列表**/
```

```
        goodsGridView=(GridView)findViewById(R.id.gridView1);
        goodsAdapter=new SimpleAdapter(getApplicationContext(),goodsArrayList,
R.layout.activity_goods_list_item,  new String[]{"Image", "goodsCardId",
"goodsName"},new int[]{R.id.itemImage,R.id.tv_goods_cardid,R.id.tv_goods_ name});
        goodsGridView.setAdapter(goodsAdapter);
        goodsGridView.setOnItemClickListener(new OnItemClickListener(){
            @SuppressWarnings({"unchecked"})
            public void onItemClick(AdapterView<?>arg0, View arg1, int arg2, long
arg3){
                HashMap<String,Object>item=(HashMap<String,Object>) arg0.
getItemAtPosition(arg2);
                final String goodsCardId= item.get("goodsCardId").toString();
                AlertDialog.Builder builder=new AlertDialog.Builder(GoodsOutport
Activity.this).setIcon(R.drawable.ic_launcher).setItems(new String[]{"出库"},
new DialogInterface. OnClickListener(){
                    public void onClick(DialogInterface dialog, int which){
                        switch(which){
                        case 0:
                            mGoodsOutportTask=new GoodsOutportTask();
                            mGoodsOutportTask.execute(goodsCardId);
                            break;}}});
                builder.create().show();}});
        mGoodsRequestTask.execute();}
    /***获取物品信息任务类**/
    private class GoodsRequestTask extends AsyncTask<String,Integer,List<GoodsJs
on>>{
        protected List<GoodsJson>doInBackground(String... params){
            return mGoodsDAO.getAll();}
        protected void onPreExecute(){super.onPreExecute();}
        protected void onProgressUpdate(Integer... values){
            super.onProgressUpdate(values);}
        protected void onPostExecute(List<GoodsJson>result){
            super.onPostExecute(result);
            if(result!=null){
                goodsArrayList.clear();
                GoodsJson goodsEntity=null;
                for(int i=0;i<result.size();i++){
                  goodsEntity= result.get(i);
                  if(goodsEntity!=null){
                    goodsHashMap=new HashMap<String, Object>();
    goodsHashMap.put("Image", R.drawable.metro_home_blacks_ scan_code);
                    goodsHashMap.put("goodsCardId", goodsEntity.cardNum);
                    goodsHashMap.put("goodsName", goodsEntity.dataName);
```

```
                    goodsHashMap.put("goodsAddress",goodsEntity.originPlace);
                    goodsHashMap.put("goodsUser", goodsEntity.staffName);
                    goodsArrayList.add(goodsHashMap);}}
                goodsAdapter.notifyDataSetChanged();}
            else{
            findViewById(R.id.view_loading_error).setVisibility(0);}}}
    private class GoodsOutportTask extends AsyncTask<String, Integer,
ControlResultJson>{/***物品出库**/
        protected ControlResultJson doInBackground(String... params){
            if(params!=null&&params.length>0)
                return mGoodsDAO.goodsOutport(params[0]);
            return null;}
        protected void onPreExecute(){super.onPreExecute();}
        protected void onProgressUpdate(Integer... values){
            super.onProgressUpdate(values);}
        protected void onPostExecute(ControlResultJson result){
            super.onPostExecute(result); String resStr=null;
            if(result!=null){
                resStr=result.flag==0?"出库成功! ":"出库失败";}
            else{resStr="出库失败";}
            mGoodsRequestTask=new GoodsRequestTask();
            mGoodsRequestTask.execute();
            Toast.makeText(GoodsOutportActivity.this, resStr, Toast.LENGTH_SHORT).
show();}}}
```

本章小结

本章通过智能仓储项目，让学生了解企业项目的整个开发流程：项目立项、需求分析、总体设计、详细设计、编码实现、系统调试。同时通过这个项目理解以前章节中学习到的知识并灵活应用到实际项目中，深入掌握界面的设计和实现方法、页面与页面之间数据的交互技术、数据库的设计和实现方法和视频数据处理技术。

强化练习

一、填空题

1. Android 四大核心组件分别是（　　）Activity、服务 Service、广播接收器 BroadcastReceiver 和内容提供者 ContentProvider。

2. 活动的四种状态分别是（　　）Running、暂停 Paused、停止 Stopped 和销毁 Destroyed。

3．活动的生命周期事件回调函数包括（　　）、onStart、onRestart、onResume、onPause、onStop 和 onDestroy。

4．Dalvik 虚拟机调试监控服务的英文简写为（　　）。

5．安卓开发工具的英文简写为（　　）。

6．可扩展标记语言的英文简写为（　　）。

7．Java 开发工具的英文简写为（　　）。

8．软件开发工具包的英文简写为（　　）。

9．Android 虚拟设备的英文简写为（　　）。

10．为了支持 Java 程序运行，需要安装（　　）。

11．统一建模语言的英文简写为（　　）。

12．设备独立像素的英文简写为（　　）。

13．线性布局、相对布局、帧布局和绝对布局是直接继承类为（　　）。

14．在 Android 平台中，所有的可视组件都是视图（　　）的子类；视图组 (ViewGroup) 是 View 类的一个重要直接子类；Android 布局管理器是 ViewGroup 类的一组重要的直接或间接子类。

15．android:layout_column 属性指定该单元格在第几（　　）显示；android:layout_span 属性指定该单元格占据的列数。

16．Android 中使用 SharedPreferences 使用（　　）的方式来存储数据。

17．在程序中可以通过 setBackgroundColor() 方法来设置布局的（　　）。

18．对按钮事件进行（　　）的方法是 setOnClickListener()。

19．Android 系统为我们提供了事件的处理机制。在处理点击事件时调用 setOnClickListener() 方法，其回调方法为（　　）。

20．线性布局的方向有两种不同的方向属性，分别为纵向和（　　）。

二、单选题

1．android.util.Log 可设置 5 种输出日志的级别，其中可显示所有类型的消息的调用方法是（　　）。

A. Log.i (TAG, strings)　　B. Log.d (TAG, strings)

C. Log.w (TAG, strings)　　D. Log.v (TAG, strings)

2．活动（Activity）运行时首先执行的生命周期事件回调函数是（　　）。

A. onCreate()　　B. onStart()　　C. onResume()　　D. onDestroy()

3．活动（Activity）销毁时最后执行的生命周期事件回调函数是（　　）。

A. onCreate()　　B. onDestroy()

C. onRestart()　　D. onPause()

4．R.java 类文件，在建立项目时由 Android 提供的工具自动生成，该工具名称是（　　）。

A. 资源打包工具　　B. Dalvik 虚拟机调试监控服务

C. 安卓开发工具　　D. 软件开发工具包

5．在 AndroidManifest.xml 中设置活动在桌面启动器 (Launcher) 中显示的名称的属性是（　　）。

A. android:name="string"　　B. android:theme="resourceortheme"

C. android:label="stringresource" D. android:icon="drawableresource"

6. 在 AndroidManifest.xml 中设置应用程序可调试的应用 (Application) 的属性是 ( )。

A. android:description="string resource" B. android:enabled=["true"| "false"]

C. android:debuggable=["true"| "false"] D. android:icon="drawable resource"

7. 下面不属于Android体系结构中的应用程序层是 ( )。

A. 短消息程序 B. 日历

C. 电话簿 D. 嵌入式数据库 SQLite

8. 资源目录res包含了Android应用项目的全部资源，命名规则正确的说法是 ( )。

A. 可以支持数字（0~9）、下画线（_）和大小写字符（a~z, A~Z）

B. 只能支持数字（0~9）和大小写字母（a~z, A~Z）

C. 只支持数字（0~9）、下画线（_）和小写字符（a~z），第一个可以是字母和下画线（_）

D. 只支持数字（0~9）、下画线（_）和小写字符（a~z），第一个必须是字母

9. Android应用程序设计主要是调用（ ）提供的API进行实现。

A. 应用程序层 B. 应用框架层

C. 应用视图层 D. 系统库层

10. Android应用项目启动最先加载的是AndroidManifest.xml，若项目中包含有多个活动 (Activity)，则决定最先加载活动的属性是（ ）。

A. android.intent.category.LAUNCHER B. android.intent.category.ICON

C. android.intent.action.MAIN D. android.intent.category.ACTIVITY

11. 在活动（Activity）的生命周期中，首先被调用的回调函数是（ ）。

A. onStart() B. onResume()

C. onRestart() D. onCreate()

12. 若要向Android工程添加字符串资源，应将其添加（ ）文件。

A. dimens.xml B. styles.xml

C. strings.xml D. value.xml

13. 在线性布局中，设置布局管理器内组件排列方式的xml属性名为（ ）。

A. android:orientation B. android:baselineAligned

C. android:divider D. android:gravity

14. 下面相对布局 Layout Params 属性中只能设置boolean值的属性为（ ）。

A. android:layout_centerHorizontal B. android:layout_toLeftOf

C. android:layout_above D. android:layout_alignLeft

15. Android用户界面布局xml文件中的帧布局标签是（ ）。

A. LinearLayout B. FrameLayout

C. RelativeLayout D. TableLayout

16. 相对布局 Layout Params 属性中确定是否在父容器中位于中央的属性名为（ ）。

A. android:layout_center Horizontal B. android:layout_centerVertical

C. android:layout_centerInParent D. android:layout_alignParentLeft

17. 相对布局 LayoutParams 属性中设置需要被隐藏的列的序号的xml属性名为（ ）。

A. android:collapsedColumns B. android:stretch Columns

C. android:shrinkableColumns　　D. android:layout_column

18. 在网格布局（GridLayout）中设置该网格列数的xml属性为（　　）。

A. android:alignmentMode　　B. android:columnCount

C. android:columnOrderPreserved　　D. android:rowCount

19. 网格布局（GridLayout）的LayoutParams属性中设置子组件在GridLayout的哪一列的xml属性名为（　　）。

A. android:layout_column　　B. android:layout_columnSpan

C. android:layout_gravity　　D. android:layout_row

20. 网格布局（GridLayout）的LayoutParams属性中设置该子组件在GridLayout纵向横跨几行的xml属性名为（　　）。

A. android:layout_column　　B. android:layout_columnSpan

C. android:layout_row　　D. android:layout_rowSpan

21. 绝对布局（AbsoluteLayout）中设置组件的y坐标值的xml属性名为（　　）。

A. android:layout_x　　B. android:layout_y

C. android:layout_width　　D. android:layout_height

22. 在Android中常用来显示程序执行进度的控件为（　　）。

A. 文本视图（TextView）　　B. 编辑框(EditText)

C. 按钮（Button）　　D. 进度条（ProgressBar）

23. 获得进度条（ProgressBar）控件当前进度值的方法为（　　）。

A. getProgress()　　B. getSecondaryProgress()

C. setMax(int max)　　D. setProgress(int progress)

24. 在Android中用于显示图片的控件为（　　）。

A. 按钮（Button）　　B. 文本视图（TextView）

C. ImageView　　D. EditText

25. 若要捕获某类视图控件的事件，需要为该控制设置（　　）。

A. 方法　　B. 监听器　　C. 回调函数　　D. 属性

26. 在Android中能简单有效地处理单击事件响应OnClickListener的控件为（　　）。

A. 按钮（Button）　　B. 文本视图（TextView）

C. ImageView　　D. EditText

27. 图像视图（ImageView）控件中使用什么属性引用图片资源（　　）。

A. android:src　　B. android:text　　C. android:id　　D. android:img

28. 控制虚拟键盘输入类型的属性为（　　）。

A. android:text　　B. android:src

C. android: inputType　　D. android:id

29. 在Android中，若要向工程中导入图片资源，应将图片放在（　　）目录。

A. res\picture　　B. res\string

C. res\icon　　D. res\drawable

30. 在Android中，若要向工程添加字符串资源，应将其添加到（　　）文件。

A. dimens.xml　　B. styles.xml　　C. strings.xml　　D. value.xml

31. ListView是常用的（　　）类型控件。
A. 按钮（Button）　　B. 图像视图（ImageView）
C. 列表　　D. 下拉列表

32. ListView与数组或List集合的多个值进行数据绑定时使用（　　）。
A. ArrayAdapter　　B. BaseAdapter
C. SimpleAdapter　　D. SimpleCursorAdapter

33. 在Android列表视图中以下表示系统自定义的，只显示一行文字的布局文件是（　　）。
A. android.R.layout.simple_list_item_0
B. android.R.layout.simple_list_item_1
C. android.layout.simple_list_item_0
D. android.layout.simple_list_item_1

34. Android中包含多种基本UI控件和高级UI控件，这些UI控件均派生于（　　）类。
A. 视图组（ViewGroup）　　B. 控件（Control）
C. 文本视图（TextView）　　D. 视图（View）

35. Android布局方式不包括下面的（　　）项。
A. 线性布局　　B. 相对布局　　C. 多维布局　　D. 单帧布局

36. Android线性布局分为两种方式：纵向和横向，设置该方式的属性为（　　）。
A. android: orientation　　B. android:layout_gravity
C. android:layout_width　　D. android:layout_height

37. 在Android应用程序开发中用户要用到一些开发工具，但其中不包括（　　）。
A. Eclipse　　B. VC++6.0　　C. ADT　　D. JDK

38. Android开发中用户经常用到打印日志的方式来进行调试，其中日志类型不包括下面的（　　）。
A. Log.v　　B. Log.c　　C. Log.e　　D. Log.d

39. 在Android程序开发时，用户经常用到文本框控件，该控件在布局文件中的标签为（　　）。
A. EditText　　B. Text　　C. TextView　　D. Label

40. 在调用对话框Dialog时，需要最后调用（　　）方法来显示对话框。
A. onLongClick()　　B. onClick()　　C. onTouch（）　　D. show()

41. 列表控件在布局文件中的标签为（　　）。
A. ListArray　　B. List　　C. List Adapter　　D. ListView

42. 在Android中，进度条对话框ProgressDialog必须要在后台程序运行完毕前，以（　　）方法来关闭所取得的焦点。
A. finish()　　B. close()　　C. dimiss()　　D. 以上都不是

43. 在使用系统自带的TabHost时需要注意：TabHost的ID必须为（　　）。
A. android:id/tabhost　　B. android:id/tabs
C. android:id/tabcontent　　D. android:id/tabframework

44. 为了在Android程序中能够自适应手机屏幕的大小，一般所选择的布局方式是（　　）。

A. 线性布局　　B. 相对布局　　C. 单帧布局　　D. 绝对布局

45. Android的基本UI控件和高级UI控件均继承自（　　）类。

A. TextView　　B. Button　　C. GridView　　D. View

46. Android UI所包括的几类布局管理器：线性布局、相对布局、帧布局、表格布局和绝对布局等，它们均继承自（　　）类。

A. Activity　　B. Service　　C. ViewGroup　　D. SurfaceView

47. ListView是Android中最常用的控件之一，ListView中一个重要的概念就是（　　），它是控件与数据源之间的桥梁。

A. 适配器Adapter　　B. 视图View　　C. 数据源　　D. 事件监听器

48. 在启动Activity的方法中，不包括以下（　　）方法。

A. startActivity　　B. startActivityFromChild

C. startActivityForResult　　D. startActivityForFragment

49. Activity的生命周期方法不包括下列（　　）方法。

A. onPause　　B. onCreate　　C. onNewIntent　　D. onRestart

50. 设置编辑框（EditText）提示信息的属性为（　　）。

A. android:inputType　　B. android:hint

C. android:digits　　D. android:text

51. Android 意图（Intent）的作用在于（　　）。

A. 连接四大组件的纽带，可实现界面间的切换，可包含动作和动作数据

B. 实现应用程序间的数据共享

C. 可保持应用在后台运行，而不会因为切换页面而消失

D. 处理一个应用程序的后台工作

52. 句柄（Hanlder）是线程（Thread）与活动（Activity）通信的桥梁，若线程处理不当，会导致机器变得越来越慢，则可调用（　　）方法销毁该线程。

A. onStop()　　B. onFinish()　　C. onDestroy()　　D. onClear()

53. 下面属于View的子类的是（　　）。

A. Activity　　B. Service　　C. ViewGroup　　D. Content Provider

54. 当活动（Activity）被消毁时，应该实现它在（　　）方法来保存其原来的状态。

A. onResume()　　B. onSaveInstance（）

C. onInstanceState（）　　D. onSaveInstanceState（）

55. 活动的状态包括（　　）哪些状态。（请选择3个答案）

A. 睡眠状态　　B. 暂停状态　　C.停止状态　　D. 运行状态

56. 下面关于句柄的说话中正确的是（　　）。（请选择2个答案）

A. 它是采用栈的方式来组织任务的　　B. 它可以属于一个新的线程

C. 它是不同线程间通信的一种机制　　D. 它避免了新线程操作用户界面（UI）组件

57. 下面（　　）是属于视图的子类组件包。（请选择2个答案）

A. TextView　　B. 服务（Service）　　C. ViewGroup　　D. 活动(Activity)

58. 在布局文件中，定义一个视图组件时，有（　　）两个属性必须写。（请选择2个答案）

A. android:id
B. android:text
C. android:layout_width
D. android:layout_height

三、多选题

1. 下列属于Activity的状态是（　　）。
   A. 运行状态
   B. 暂停状态
   C. 停止状态
   D. 睡眠状态
2. 关于Handler的说话正确的是（　　）。
   A. 它实现不同线程间通信的一种机制
   B. 它避免了新线程操作UI组件
   C. 它采用栈的方式来组织任务的
   D. 它可以属于一个新的线程
3. 关于广播的作用，正确的说法是（　　）。
   A. 它是用接收系统发布的一些消息的
   B. 它可以帮助service修改用户界面
   C. 它可以启动一个Activity
   D. 它可以启动一个Service
4. 下面属于View的子类的是（　　）。
   A. Activity
   B. Service
   C. ViewGroup
   D. TextView
5. 在main.xml中，定义一个组件时，有两个属性必须写（　　）。
   A. android:layout_width
   B. android:layout_height
   C. android:id="@+id/start"
   D. android:text
6. 请找出你学过的适配器类（　　）。
   A. SimpleAdapter
   B. SimpleArrayAdapter
   C. SimpleCursorAdapter
   D. SimpleCursorsAdapter
7. 关于主题的说法，正确的是（　　）。
   A. 它是属性集合
   B. 它可以在程序中来设置
   C. 它通常用于一个Activity或所有Activity上
   D. 它可以用于单个TextView上
8. 意图可分为（　　）。
   A. 显式意图　B. 隐式意图　C. 组件意图　D. 类意图
9. 解析xml的方式有（　　）。
   A. 字符器类型
   B. 流方式
   C. 文档对象模型(DOM)
   D. SAX(Simple API for XML)
10. Android 通过 startService 的方式开启服务，关于Service生命周期的onCreate()和onStart()说法正确的是（　　）。
    A. 当第一次启动时先后调用 onCreate() 和 onStart() 方法
    B. 当第一次启动时只会调用 onCreate() 方法
    C. 如果 Service 已经启动，将先后调用 onCreate() 和 onStart() 方法

D. 如果 Service 已经启动，只会执行 onStart()方法，不在执行 onCreate()方法

四、简答题和编程题

1. 单帧布局（FrameLayout）的特点是什么？
2. Android软件框架结构自上而下可分为哪些层？
3. Android应用程序的四大组件是什么？
4. Android 系统提供的数据存储方式有几种？分别是什么？
5. Android 中常用的5种布局分别是什么？
6. 什么是 Intent? 其作用是什么？
7. Android的动画由哪几种类型组成？
8. 实现SharedPreferences存储的主要操作步骤包括哪些？

参 考 文 献

[1] 杨丰盛. Android 应用开发揭秘 [M]. 北京:机械工业出版社，2010.

[2] 李宁. Android 开发权威指南 [M]. 北京:人民邮电出版社，2011.

[3] 盖索林. Google Android 开发入门指南 [M]. 2 版 . 北京:人民邮电出版社，2009.

[4] 梅尔. Android 2 高级编程（第 2 版）[M]. 王超，译. 北京:清华大学出版社，2010.

[5] 余志龙，王世江. Google Android SDK 开发范例大全 [M]. 2 版. 北京:人民邮电出版社，2010.

[6] 李宁. Android/OPhone 开发完全讲义 [M]. 北京:水利水电出版社，2010.

[7] 李刚. 疯狂 Android 讲义 [M]. 北京:电子工业出版社，2011.

[8] 汪永松. Android 平台开发之旅 [M]. 北京:机械工业出版社，2010.

[9] E2EColud 工作室. 深入浅出 Google Android[M]. 北京:人民邮电出版社，2009.

[10] 梅尔. Android 高级编程 [M]. 王鹏杰，霍建同，译. 北京:清华大学出版社，2010.

[11] 李华忠，梁永生，刘涛. Android 应用程序设计教程 [M]. 北京:人民邮电出版社，2013.